T0191923

Environmental Footprints and Eco-design of Products and Processes

Series Editor

Subramanian Senthilkannan Muthu, Head of Sustainability - SgT Group and API, Hong Kong, Kowloon, Hong Kong

Indexed by Scopus

This series aims to broadly cover all the aspects related to environmental assessment of products, development of environmental and ecological indicators and eco-design of various products and processes. Below are the areas fall under the aims and scope of this series, but not limited to: Environmental Life Cycle Assessment; Social Life Cycle Assessment; Organizational and Product Carbon Footprints; Ecological, Energy and Water Footprints; Life cycle costing; Environmental and sustainable indicators; Environmental impact assessment methods and tools; Eco-design (sustainable design) aspects and tools; Biodegradation studies; Recycling; Solid waste management; Environmental and social audits; Green Purchasing and tools; Product environmental footprints; Environmental management standards and regulations; Eco-labels; Green Claims and green washing; Assessment of sustainability aspects.

More information about this series at https://link.springer.com/bookseries/13340

Eric Lichtfouse ·
Subramanian Senthilkannan Muthu · Ali Khadir
Editors

Inorganic-Organic Composites for Water and Wastewater Treatment

Volume 2

Editors
Eric Lichtfouse
Aix-Marseille University
Marseille, France

Subramanian Senthilkannan Muthu
SgT Group & API
Hong Kong, Kowloon, Hong Kong

Ali Khadir
Islamic Azad University of Shahre Rey
Branch
Tehran, Iran

ISSN 2345-7651 ISSN 2345-766X (electronic)
Environmental Footprints and Eco-design of Products and Processes
ISBN 978-981-16-5930-0 ISBN 978-981-16-5928-7 (eBook)
https://doi.org/10.1007/978-981-16-5928-7

This Springer imprint is published by the registered company Springer Nature Singapore Pte Ltd.
The registered company address is: 152 Beach Road, #21-01/04 Gateway East, Singapore 189721, Singapore

Contents

About the Editors

Eric Lichtfouse is a research scientist at Aix-Marseille University and an invited professor at Xi'an Jiaotong University. His research interests include climate change, carbon, pollution and organic compounds in air, water, soils and sediments. He his teaching biogeochemistry and scientific writing. He is the chief editor of the journal Environmental Chemistry Letters and the book series Sustainable Agriculture Reviews and Environmental Chemistry for a Sustainable World.

Subramanian Senthilkannan Muthu currently works for SgT Group as the Head of Sustainability, and is based out of Hong Kong. He earned his PhD from The Hong Kong Polytechnic University, and is a renowned expert in the areas of Environmental Sustainability in Textiles & Clothing Supply Chain, Product Life Cycle Assessment (LCA) and Product Carbon Footprint Assessment (PCF) in various industrial sectors. He has 5 years of industrial experience in textile manufacturing, research and development and textile testing and over a decade's of experience in LCA, carbon and ecological footprints assessment of various consumer products. He has published more than 100 research publications, written numerous book chapters and authored/edited over 100 books in the areas of Carbon Footprint, Recycling, Environmental Assessment and Environmental Sustainability.

Ali Khadir is an environmental engineer and a member of the Young Researcher and Elite Club, Islamic Azad University of Shahre Rey Branch, Tehran, Iran. He has published several articles and book chapters in reputed international publishers, including Elsevier, Springer, Taylor & Francis and Wiley. His articles have been published in journals with IF greater than 4, including the Journal of Environmental Chemical Engineering, International Journal of Biological Macromolecules and Journal of Water Process Engineering. He also has been the reviewer of journals and international conferences. His research interests centre on emerging pollutants, dyes and pharmaceuticals in aquatic media, advanced water and wastewater remediation techniques and technology.

Metal Organic Frameworks to Remove Arsenic Adsorption from Wastewater

Sruthi Rajasekaran, K. R. Sunaja Devi, D. Pinheiro, M. K. Mohan, and P. Iyyappa Rajan

Abstract Water is an integral part of life on earth. Rapid industrialization, urbanization, and population explosion have all contributed to the pollution of ground and surface water with, among other things, heavy metals. This has led to an acute shortage of clean drinking water. Arsenic is one of the most toxic heavy metals found in water, posing a serious threat to the environment, human beings, and aquatic life. Over the years, a considerable amount of research has been directed toward the elimination of arsenic from water via sustainable methodologies. Metal organic frameworks are a class of materials possessing exceptional features like chemical stability, high porosity, multiple functional groups, and large surface areas. These properties can be effectively channelized to make metal organic frameworks excellent adsorbents for the removal of arsenic from contaminated water and make it drinkable. We have reviewed herein, the problems of heavy metal contamination, specifically the different forms of arsenic that pollute water. The importance of metal organic frameworks and the progress made in the synthesis of materials having a metal oxide framework have been discussed. Significant properties like adsorption and mechanistic aspects of adsorption through metal organic frameworks have been described. Furthermore, the characterization of the electronic and geometric aspects

S. Rajasekaran · K. R. S. Devi (✉)
Department of Chemistry, CHRIST (deemed to be University), Bangalore, Karnataka 560029, India
e-mail: sunajadevi.kr@christuniversity.in

S. Rajasekaran
e-mail: sruthi.rajasekaran@res.christuniversity.in

D. Pinheiro · M. K. Mohan
Department of Science and Humanities School of Engineering and Technology, CHRIST (deemed to be University)) Kengeri Campus, Bangalore, Karnataka 560074, India
e-mail: dephan.pinheiro@christuniversity.in

M. K. Mohan
e-mail: mothikrishna.mohan@christuniversity.in

P. Iyyappa Rajan
Asia Pacific Center for Theoretical Physics, POSTECH Campus, Pohang 37673, Republic of Korea

E. Lichtfouse et al. (eds.), *Inorganic-Organic Composites for Water and Wastewater Treatment*, Environmental Footprints and Eco-design of Products and Processes,
https://doi.org/10.1007/978-981-16-5928-7_1

of metal organic frameworks using density functional theory has been reviewed. Insight into proper scaling up and development of metal organic frameworks for practical applications have also been suggested.

Keywords Metal organic frameworks · Arsenic pollutants · Adsorption · Wastewater · Heavy Metals · Water contamination · Environmental chemistry

1 Introduction

The importance of water for human beings and indeed for life on earth cannot be overstated. During the past decades, rapid industrialization and urbanization have polluted our ecosystem, notably our water bodies. The above reasons, coupled with the explosion in the human population have led to an increased demand for pure drinking water. The wastewater discharged from industrial wastes often carries a large concentration of heavy metals often as high as 24.2 ppb [39, 50, 83]. Several methods, biological, chemical, and physical have been used for the remediation of heavy metal pollution of water. However, many of these techniques release secondary pollutants into water and are therefore not completely environment friendly [122]. Thus, there is a wide scope for identifying quick, cost-effective, and cleaner methods to purify wastewater from heavy metals. This can contribute to a cleaner environment and also help meet the demand for clean drinking water [31]. Of the several methods employed for the remediation of pollutants, many are based on adsorption to achieve the elimination of pollutants from the water bodies [84].

Adsorption is a process that is used for the removal of a substance from gaseous or liquid solutions, where a solid is used as a medium for the removal [70]. The van der Waals and electrostatic forces between the adsorbate molecule and adsorbent surface are the prime drivers for the adsorption process [91]. The concentration of adsorbent and pH play a vital role in any adsorption process. Adsorption is an effective technique for the removal of arsenic (As) because of its excellent removal efficiency, easy operation, and absence of sludge formation [41, 67]. Arsenate adsorption is favored at low pH values, whereas arsenite requires higher pH conditions. The nature of adsorbents is another crucial factor that determines the effectiveness of arsenic removal through adsorption. Widely used sorbents are activated coal, red mud, coal, fly ash, and metal organic frameworks.

Metal organic frameworks (MOFs) are materials, made up of porous coordination polymers, with easily tailorable pore sizes and possessing large surface areas. The pores in the constructed MOFs play a major role in their effectiveness in the target application. Some of the highly porous MOFs, like other porous materials having long-range order, exhibit excellent selectivity in adsorption. MOFs contain a metal ion or cluster and an organic linker, which are well known for exhibiting exceptional performances in diverse applications such as carbon dioxide capture, gas storage, drug delivery, catalysis, and environmental protection [23, 32, 123]. MOFs have also been [26] extensively used in the elimination of heavy metal ions from wastewater.

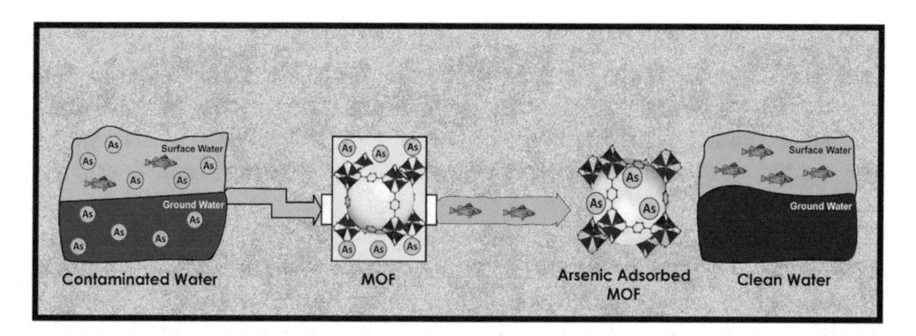

Fig. 1 Removal of arsenic using metal organic framework through adsorption

Considerable efforts have gone into the design and preparation of MOFs with new structures and functions besides exploring potential applications in areas like gas adsorption or separation [60], sensors [60], drug delivery [55], magnetic materials [18], and optical devices [30].

Gas separation is an important process in industries where microporous materials with high adsorption capacity and selectivity are required to achieve efficient gas separation. Gases like propylene–propane mixture are used for gas separation and consume 90% less energy compared with the distillation process [6, 60]. Scrubbing agents like calcium oxide and zeolites have also been used for applications in sensing. Sensors have been developed by using MOFs KAUST-7 and KAUST-8 coated with quartz crystal microbalance transducers, which have the ability to detect sulfur dioxide gas at low concentrations. The structure–property relationship in adsorbents can be tuned to detect and selectively capture toxic molecules [6, 100]. MOFs are also potential candidates in drug delivery as they can be engineered to have key properties. UiO-66 is a good example of MOFs used in drug delivery and anticancer treatment where metal clusters occupy the co-ordination sites of MOFs [55].

MOFs with their high electric conductivity and excellent magnetic ordering enable novel functionalities in spintronic applications besides imparting ferromagnetic and semiconducting properties to them [18]. In addition to the applications mentioned above, MOFs have been widely used in the elimination of pollutants from wastewater. This review provides a detailed description of the removal of arsenic from polluted as well as groundwater with the help of MOFs and modified MOF materials. Figure 1 depicts the graphical representation of adsorption for arsenic removal using MOFs.

2 Heavy Metals in Water and the Impact on Health

Heavy metals are inherent constituents of the lithosphere and are among the oldest toxic substances known to humans [38]. They make up a major percentage of pollutants in the effluents emerging from industries [89] and their presence imposes serious side effects on living organisms [38, 83, 107]. As per the Environmental Protection Agency of US Federal Government, metals like beryllium, cadmium, chromium, arsenic, and others are considered to be priority pollutants for their extensive usage, toxicity, and contamination of water through sewage, sludge, pesticides, municipal wastes, and industrial effluents. Some of them are carcinogenic and even fatal at high concentrations.

During the past few decades, the enormous growth of chemical industries has led to an increased amount of waste released into the environment and the subsequent accumulation of heavy metals [85, 97]. Heavy metals like mercury, lead, cadmium, molybdenum, arsenic, and thallium ions are the major culprits polluting drinking water where the last two are the most toxic among them [87]. These heavy metals enter the atmosphere and water mainly through volcanic activity, mining operations, electroplating, plumbing fixtures, batteries, paint, oils, fertilizers, and nuclear power plants. The existence of heavy metals in water leads to harmful effects on many organs of the human body and causes diseases like cancer, infertility, arsenicosis, paralysis, hair loss, and mutagenesis. The permitted limit of heavy metals in drinking water ranges from 0.1 to 100 ppb depending on the metal [21, 33, 124]. Earlier, adsorbents like zeolites, carbonaceous, and polymer-based chitosan were employed for remediating water from heavy metals [97]. Carbon-based materials offer a lot of scope as adsorbents for the removal of toxic metals from wastewater.

3 Arsenic as a Contaminant

Arsenic is a toxic metalloid and a carcinogen [80]. It is the 20th most abundant element on earth and is found in seawater and the human body [104]. Arsenic combines with oxygen, chlorine, and sulfur to form a variety of inorganic species. The name 'King of Poison' is given to arsenic by World Health Organization [90]. Primarily, arsenic contamination is caused by the usage of phosphate fertilizers, insecticides, smelting, and mining operations that release arsenic into groundwater and soil leading to severe environmental issues [112]. Water contaminated with arsenic poses a major threat to human health and aquatic organisms in numerous regions of the world [4]. Both ground and surface water contain arsenic contamination that in turn affects human health and aquatic organisms worldwide.

In natural water bodies, arsenic can exist in two forms, As(V) and As(III) of which the latter is more hazardous [77, 84]. Major biochemical processes will be hindered when As(III) binds with thiol groups altering the immune system leading to various gastrointestinal diseases [45]. Generally, As(V) does not bind to the sulfhydryl group and is, therefore, less toxic [11]. Continuous exposure to arsenic leads to acute diseases like skin lesions and cancer in addition to affecting the central nervous system [77]. The toxin level for arsenic in potable water is $0.01 mgL^{-1}$, but in some cases, it can even go up to 0.1–2 mgL^{-1} in groundwater which can be very harmful [9]. Oxidation, coagulation, and membrane filtration are some of the viable methods that have been practiced to remove arsenic [104].

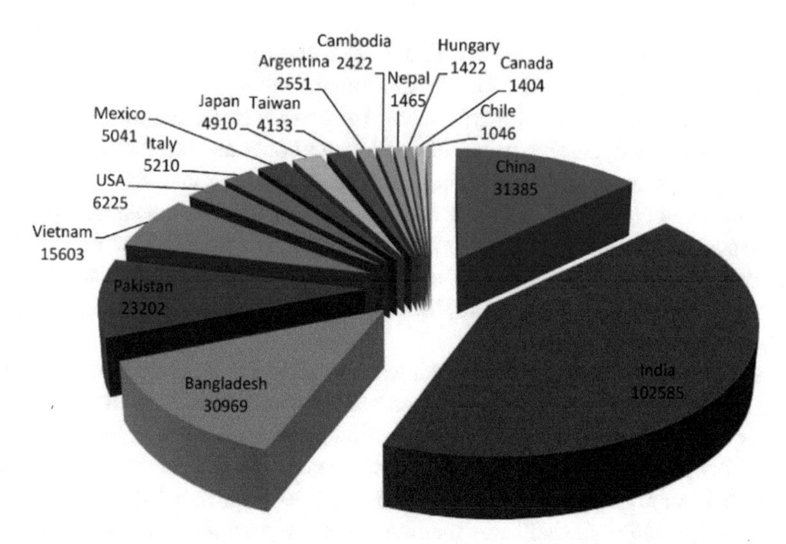

Fig. 2 Population exposed to arsenic contamination in selected countries. The maximum affected countries are India and China. Reprinted with the permission of Elsevier from [40]

The natural weathering processes like soil erosion and leaching largely contribute to the accumulation of arsenic, and, because of its high mobility, re-enters the water bodies to adversely impact soil fertility. It was reported that the concentration of arsenic is high in shallow groundwater, which affects millions of people [35, 67]. Reverse osmosis plants were reportedly installed around various parts of Argentina to remediate the high amounts of arsenic found in the groundwater for the public's drinking water [62]. Figure 2 gives the data for the presence of arsenic in different countries, in which the worst affected countries are India and China. High arsenic content is found in semi-arid regions, and the studies show that utilization of more shallow water is the major cause of the problem. The regions that have high arsenic concentrations contain oxidized groundwater leading to waterborne disease, which, in turn, affects human life [40]. According to the World Health Organisation, arsenic is a Class I carcinogenic. Thus, it is crucial to remove arsenic ions from the environment using all possible methods.

4 Different Forms of Arsenic in Water

Water-soluble arsenic is predominantly found in the inorganic form in natural water though it is not uncommon to find arsenic in the organic form. The chemical behavior, distribution between organic and inorganic forms of arsenic as well as its diffusion are all pH dependent and hence can vary with the pH of the water bodies [39, 80]. The neutral form (H_3AsO_3) of As(III), with higher toxicity, is generally seen when the pH of the solution is less than 9. Arsenic species are also found in natural sources like rocks, volcanic activity, and produced during metallurgical operations like mining, roasting, etc. [103]. The dissolution of minerals, ores, soil, and sediments enables arsenic to enter the water, leading to its greater concentration in groundwater compared with surface water. Natural water contains both As(III) and As(V) where As(III) is stable under reducing aqueous conditions, and As(V) is stable in oxygenated water [56]. Figure 3 illustrates the different forms of arsenic in water [37]. The various techniques employed for the identification of arsenic in water are inductively coupled plasma spectrometry, atomic absorption spectroscopy, atomic emission spectrometry, mass spectroscopy, high-performance liquid chromatography, and ion chromatography [37, 81].

Due to the acute shortage of drinking water worldwide, it is important that heavy metals and trace elements be removed from both land and contaminated water bodies. The raw materials used for preparing arsenic removing adsorbents are commonly sourced from activated carbon, agricultural waste like rice husk, coconut shell, potato peel, alumina, granular ferric hydroxide, zeolites, and synthetic resins [9, 56, 67]. Many industries still use methods such as precipitation, coagulation, ion exchange, and flocculation for the removal of arsenic, and most of these techniques are not successful in completely eliminating arsenic. Studies show that complexation between hydroxyl iron oxides, As(III), and Sb(III) on the surface play a primary role during adsorption [121]. The hydroxyl groups present on the surface of chitosan have enabled the formation of hydrogen bonds, which felicitates the adsorption of As(III) and Sb(III). Modification of chitosan with amino groups increases the surface groups leading to more active sites and hence even better adsorption.

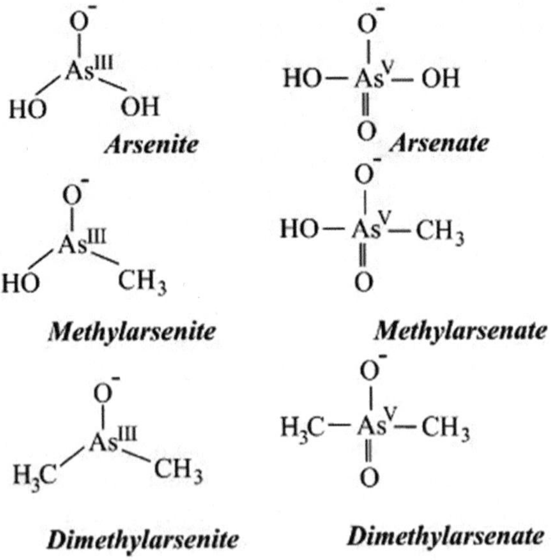

Fig. 3 Different forms of arsenic species with methyl and dimethyl groups are attached to arsenate and arsenite in water. Reprinted with the permission of Elsevier from [37]

Arsenite — $HO-As^{III}(O^-)-OH$

Arsenate — $HO-As^{V}(O^-)(=O)-OH$

Methylarsenite — $HO-As^{III}(O^-)-CH_3$

Methylarsenate — $HO-As^{V}(O^-)(=O)-CH_3$

Dimethylarsenite — $H_3C-As^{III}(O^-)-CH_3$

Dimethylarsenate — $H_3C-As^{V}(O^-)(=O)-CH_3$

5 Conventional Strategies for the Removal of Arsenic

The composition and nature of arsenic found in contaminated water and the chemistry behind it are key factors that govern the elimination of arsenic from water [94]. Most of the arsenic removal strategies are efficient if it remains predominantly non-charged at pH less than 9.2 [70]. Most technologies adopt a two-step approach. The first step involves oxidation, where arsenite is oxidized to arsenate, and the second step involves the actual removal of arsenate [78]. The most important strategies for the removal of arsenic include oxidation, coagulation–flocculation, and membrane filtration technology.

The oxidation of As(III) to As(V) is a critical step in the remediation of arsenic as most of the soluble arsenic under neutral pH exists in the arsenite form. Apart from atmospheric oxygen, other chemicals and microorganisms are also employed to oxidize arsenite in water [70]. Costa et al. performed a systematic process for the elimination of arsenic with the help of UV-assisted catalytic H_2O_2 oxidation followed by adsorption using ilmenite ($FeTiO_3$) a natural photocatalyst [13]. The coagulation–flocculation method is another widely used method for the elimination of arsenic from water [70]. The negative charge of the colloids is reduced by the positively charged coagulants.

Flocculation involves the inclusion of anionic flocculants to cause a link between the larger particles formed, leading to the development of flocs. Here, aided by the added chemicals, arsenic is converted into an insoluble solid, which later precipitates as sediment [91]. The soluble arsenic species can sometimes be co-precipitated using metal hydroxides, and the solid residue can then be separated either by filtration or sedimentation. The usage of different coagulants like ferric chloride and alum showed that the extent of arsenic removal depends considerably on the pH [68]. Ge J and others studied the removal of arsenic from wastewater by coagulation technology and investigated the adsorption efficiency under different conditions such as the effects of suspended solids in the mixed liquor, pH, and the effect of orthophosphate [27].

Membrane-based filtration is a widely used method for obtaining drinking water. Membranes are fine pliable sheets made of synthetic materials and act as a barrier preventing the impurities in water to pass through [70]. Generally, pressure-driven membrane filtration can be categorized into two types, microfiltration and ultrafiltration which use low pressures whereas reverse osmosis and nanofiltration use high pressures. For the removal of arsenic from polluted water using microfiltration, the pore size of the membrane used is between 0.1 and 10 μm. On the other hand, ultrafiltration is not an effective technique due to the larger size of the membrane pore. Therefore, to make the membrane filtration technique effective, surfactant-based separation techniques such as micellar enhanced ultrafiltration can be employed. Similarly, both nanofiltration and reverse osmosis are used for the removal of dissolved compounds with molecular weights larger than 300 $gmol^{-1}$. Thus, these two methods can be effectively used to remediate the dissolved arsenic from water [93]. Yang X and others reported the usage of ceramic filters modified by nano-CeO_2 to remove As(V) from water in remote areas in a cost-effective manner [118].

6 Metal Organic Frameworks

Metal organic frameworks (MOFs) are a group of porous crystalline materials made up of metal ions, clusters, and organic ligands. The pores in the MOFs play a vital role in determining the properties of the structural framework. The size and shape of the pores in MOFs can be adjusted by making changes in the ligands and metal ions. [84]. Unlike other solids, MOFs maintain their structure and crystalline nature upon expansion of the organic linkers and inorganic secondary building units, which broadens the scope of their applications [25].

The frequently used metals for the synthesis of MOFs are Co(II), Ag(I), Cu(II), Mg(II), Fe(III), Al(III), which result in different structural geometries such as trigonal bipyramidal, pyramidal, hexagonal, tetrahedral, and octahedral. Organic linkers like carbonates, sulphonates, amines, and phosphates are important components in the synthesis of MOFs. The early MOFs synthesized were sensitive to water due to the lability of the ligand–metal bond, and this aspect restricted the applications of MOFs. The most important property of MOFs as adsorbents is the surface area, which can range from 1000 to 10,000 m^2g^{-1}. MOFs, possessing desirable properties like high surface area of upto 6000 m^2g, porosity, tunable pore size, rigidity, structural flexibility, and thermal stability without doubt, offer a lot of promise as a material for efficient water treatment [110]. Figure 4 shows the schematic structure of MOF [43].

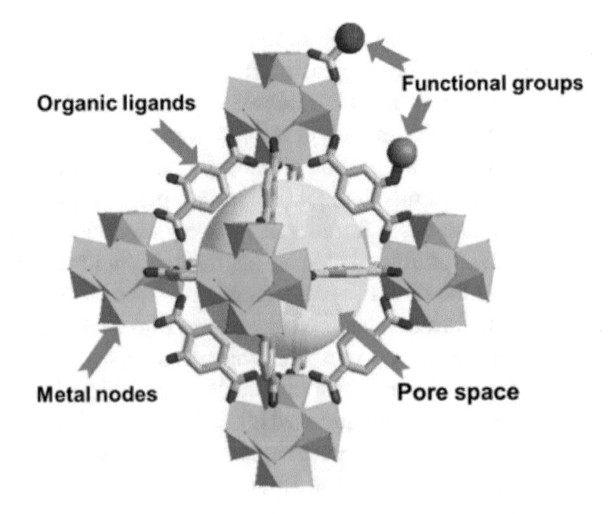

Fig. 4 Structure of a metal organic framework comprising of organic ligands, functional groups, pore space, and metal ions. Reprinted with the permission of Elsevier from [43]

7 General Methods of Preparation of Metal Organic Framework

Several techniques have been adopted for the synthesis of MOFs. For liquid-phase syntheses, metal salts and ligand solutions are added separately to a reaction vial. Besides the liquid phase, solid-phase synthetic methods too have been carried out. Some of the methods are solvothermal [111], microwave [52], sonochemical [28], slow evaporation [119], mechanochemical [98], and electrochemical methods [76].

Slow evaporation is a traditional method used for the synthesis of MOFs, which do not require any external energy supply. In this method, the starting solution is concentrated, while the solvents are at room temperatures. However, this is a tedious and time-consuming technique. Sometimes different combinations of solvents are also used to increase the solubility and rate of the reaction [22, 86, 119]. In solvothermal synthesis, the reactions are performed in closed vessels or Teflon-lined autoclaves at temperatures beyond the boiling point of the solvent [111]. High boiling solvents are usually used, and when solubility is a problem, the mixture of solvents is also used. MOFs prepared by this method have been used for sensing, adsorption, and gas separation processes [19, 71]. Microwave-assisted syntheses or microwave-assisted solvothermal syntheses are rapid methods for the synthesis of MOF, which involves microwave heating of a solution for a fixed time period [46, 52, 105].

Electrochemical synthesis is a viable method for the construction of large crystals of MOF by varying pH under mild conditions [76]. It does not require metal salts, which is a major advantage and is widely used in the industrial process. MOFs generally synthesized via the above technique are Cu-BTC-MOF, Ni@C, IRMOF-3, MIL-53(Al), and many others [1, 114]. Synthesis of Cu-BTC- MOF carbon nanotube was carried out via electrochemical method and has paved the way for non-enzymatic determination of glucose [1]. Mechanochemical synthesis is a solvent-free method for the synthesis of MOFs in which chemical reactions are performed using mechanical force. It is efficiently used for the rapid synthesis of MOFs where a small amount of solvent is mixed with a solid mixture. This is used for the construction of bonds through a simple, economical, and environmentally friendly mechanochemical method [64, 98].

Sonochemical synthesis is a method where chemical changes are brought about by applying ultrasonic radiations [7, 28]. The sonochemical method can generate a homogenous nucleation center in lesser time. Graphene nanosheet with amine-functionalized Cu terephthalate MOF (Cu-BDC-NH_2 @GO) was synthesized through this technique and used for hydrogen adsorption studies [14].

8 Applications of Metal Organic Frameworks

The large surface area, porosity, unique functional group are some of the properties of MOFs that have found use in applications like gas separation, storage, catalysis, sensors, drug delivery, and water purification [3, 17, 44, 47, 54, 63, 66]. Figure 5

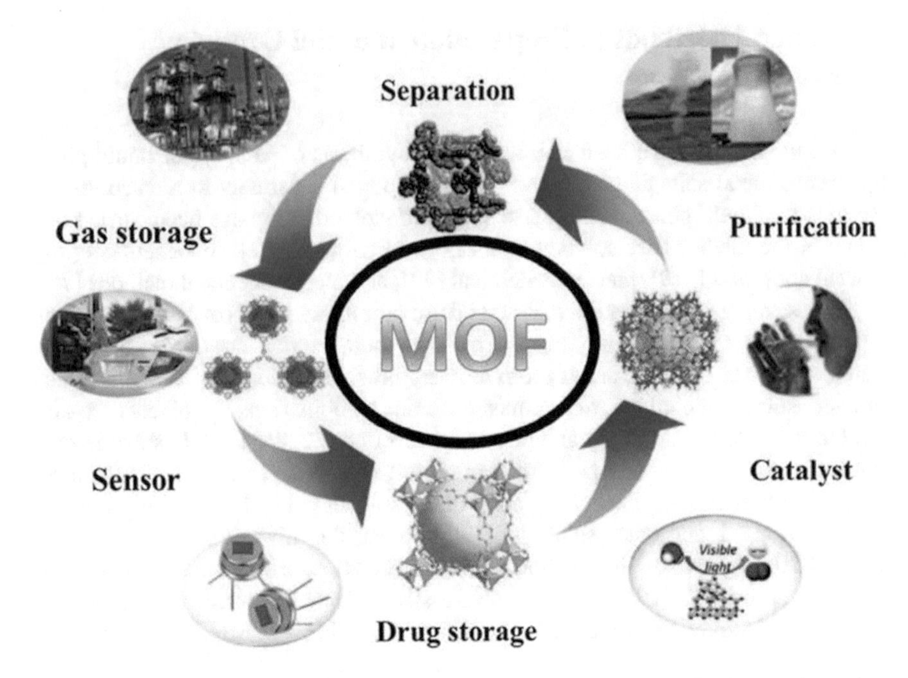

Fig. 5 Important applications of metal organic framework such as water purification, catalysts, drug storage, sensing, gas storage, and separation of gases. Reprinted with the permission of Elsevier from [16]

illustrates the various applications of MOFs in different areas [16]. MOFs have excellent permeation and separation properties and are used for gas separation with excellent results. The gases preferred are mixtures of CO_2 or H_2 and CO_2 or N_2. Here CO_2 is a good adsorbate that suppresses less adsorbing permeances of H_2 or N_2 [47]. The main gases used for storage applications by MOF are methane, natural gas, biogas, and to a lesser extent NO and CO_2. The high porosity of MOF materials is the main advantage in methane storage through adsorption [3, 17]. Gas sensors are of great interest, as the material can be used in the detection of toxic gases. ZIF-8 based nanocomposite sensors show a good response to H_2 with ZnO nanowire sensors [63].

MOFs, with their large surface area and high pore volume, have found extensive application in drug delivery and health care. MOF-74 materials have been used as a storage material and in the drug delivery of anticancer agents like methotrexate and 5-fluoroacil [3, 54]. Co-MOF-74 proved to be a good catalyst in the cycloaddition of CO_2 to styrene oxide under benign conditions without any deterioration in the structural properties [3, 17]. NH_2-MIL-53 cross-linked with PVDF membrane was used for the purification of H_2 with excellent results with a huge potential in industrial applications [66].

9 Metal Organic Framework as an Adsorbent

Adsorption plays a major part in the elimination of contaminants from wastewater. Arsenate species have high mobility in water, which, in turn, affects nature and human beings. The porous nature of MOFs result in large surface areas, enabling the material's capacity for adsorption [10]. The first MOF ($CoC_6H_3(COOH_{1/3})_3(NC_6H_5)_2.2/3$ NC_6H_5) was reported in 1995 by Yaghi O. M et al. and was shown to have higher surface areas compared with zeolites and carbonaceous materials [115]. Substantial research has been carried out in the field of water purification to show that adsorption is a favorable technique for the removal of pollutants from wastewater as it involves a simple technology at low cost, requiring only adsorbents. For the arsenic removal, natural adsorbents were used in the earlier days but when comparative studies were conducted between natural and laboratory synthesized adsorbents, the laboratory synthesized adsorbents were found to be significantly superior [67]. Nowadays, MOFs are used extensively to treat contaminants from water through adsorption and are effective in eliminating heavy metals [104].

MOFs are materials that are formed from the nodes through coordination bonds, an area where several adsorption studies are carried out by researchers [48]. While using the adsorption technique, it is difficult to separate the nanoparticles from water post-treatment. This problem was overcome when composites of UiO-66 along with activated carbon were used with excellent outcomes [95]. Several direct and indirect factors play a part in arsenic elimination, and optimization of these factors can lead us to excellent results in the treatment of arsenic. Da Pang and coworkers raised the possibility of using cotton fiber decorated with MOF MIL-88A(Fe) for removing both organic and inorganic arsenic pollutants from wastewater [73]. MIL-88A(Fe) synthesized by post-synthetic method (MC-1) was coated onto the cotton fiber. Figure 6 shows the adsorption capacity of this MOF toward arsenite and arsenate species at different pH conditions ranging from 4.0 to 12.0 in which the adsorption decreases with an increase in pH [73]. Thus adsorption characteristics of arsenic solution are

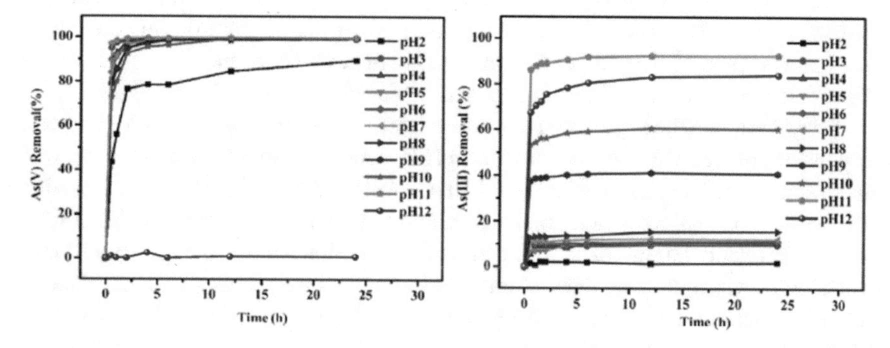

Fig. 6 Metal organic framework MIL-88A(Fe) used for removing As(V) and As(III) under different pH conditions. The maximum removal efficiency of 92.6% for both As(V) and As(III) was achieved at pH 11. Reprinted with the permission of Elsevier from Pang et al. [73]

greatly influenced by the pH as the surface properties and anionic species of the adsorbent are dependent on it.

The removal efficiencies of As(III) were found to be considerably less under acidic and mild alkaline conditions. But at higher alkaline conditions of around pH 11, the removal efficiency of MIL-88 A (Fe) was 92.6%. Furthermore, studies have revealed that electrostatic forces are not major factors determining the adsorption of MIL-88 (Fe). The key factor for treating wastewater and elimination of arsenic from contaminated water can be explained via adsorption kinetics. The adsorption kinetics and isotherm can be explained through pseudo-second-order kinetic model and Langmuir isotherm model, respectively. The maximum adsorption capacities of MIL-88A(Fe) toward As(III) and As(V) were 126.5 and 164.0 mgg^{-1}, respectively.

The synthesis of ZIF-8 was done via solvothermal method at 25° C and used in the elimination of arsenic from water [42]. The adsorbent here is ZIF-8 nanoparticles and it has a surface area of 1063.5 m^2g^{-1}. The Langmuir model fits well for the adsorption and the kinetics which showed second order. The Langmuir model showed that the highest adsorption capabilities of As(III) and As(V) were found to be 49.9 mgg^{-1} and 60.03 mgg^{-1}, respectively. The adsorptive capacities of MOF ZIF-8 at low concentrations of arsenic and its removal efficiency were also determined. For the ZIF-8, As(V) concentration decreases rapidly to 2.8 μgL^{-1} from 100 μgL^{-1} for 0.06 gL^{-1} of ZIF-8. From the initial concentration, 100 μgL^{-1} As(III) concentration can only be reduced to 73 μgL^{-1} even at a high dose of 0.2 gL^{-1}. This MOF is found to be stable at both neutral and basic conditions. While SO_4^{2-} and NO_3^- did not exert a considerable influence on the adsorption, PO_4^{3-} and CO_3^- were found to significantly inhibit the adsorption of arsenic species. Here, arsenic forms a complex with hydroxyl group on the adsorbent (ZIF-8) leading to better removal of arsenic through adsorption mechanism. Figure 7 explains the adsorption isotherm of As(III) and As(V) using ZIF-8 [42].

Synthesis of MOF MIL-53(Fe) was carried out through solvothermal method at room temperature [106]. Kinetics and adsorption isotherm-based studies were carried out to evaluate the efficacy of MIL-53(Fe) to remove As(V). The capacity to adsorb As(V) at different initial concentrations (5, 10, 15 mgL^{-1}) was investigated. The results showed an increase in the adsorption rate during the first 60 min attributed to a large number of free adsorptive sites and high As(V) concentration gradient. At higher concentrations, the equilibrium was achieved only after 90 to 120 min.

The pH is again the key factor in arsenic adsorption, and maximum As(V) removal efficiency rate of 99% was achieved at a pH 5, which fell to 87% when the pH was increased to 11. The best efficiency of MIL-53 (Fe) to adsorb As(V) was achieved at pH less than 6.9, whereas in an aqueous solution, As(V) is dominated by $H_2AsO_4^{3-}$, an anionic ligand that interacts with centered Fe^{3+} cations in the MIL-53 (Fe) framework via Lewis acid–base interactions. Additionally, in aqueous solutions, electrostatic interactions take place between Fe^{3+} cation and $H_2AsO_4^-$. The above two interactions combine to provide an improvement in the As(V) adsorption capacity of MIL-53 (Fe). Here, the MOF showed an adsorption capacity of 21.27 mgg^{-1} on As(V) in an aqueous solution compared with As(III).

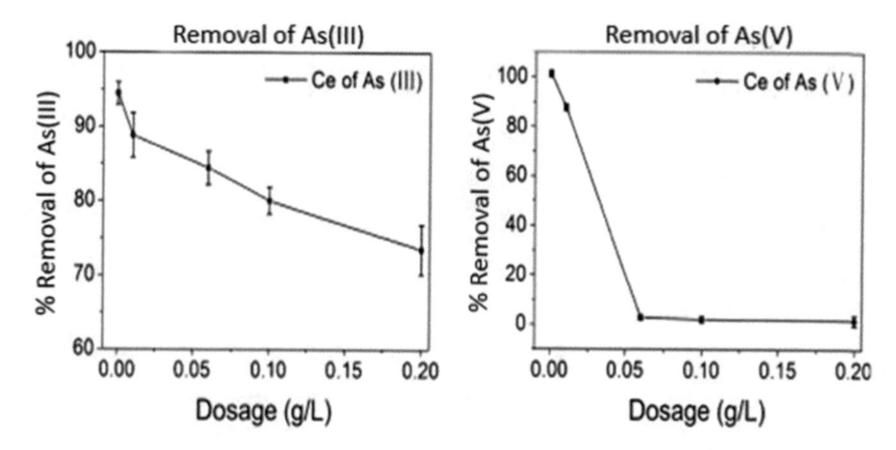

Fig. 7 Different sorbent dosages of ZIF-8 nanoparticles used for the removal of As(V) and As(III). As(III) concentration decreases to a lower value at a high dosage of 0.2 gL^{-1} of ZIF-8, whereas As(V) concentration decreases rapidly at a lower dosage of 0.05 gL^{-1} of ZIF-8. Reprinted with the permission of Elsevier from [42]

Folens K and others used a selective MOF adsorbent Fe_3O_4@MIL-101, synthesized via solvothermal technique for the elimination of arsenic species in water. Here MIL-101(Cr) acts as a host for Fe_3O_4 nanoparticles [24]. The adsorbent used is a hybrid nanomaterial of MIL-101, which shows great affinity toward both forms of arsenic. In this study, the As(V) adsorption efficiency was 99 and 97.4% for Fe_3O_4@MIL-101 and parent MOF, respectively. However, for As(III), the hybrid MOF exhibited a good removal efficiency of 94.7% compared with the parent MOF (22.8%). Adsorption capacities were found to be 121.5 for As(III) and 80.0 mgg^{-1} for As(V) at pH 7. Generally, water contains other ions like Ca^{2+}, Mg^{2+}, PO_4^{3-}, etc., and there is a chance of interference from these ions during adsorption, resulting in a decreased adsorption efficiency. However, these ions didn't show any interference to affect the removal efficiency and selectivity of arsenic for this hybrid MOF. This study explains the use of MOFs as hosts to make hybrid material having a high affinity for arsenic even when other ions in water bodies are present.

Fe-based MOF (MOF MIL-88A) micro rods were synthesized hydrothermally by Wu. H et al. and used it for the elimination of arsenic from water [113]. The arsenic removal efficiency was found to be inversely related to the initial concentration of the arsenic solution. The arsenic removal efficiency of MOF MIL-88A was studied for different concentrations of As(V) and is shown in Fig. 8 [113]. A pH study showed a maximum adsorption capacity of 145 mgg^{-1} at pH 5. The arsenic removal ratio reaches 98.72 and 94.60% within 20 min at initial concentrations of 10 and 20 mgL^{-1}, respectively, thus proving to be an excellent adsorbent at low concentrations of arsenic. The results indicate that arsenic removal depends on pH. Lewis acid–base interactions take place between cationic and anionic species, i.e. Fe^{3+} and H_2AsO^{4-}. Here, under acidic pH, arsenic is present as an anionic species. The arsenic adsorption capacity of MIL-88 A is high compared with other adsorbents due to the presence

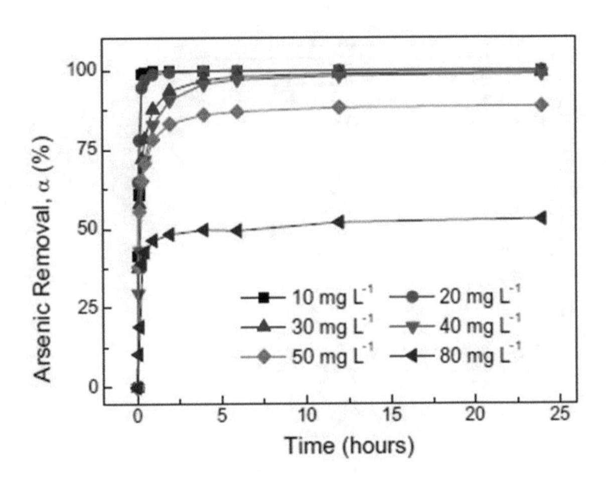

Fig. 8 Kinetic data for arsenic removal using various initial concentrations of metal organic framework MIL-88A. The maximum removal efficiency of 98.72 and 94.60% is attained at 10 and 20 mgL^{-1}, respectively. Reprinted with the permission of Springer nature from Wu et al. [113]

of a large number of OH^- groups on the MIL 88A surface which can exchange with $H_2AsO_4^-$ in aqueous solution. This is useful for ion transportation into its inner regions leading to a greater ligand exchange between $H_2AsO_4^-$ and OH^- ions.

Arsenic removal from water was carried out using MOF-53(Al) synthesized by microwave-assisted technique [57]. Batch-wise experiments for adsorption kinetics studies exhibit an initial rate of 80% after 11 h when the pH was 6–9. MIL-53(Al) shows a rise in the adsorption capacity of 90% at a pH of 8. According to Langmuir adsorption isotherm study, the removal capacity was 105.6 mgg^{-1}. The presence of co-existing anions like Cl^-, F^-, NO_3^-, and SO_4^{2-} does not interfere in the adsorption efficiency of MOF-53(Al). Meanwhile, the presence of PO_4^{3-} showed a lower adsorption capacity because of the competition with As(V) for binding on the MOF. Huang Z et al. carried out arsenic elimination from water using regenerative adsorbent Zn(II) imidazole framework (Zn-MOF) via solvothermal technique and exhibited high performance and potential application in wastewater treatment, which is shown in Fig. 9 [36]. Batch-wise experiments were carried for the elimination of heavy metals at different pH (1–6). The removal rate of Zn-MOF for hazardous metal ions is less at lower pH, and the maximum uptake was 718 mgg^{-1} for AsO_4^{3-} (50 mgL^{-1}). The thermodynamic studies showed that the reaction is endothermic and can be reused for six cycles.

The hydrothermal technique was used to synthesize $CoFe_2O_4$ and $CoFe_2O_4$@MIL-100(Fe), and the application of mesoporous hybrid nanoparticles on arsenic removal was carried out. Here, simultaneous removal of inorganic arsenate and arsenite was done [117]. Here, inorganic arsenic, helped by its nanoscale and microporous character, shows a high adsorption capacity at a fast rate on the prepared MOF. A higher adsorption capacity of 114.8 and 143.6 mgg^{-1} for As(V) and As(III), respectively, after 2 min was observed for an initial concentration of 0.1 mgL^{-1}. The model platform for As(V) and As(III) uptake by UiO-66 derivative is interesting and provides an option to tune the properties of the p-benzene dicarboxylate derivative.

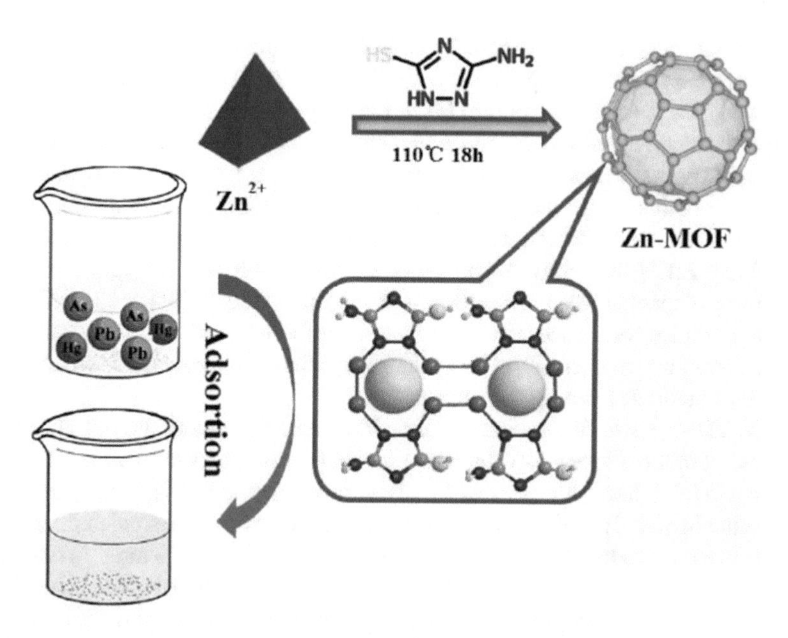

Fig. 9 Regeneration of the zinc metal organic framework adsorbent used in the removal of arsenic. Zinc metal and an organic linker are heated at 110 °C for 18 h to form the framework. Reprinted with the permission of ACS from [36]

MOF is synthesized by a post-synthetic method [2]. The results show that more than 90% As(V) oxyanion from a 5 ppm solution was removed after 30 min with the ratio of As(V): Zr_6 as 0.29:1 and overall removal was evaluated after 3 h. For a demonstration of As(III) uptake, neutral As(III) was captured using thiolated ligand sites where acetic acid and hydrochloric acid were the modulators. The As(III) uptake amount of 40 mgg^{-1} for 1:1 ratio of As(III) and Zr_6 was twice that obtained from AcOH-UiO-66$(SH)_2$. These results indicate that UiO MOFs can selectively capture neutral As(III) and anionic As(V) in which both recognition motifs can be incorporated for dual purposes.

Zr MOF (UiO-66) synthesized by solvothermal procedure showed high efficiency in removing arsenic by adsorption [34]. Here the As(V) and As(III) removal process was performed batch-wise to study the kinetics of adsorption. The removal of arsenic is an exothermic process as verified by chemisorption reactions. The effect of pH and the presence of co-existing ions like CO_3^{2-}, NO_3^-, Br^-, Cl^- on arsenic removal were studied. Better results were obtained at a pH of 9, and there was no remarkable effect on adsorption due to co-existing ions. The adsorption capacity of the prepared MOF was not affected even in the presence of co-existing anions. The BET surface areas for UiO-66 and UiO-66-(NH_2) were found to be 485.9 and 113.4 m^2g^{-1}, respectively. The arsenic removal efficiency was 91.83% for UiO-66, which was higher than UiO-66-NH_2 (73.47%).

A similar water-stable UiO-66 MOF was also designed by Wan P and others using phase inversion technique and studied for the arsenate removal in aqueous solution with good results [108]. Zirconium MOF UiO-66 was synthesized by Wang C et al. using a microwave-assisted method, and it showed excellent arsenic removal properties [109]. The As(V) removal was 303 mgg^{-1}. Elsewhere, zirconium organic framework was obtained from biomass-derived porous graphite nanocomposites carried out by in situ method, which was also evaluated for the adsorption of arsenic from water [72]. $ZrCl_4$ was used for the synthesis of UiO-66, and the concentration of both forms of arsenic was found reduced to less than 10 μg/L. Water stable Zr-MOF 66 nanoparticles were successfully prepared for the removal of trace quantities of arsenate from wastewater [99]. The MOF was synthesized through solvothermal and sonication methods from $ZrCl_4$ solution.

MOF ZIF-L, a two-dimensional zeolitic imidazole framework, was synthesized at room temperature through solvothermal process and used for the efficient removal of arsenic and other hazardous wastes like dyes, aromatics, etc. [69]. Adsorption study for arsenite included pH effect study, kinetics, and isotherms. The studies showed that pH 10 gave the maximum adsorption capacity for arsenite (43.74 mgg^{-1}) compared with pH 8 (15.50 mgg^{-1}). Another MOF named as La-MOF 900 derived from porous carbon and covered with La_2O_3 was synthesized via hydrothermal method [8]. The efficiency of As(V) removal using this MOF was 87%.

The oxidation of arsenite species to arsenate was attributed to the presence of porous carbon in La-MOF-900 while La_2O_3 components affected the removal of As(V) through adsorption. A hydrothermal or pyrolysis method was employed to synthesize a hollow sphere structured composite Ni/Ni@C400 and achieved an As(V) removal of 454.9 mgg^{-1} [65] The synthesis of lanthanum-based MOFs was studied by Prabhu and coworkers using solvothermal technique [79]. Different linkers like La-benzoic acid, La benzene dicarboxylic acid, La benzene tricarboxylic acid were used with $La(NO_3)_2.6H_20$ at 120 °C for 24 h. The As(V) removal efficiency was found to be 10 mgg^{-1}.

Li Z et al. used acetate modified yttrium-based metal organic framework MOF-76(Y) and MOF-76(Y) (Ac) and studied the adsorption behavior on arsenic [59]. The synthesis was carried out through microwave-assisted method, and the adsorption reaction was carried out at pH 10 using 1.12 mmol of MOF-76 (Y) (Ac) and 0.084 g of MOF-76(Y) at 80 °C for 24 h. The removal rate of As(V) was found to be 201.46 mgg^{-1}, and the results indicate the potential of MOF 76 (Y) (Ac) to remove highly concentrated arsenic content from wastewater.

10 Mechanistic Insight on the Adsorption of Arsenic by Metal Organic Framework

To have better clarity on the arsenic removal, it is important to have a good understanding of the nature of interactions taking place between arsenic and the adsorbent

[5, 102]. Generally, there are two processes that occur during adsorption, adsorption on the surface, and adsorption inside the pores of the adsorbent. In porous materials, both processes can take place whereas for crystalline adsorbents, adsorption on the surface alone takes place [15, 20].

Adsorption of toxic elements on MOFs can occur either in the modified or unmodified forms based on surface area and porosity. Physisorption and chemisorption are both involved in the adsorption of toxic metals on MOFs. Figure 10 represents the general mechanism of adsorption, which includes electrostatic and coordination interactions [82]. Generally, the mechanism for adsorption using MOFs can be explained based on acid–base interaction, π–π interaction [61], ion exchange [88], and coordination [12]. Previous literature shows that unmodified MOFs generally have a lower adsorption capacity compared with pristine or composite modified MOFs. Different characteristics of MOFs are influenced by the adsorption capacity of both adsorbent and adsorbate. Hence, the original mechanism for adsorption is comparatively complex and differs for various adsorbents. Thus a molecular level investigation is essential to understand the exact mechanism [82].

The adsorption mechanism is homogenous for $CoFe_2O_4$ and $CoFe_2O_4$@MIL-100(Fe) toward both forms of arsenic [117]. The FTIR spectrum revealed the stretching frequencies of Fe-O-As group. In this mechanism, the monodentate attachment of the deprotonated As(V) and As(III) is explained using X-ray absorption spectroscopy data. The –OH groups on the hybrid surface were exchanged by the

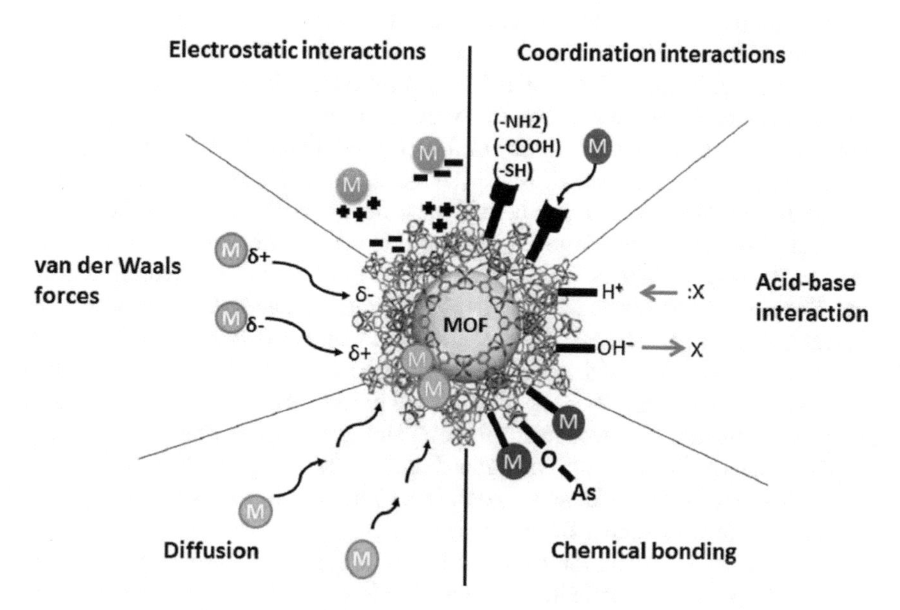

Fig. 10 The general mechanism for the adsorption process using metal organic framework. The adsorption generally occurs through diffusion, chemical bonding, acid–base interaction, coordination interaction, electrostatic interaction, and van der Waals forces. Reprinted with the permission of RSC Advances from [82]

deprotonated inorganic arsenic species through an exchange of hydroxyl ions. The distribution coefficient of inorganic arsenic species has a telling effect on the efficiency and effectiveness of adsorption of inorganic arsenic. Equation (1–3) gives the stepwise deprotonation of As(V).

$$H_3AsO_4 \rightarrow H_2AsO_4^- + H^+ \tag{1}$$

$$H_2AsO_4^- \rightarrow HAsO_4^{2-} + H^+ \tag{2}$$

$$HAsO_4^{2-} \rightarrow HAsO_4^{3-} + H^+ \tag{3}$$

Through hydrothermal exchange with $H_2AsO_4^-$ and $HAsO_4^{2-}$, arsenate is more to be adsorbed on hybrid. Since H_3AsO_3 is a tribasic weak acid, arsenite shows a lower hydroxyl exchange capability than arsenate. A natural H_3AsO_3 was adsorbed on the adsorbent via hydrogen bonding to give a structure where the surface of $CoFe_2O_4$ was covered with hydroxyl groups. Changes were apparent in the surface charge when different groups like $–FeOH_2^+$, –FeOH to $–Fe(OH)_2^-$ and even $–Fe(OH)_3^{2-}$ were present with respect to change in pH. A sharp fall in the zeta potential value of $CoFe_2O_4$@MIL-100(Fe) immediately after the adsorption of arsenate species is convincing evidence for strong specific adsorption and the formation of inner complexes on the surface of the adsorbent. The mesoporous nature of the hybrid adsorbent enables the adsorbed arsenite to penetrate deep into the pores, thus increasing the efficiency of the MOF. Therefore, this hybrid adsorbent was used to evaluate the amount of inorganic arsenic present in natural water sources. Figure 11 explains the mechanism of adsorption of As(III) on $CoFe_2O_4$@ MIL-100(Fe) [117].

Audu and coworkers used Zr MOF (UiO-66) and its analogs to demonstrate a capture-and-release mechanism where organic linkers were tuned to isolate heavy metal ions. The metal cluster nodes were used to capture the anionic arsenate ions whereas the organic linkers were functionalized to bind with the neutral arsenite ions. The complementary action was, thus, used to remediate water from both types of arsenic species [2]. The MOF was synthesized with hexazirconium oxo hydroxo groups as cluster nodes, and the thiol groups were the linkers. The high stability of this MOF helps in its reusability and application in the removal of As(III) from anaerobic groundwater stream. The mechanism proposes strong interactions between the nodes of UiO-66 and $[As(V)O_4H_{3-n}]^{n-}$ oxyanion based on the strong coordination between the $Zr_6O_4(OH)_4$ cluster nodes and phosphonates/phosphates, which are isostructural to arsenates and have similar Bronsted basicities.

Moreover, the incorporation of -SH group containing 1,4 benzene dicarboxylic acid ligands (p-dithiol terepthalic acid) into UiO-66 enabled binding with the neutral $[As^{III}(OH)_3]_n$ species, quite similar to the arsenophilicity exhibited by sulfur-bearing enzymes and chelators with -SH groups. The efficiency of these paired binding features is excellent when these sites are made easily accessible by either increasing the pore size of MOF or by decreasing the particle size of MOF nanocrystal. Figure 12

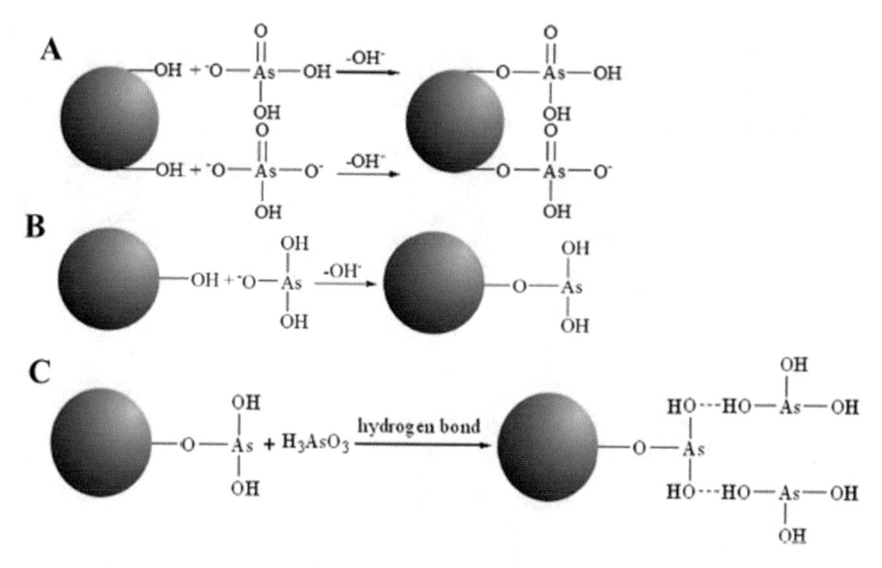

Fig. 11 The mechanism of the substitution of –OH group on $CoFe_2O_4$@MIL-100(Fe). **A** and **B** show the arsenic species adsorbed on the hybrid metal organic framework where substitution of –OH group has taken place by deprotonation of inorganic species. **C** shows the adsorption of H_3AsO_3 on the hybrid via hydrogen bonding to form a multilayer structure for As(III). Reprinted with the permission of Springer nature from [117]

represents a schematic representation of anionic arsenate binding at the nodes and neutral arsenite binding with the linkers [2]. The results show the binding of both the arsenic species into $Zr_6(O)_4(OH)_4$ to be reversible where it is designed as a regenerable/reusable adsorbent for capturing pollutants or toxic agents in aqueous condition.

Adsorption study of arsenic on Zr MOF shows that the bond between Zr and O plays a major role in the arsenic removal from wastewater. Spectroscopic studies like X-ray absorption near edge structure (XANES) were used to establish the oxidation state of arsenic anchored on UiO-66. The As K-edge XANES results indicate that the As(III)-UiO-66 adsorption edge is same and in good agreement with arsenite and arsenate spectra. The first As-O coordination was restored with oxygen atom in both As(III)-UiO-66 and As(V)-UiO-66. The coordination shell of As(III)-UiO-66 in Zr-As is surrounded by Zr atoms with the coordination number of 1.61 confirming the presence of bidentate binuclear complexes. On the other hand, As(V) predominantly forms monodentate mononuclear complexes, as supported by other studies. In the present study, the As(V) adsorption takes place on zirconyl, whereas 9.10 Zr atom donates As-Zr shell of As(V) UiO-66 at 2.95 Å. These data do not agree with monodentate mononuclear complex at (3.5 Å). Therefore, the bidentate mononuclear complexes are formed from the contribution of As-Zr. Figure 13 explains the binding modes of arsenite (a) and arsenate (b) to the hexanuclear Zr cluster of UiO-66 [34].

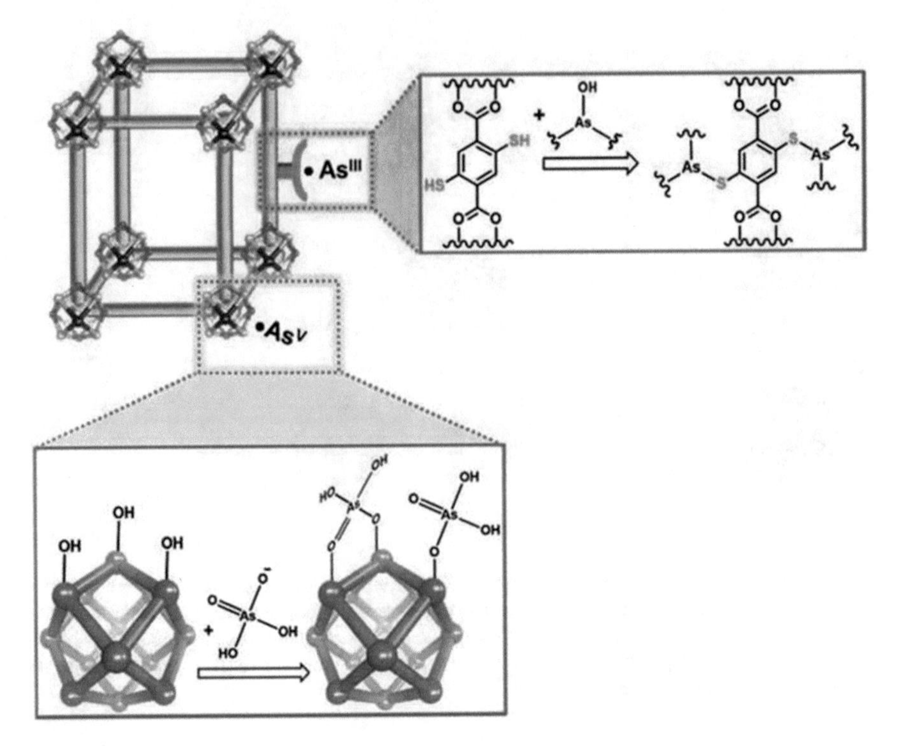

Fig. 12 The mechanism of arsenate and arsenite on modified metal organic framework coordinate showing the anionic As(V) groups at nodes, and As(III) acting as a binding agent with the linker. Reprinted with the permission of RSC Advances from [2]

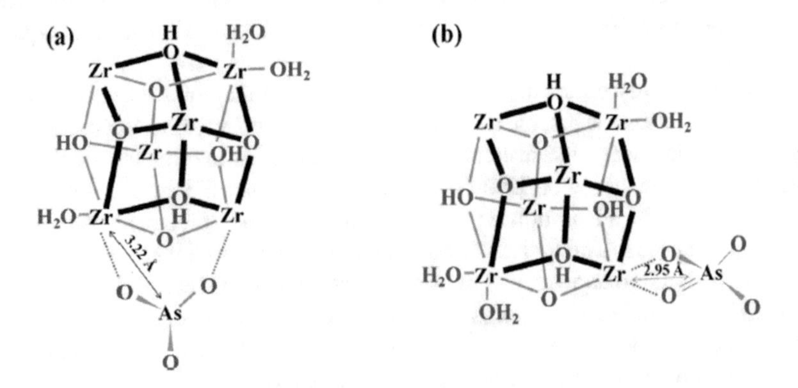

Fig. 13 Binding modes of arsenite and arsenate species to the Zr cluster of UiO-66 showing (a) the coordination of As-O restored with oxygen atoms and (b) As(V) predominantly forming a monodentate mononuclear complex. Reprinted with the permission of Elsevier from [34]

MOF ZIF-L is another promising and cost-effective adsorbent for the removal of arsenite from wastewater [69]. In the proposed mechanism, arsenite adsorption on ZIF-L involves a water-assisted breaking of a part of Zn-N in ZIF-L's structure to form activated hydroxyl Zn(Zn-OH) and protonated nitrogen atoms such as $C = NH^+$ and $C\text{-}NH_2^+$. These sites provide the positive charges to electrostatically attract the negatively charged As(III) ions. In the presence of arsenite solution, ZIF-–L will form Zn-O-As bonds, which is the main transition group for arsenite adsorption. Thus, the proposed mechanism suggests that the arsenite adsorption on ZIF L occurs both physically and chemically based on the hydroxyl substitution and electrostatic interactions.

MOF UiO-66, which is stable in water, has been synthesized as an adsorbent for removing arsenic from water [109]. The study reports the outstanding results as an adsorbent for water remediation and also provides insight into the application of MOFs. Fourier transform infrared spectroscopy analysis showed a significant shift in the band at 813 cm^{-1} pertaining to the Zr-O-As group showing the binding of arsenic on the adsorbent UiO-66. Another peak at 865 cm^{-1} and 660 cm^{-1} shows As-O and As-OH bond confirming the presence of arsenic complexes inside the UiO-66 framework aided by the formation of Zr-O-As coordination bond. In the UiO-6 unit cell, there are two types of linkages, namely, Zr-O(μ3)Zr that connects the Zr centers, and Zr-O-C that links Zr and benzene dicarboxylate. Studies reveal that the hydroxyl groups on the surface of metal oxide are mainly responsible for the adsorption of arsenic. Thus, the adsorption takes place primarily on the μ3-O sites of UiO-66, especially the protonated oxygen connected to Zr in a unit of Zr_6 cluster to attract equivalent arsenate species. The adsorption can also take place by interchanging some benzene dicarboxylate ligands leading to an arsenic complex in the framework. Methods such as reducing particle size and preparing hierarchically ordered materials/core shells have been adopted to increase the adsorbent's surface area and thus its efficiency. Figure 14 gives an illustration of adsorption mechanism of arsenate onto UiO-66 [109].

A cost-effective adsorbent, Fe-Co-based MOF-74 was successfully produced via solvothermal process and used for remediation of arsenic in water [96]. The hydroxyl and metal–oxygen bonds play a major role in the adsorption process besides the usual electrostatic interactions. Changes in the position and intensity of the peak in the infra-red spectrum reveal the interactions between the arsenic species and the functional groups on the adsorbent. The decrease in the intensity of the spectral peaks shows the participation of the functional groups like C-O, C = O, M-OH, and M-O in the MOF. After the adsorption of arsenic, a new peak appears around 800 cm^{-1} indicating As-O stretching vibration due to the surface adsorption of arsenic.

X-ray photoelectron spectroscopy studies carried out before and after adsorption of the arsenic species helped to determine the surface composition and to understand the mechanism of adsorption. The spectra reveal the binding energy of As(III) and As(V) at 44.3 and 45.2 eV, respectively. After the uptake of arsenic, the peak of C–OH was absent due to the substitution of hydroxyl group with $H_2AsO_3^-$ and $H_2AsO_4^-$. The peaks of M-O and M-OH drastically increased from 18.9 to 34.9% and 13.3 to 33.1%, respectively, post-adsorption. A similar result was also seen for As(V). These

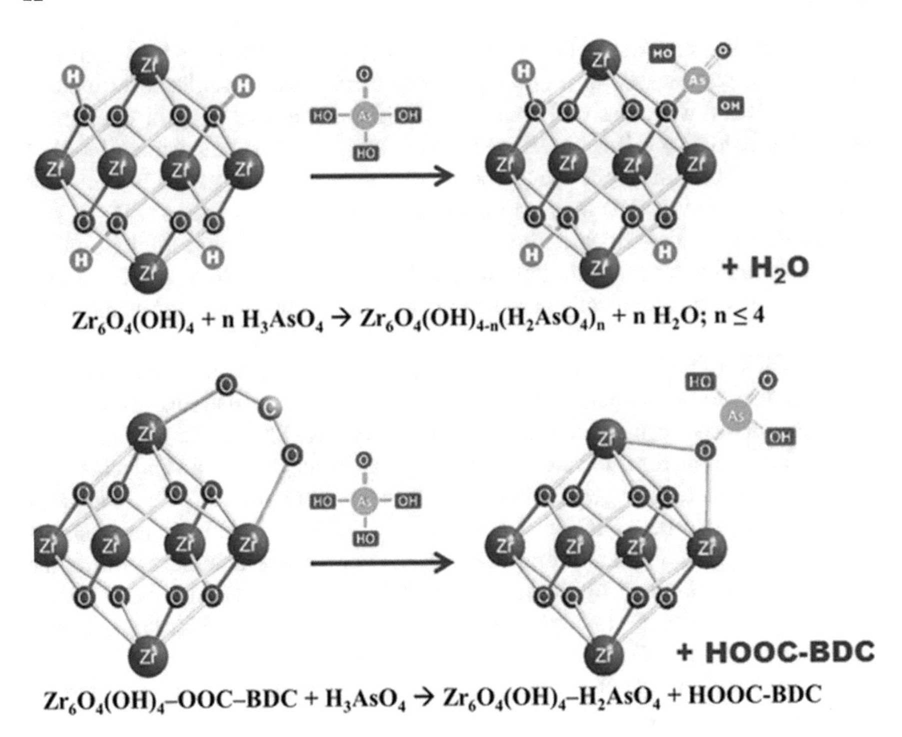

Fig. 14 The mechanism proposed for the adsorption of arsenate onto UiO-66. The adsorption of arsenate onto UiO-66 takes place through coordination at hydroxyl group and the organic linker (benzene dicarboxylate). For clarity, H atoms are not shown in the cluster, and OOC is shown as a part of the ligand that is bridged to another Zr_6 cluster. Reprinted with the permission of Springer nature from [109]

results explain the formation of a newly formed chemical bond between the metal and oxygen, and the binding energies of both Fe and Co saw a shift after arsenic adsorption due to the chemical interaction between cobalt, iron, and arsenic. Thus, the key active sites like C-OH, M-O, and M-OH are involved in the adsorption process in Fe_2Co MOF-74 making it an excellent adsorbent for the removal of arsenic ions from water.

While the study of porous adsorbents to remove heavy metal contamination is of great interest and significance, adsorbents capable of efficiently removing arsenic are still scarce [120]. Crystalline Zn-MOF-74 has been effectively used for the removal of arsenic. Extensive characterizations have revealed the adsorption of NaH_2AsO_4 and NaH_2AsO_3. The IR spectra show a stretching frequency of As-O bond for pristine Zn-MOF-74 at 881 cm^{-1} and 883 cm^{-1} indicating the presence of both the adsorbed As(V) and As(III) species. In the Zn-MOF-74 sample, ZnO bond shows a redshift from 807 cm^{-1} to 811 cm^{-1} for As(V) adsorbed sample and 813 cm^{-1} for As(III) adsorbed samples due to the presence of new coordination interactions such as Zn-O-As. After loading arsenic, there is a slight shift in the XPS peak, and the binding energies have decreased owing to the formation of Zn-O-As bond reverting to the original Zn-O-(H_2O) bond. Therefore, the adsorption mechanism obtained by IR and XPS studies reveals the Zn-O-As bond formation between the open metal sites $H_2AsO_4^-$ or $H_2AsO_3^-$ paving a new mechanism for the removal of arsenic. Table 1 gives the details of the different methods adopted for the MOF synthesis and their efficiency in arsenic removal.

11 Theoretical Study of Metal Organic Framework and Adsorption

Density functional theory (DFT) is a handy tool to theoretically examine the electronic and geometric properties of a broad range of molecules and complexes [51]. However, theoretical investigations supporting the adsorption capabilities of MOFs are not many [29, 75]. Tarboush and co-workers performed a density functional theory study to verify the results obtained from the experiment and also to explain the mechanism of adsorption on the surface of metals [99]. The calculations were completed by Plane-Wave self-consistent field code using Quantum Expresso Software package for exchange co-relation energy with the Perdew Burke-Ernzerh of functional theory. The lattice parameter was fixed at $1 \times 1 \times 1$ supercell and Monkhorst pack k-point grid was set at $1 \times 1 \times 4$. Adsorption sites were investigated, and zinc atom was determined to be the primary adsorption site, and the secondary adsorption was found to be electrostatically bound to primary adsorbate. In order to optimize nuclear coordinate, Broyden Fletcher Goldfarb Shanno algorithm was used. Crystal structure for MOF Zn-74 was determined, and the binding energies were calculated using Eq. (4).

$$\Delta E = E_{MOF-74/A} - (E_{MOF-74} + E_A) \quad (4)$$

where E_A and E_{MOF-74} are the energies of adsorbent and adsorbate, respectively, and $E_{MOF-74/A}$ represents the energy of the system after adsorption. Density functional theory calculations for primary and secondary adsorption energies were calculated, and the presence of multiple layers of water surrounding the open metal sites by the arsenate entity was established. MOF Zn-74 shows a possible application for the purification of industrial wastewater containing arsenate species using the above studies.

Li et al. have investigated the As (V) removal efficiency of zinc-metal organic framework (Zn-MOF-74) from the aqueous medium [58]. The energy level diagram obtained from the density functional theory studies clearly confirms the adsorption of water and As(V) on Zn-MOF-74. The thermodynamic feasibility of adsorption was well studied using density functional theory calculations. Various parameters were optimized to support the high adsorption efficiency of the prepared system. Normally adsorption is exothermic, but the thermodynamics of adsorption in this particular study followed an endothermic pattern, which occurred mainly because of the replacement of adsorbed water molecules in the pores of MOF by arsenate

Table 1 The different method used for metal organic framework preparation and its application in arsenic removal

MOFs	Method of preparation	Reaction conditions	Adsorption rate for removal (mgg^{-1})		Ref
			As(III)	As(V)	
ZIF-8	Solvothermal	20 mg –MOF t = 24 h	49.49	60.03	[42]
MOF-53(Al)	Microwave assisted	1.245 g—Terepthalic acid 4.220 g—$Al(NO_3)_2.9\ H_2O$ pH = 8, T = 130 °C, t = 24 h		105.6	[57]
MIL-53- (Fe)	Solvothermal	pH = 5, T = 298 K		21.27	[106]
Zr—UiO -66	Microwave-assisted	pH = 2, T = 120 °C, t = 48 h		303	[109]
CoFe2O4@MIL-100(Fe) hybrid	Hydrothermal route	148.7 mg—$CoCl_2.6H_2O$	143.6	114.8	[117]
MIL-88A- microrods	Hydrothermal	0.5406 g of $FeCl_3.6H_2O$ pH—3–7, T = 70 °C, t = 12 h		105.6	[113]
2D ZIF- L	Solvothermal	T = 70 °C, pH = 10	43.74		[69]
Zr- MOF	Solvothermal	pH = 9.2	205.0	68.21	[34]
Zn—MOF-74	Solvothermal	0.19 g - $Zn(NO_3)_2.6H_2O$ T = 120 °C, t = 24 h		99	[99]
ZrO_2- MOF	Electrospinning	T = 900 °C, t = 2 h	28	106	[92]
La(1,3,5) BDC	Solvothermal	743 mg—$La(NO_3)_3.6H_2O$ pH = 2, T = 120° C, t = 24 h		10	[79]
UiO-66	Phase inversion	$ZrCl_2$—2.37 g pH = 4.7, T = 120° C, t = 48 h		267	[108]
La- MOF-900	Hydrothermal	1.082 g $La(NO_3)_3.6H_2O$ pH = 7	As(V)—87%		[8]

(continued)

Table 1 (continued)

MOFs	Method of preparation	Reaction conditions	Adsorption rate for removal (mgg^{-1})		Ref
			As(III)	As(V)	
NiO/Ni@C	Pyrolyzing/hydrothermal route	$Ni(NO_3)_2.6H_2O = 216$ mg pH = 3, T = 150° C, t = 15 h		454.94	[65]
Yt—MOF -76	Microwave assisted	MOF-76(Ac)- 0.428 g $Y(NO_3)_3.6\ H_2O$—0.084 g pH = 10, T = 80° C, t = 24 h	As(V)—201.46		[59]
MIL-88 Fe (A)	Post-synthetic and in situ method	1.0 g of MOF, t = 8 h	49.49	60.03	[73]
Zr-MOF	In situ method	$ZrCl_2$—0.2330 g pH = 7, T = 120° C, t = 24 h	< 10 µg/L	< 10 µg/L	[72]
Z(II) imidazole	Post-modification method	10 mg—$Zn(NO_3)_2.4H_2O$ pH = 6, T = 120^0 C, t = 24 h		718	[36]
Fe_3O_4@MIL-101	In-situ synthesis	pH = 7	121 0.5	80	[24]

ions. The observed experimental results were well supported theoretically by density functional theory calculations. Adsorption energies and geometry calculations too were performed using density functional theory studies.

Shao P et al. reported enhanced adsorption of zirconia using theoretical study and quantum calculation leading to a profound understanding of the structural features and adsorption of As(III) or As(V) [92]. The defect densities of the synthesized ZrO_2 were analyzed using density functional theory calculations. This model was constructed by tweaking the crystal structure and the lattice parameter of cubic ZrO_2 is a = b = c = 5.07 A° and $\alpha = \beta = \gamma = 90°$. Later the supercell with six atomic layers was set at (1 × 2 × 3), which was separately constructed by vacuum layer of 1.44 A°. Spin polarization density functional theory studies were based on generalized gradient approximation—Vosko, Wilk, and Nusair—Beckee Perdew (GGA-VWN-BP), and adsorption energy (E_{ad}) was calculated. Adsorption is spontaneous when E_{ad} value is negative, and non-spontaneous when the E_{ad} value is positive. Figure 15 represents the density functional theory calculation of As(III) adsorption [92]. A higher level of lattice defect is observed in UiO-66-SH-A with a higher adsorption capacity, which is proven from theoretical results.

p-arsanilic acid is an organoarsenic pollutant comprising of both organic and inorganic groups. An adsorbent with high adsorbing capacity is required to effect

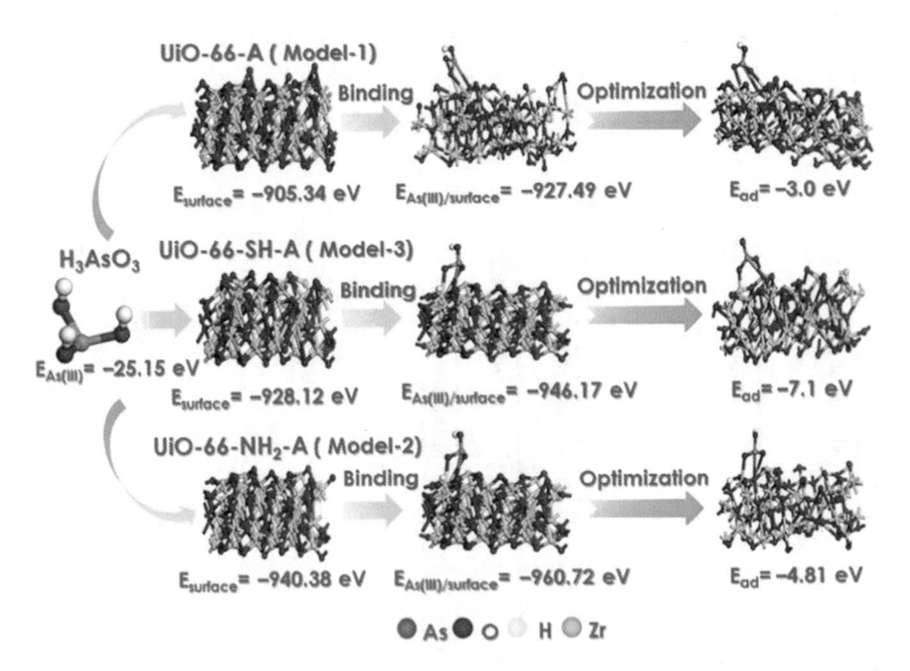

Fig. 15 Density functional theory calculations on the adsorption of As(III) on the ZrO_2 with different lattice defects. The As(III) adsorption was carried out using three models, all of them showing the adsorption of As(III) onto UiO-66. The binding and optimization take place on the surface, and the adsorbed values are found to be decreasing with different lattice defects. Reprinted with the permission of ACS from [92]

the removal of p-arsanilic acid. Tian C and others synthesized an amine-modified UiO-67 MOF in which the affinity for the material toward p-arsanilic acid was twice that of pristine UiO-67 [101]. By using Vienna ab initio simulation package (VASP), DFT calculations are carried out. Plane cut-off energy was set at 500 eV, which was examined and found suitable for UiO 67.

The scalar relativistic effects were incorporated into the effective core potential via explicit mass velocity and Darwin corrections. Hellmann–Feynman forces were used for ion relaxation, which followed conjugate–gradient algorithm. Geometry optimization was carried out until the total energy converged to within 1×10^{-4} eV when the forces were less than 0.02 eV A^{-1}. Using the Poisson–Boltzmann, solvation corrections were calculated and dielectric constant for water was $\epsilon = 78.4$. The binding energy of p-arsanilic acid with UiO-67 was calculated using Eq. (5).

$$E_{ads} = E_{sc} - E_{sc}^{'} \tag{5}$$

where E_{ads} is the energy of adsorption, E_{sc} is the energy of the optimized binding complex, and $E_{sc}^{'}$ corresponds to the energy of the system when kept apart. The density functional theory calculations showed excellent results in the adsorption of p-arsanilic acid from water.

Density functional theory calculations were also used to study the binding mechanism of arsenate species with the MOFs and the exact nature of the interactions between the adsorbent(UiO-66) and the arsenate species[108]. The structure of the unit cell represents one cluster model with 12 coordinated zirconia nodes. Energy calculation, structure of arsenate molecules, and cluster model were upgraded geometrically using generalized gradient approximation. It was found that the adsorption occurs on solid or water interface. All forms of arsenic, viz., H_3AsO_4, $H_2AsO_4^-$, $HAsO_4^{2-}$, and AsO_4^{3-} were placed in a cluster model to learn the structure of complexes. The binding energy of each complex (E_b) is given by Eq. (6).

$$E_b = E_{UiO\text{-}66\text{-}As} - (E_{UiO\text{-}66} + E_{As}) \tag{6}$$

where $E_{UiO\text{-}66\text{-}As}$ is the total energy of adsorption complexes in equilibrium state, $E_{UiO\text{-}66}$ is the independent total energy of cluster model of UiO-66 and E_{As} is the independent energy of arsenate molecule. The adsorption process of arsenate molecule with regard to UiO-66 sites shows a negative E_b value, which is an exothermic reaction, and for arsenic uptake of oxygen in Zr-O-C, it is the most favored one. Moreover, higher binding energy shows a large adsorption process. Density functional theory simulation shows sorption can occur on various sites on UiO-66 with oxygen in Zr-O-C, which is the most beneficial site for adsorption.

Adsorption of primary and secondary water molecules of Zn-MOF-74 was carried out and their open site energies were calculated to be -58 kJ mol^{-1} and -50 kJ mol^{-1}. The results obtained are in concurrence with the disclosed results [58]. Adsorptive removal of As(V) using MOF BUC-17 was reported recently. The thermodynamic parameters calculated unambiguously proved the spontaneous and exothermal nature of the adsorption process [74]. In general, density functional theory calculations are mainly used to study adsorption phenomenon at the atomic level [116]. Density functional theory calculations can be applied in the area of heavy metal using MOFs for remediation process to understand the favorable adsorption sites for heavy metals in the guest system and its most favorable configuration, exhibited by the guest system, adsorption energetics, and binding energy calculations. Density functional theory calculations are also useful in analyzing the pore size distribution of MOFs [53].

12 Conclusion

Heavy metals like arsenic are perpetual contaminants with severe effects on human health. According to the World Health Organization, arsenic contamination in water is a critical issue and can lead to various health problems at high concentrations. Arsenic removal is a challenging task, and the problem has been addressed using multiple techniques. While all methods have positive and negative aspects to them, adsorption technology, with its many advantages, is a major tool for controlling arsenic pollution. MOF-based materials, with their unique properties and large surface areas, show excellent performance for the removal of arsenic from wastewater through adsorption. The scope of utilizing MOFs for the removal of arsenic from wastewater has been substantially expanded due to the advancements made in the synthesis of novel MOFs. This review provides an insight into the removal of arsenic from wastewater, with different MOFs, thus alleviating, to some extent, the acute shortage of drinking water.

Going forward, MOFs will become a targeted area of study for the elimination of many hazardous substances such as arsenic and other heavy metals from contaminated water. The outstanding adsorptive capabilities of MOFs motivate the synthesis and development of newer materials with wider applications. Computational evaluation of MOFs is also being used nowadays to analyze and obtain the best results in the discovery of the materials for the elimination of heavy metals from water sources. While iron and zirconium-based MOF adsorbents have been extensively examined for the arsenic removal process, the years to come will see many more novel materials developed in the field of adsorption technology.

Acknowledgments The authors are thankful to CHRIST (Deemed to be University), Bangalore.

References

1. Arul P, Gowthaman NSK, John SA, Tominaga M (2020) Tunable electrochemical synthesis of 3D nucleated microparticles like Cu-BTC MOF-carbon nanotubes composite: Enzyme

free ultrasensitive determination of glucose in a complex biological fluid. Electrochim Acta 354:136673. https://doi.org/10.1016/j.electacta.2020.136673
2. Audu CO, Nguyen GT, Chang CY, Katz MJ, Mao L, Farha OK, Hupp JT, Nguyen ST (2016) The dual capture of AsV and AsIII by UiO-66 and analogues. Chem Sci 7(10):6492–6498. https://doi.org/10.1039/c6sc00490c
3. Bhadra BN, Vinu A, Serre C, Jhung SH (2019) MOF-derived carbonaceous materials enriched with nitrogen: Preparation and applications in adsorption and catalysis. Mater 25:88–111. https://doi.org/10.1016/j.mattod.2018.10.016
4. Biswas R, Sarkar A (2020) A two-step approach for arsenic removal by exploiting an autochthonous Delftia sp. BAs29 and neutralized red mud. Environ Sci Pollut 25(21):20792–20801. doi: https://doi.org/10.1007/s11356-020-10665-8
5. Bujňáková Z, Baláž P, Zorkovská A, Sayagués MJ, Kováč J, Timko M (2013) Arsenic sorption by nanocrystalline magnetite: An example of environmentally promising interface with geosphere. J Hazard Mater 262:1204–1212. https://doi.org/10.1016/j.jhazmat.2013.03.007
6. Cadiau A, Adil K, Bhatt PM, Belmabkhout Y, Eddaoudi M (2016) A metal-organic framework-based splitter for separating propylene from propane. Science 353(6295):137–140. https://doi.org/10.1126/science.aaf6323
7. Carson CG, Brown AJ, Sholl DS, Nair S (2011) Sonochemical synthesis and characterization of submicrometer crystals of the metal-organic framework $Cu[(hfipbb)(H_2hfipbb)_{0.5}]$. Cryst Growth Des 11(10): 4505–4510. doi: https://doi.org/10.1021/cg200728b
8. Chen C, Xu L, Huo JB, Gupta K, Fu ML (2020) Simultaneous removal of butylparaben and arsenite by MOF-derived porous carbon coated lanthanum oxide: combination of persulfate activation and adsorption. Chem Eng Trans 391. https://doi.org/10.1016/j.cej.2019.123552
9. Chen L, Xin H, Fang Y, Zhang C, Zhang F, Cao X, Zhang C, Li X (2014) Application of metal oxide heterostructures in arsenic removal from contaminated water. J Nanomater 793610. https://doi.org/10.1155/2014/793610
10. Chen W, Yan X (2020) Progress in achieving high performance piezoresistive and capacitive flexible pressure sensors. A review J Mater Sci Technol 43:175–188. https://doi.org/10.1016/j.jmst.2019.11.010
11. Chen YN, Chai LY, Shu YD (2008) Study of arsenic(V) adsorption on bone char from aqueous solution. J Hazard Mater 160(1):168–172. https://doi.org/10.1016/j.jhazmat.2008.02.120
12. Cheng K, Wu Y, Zhang B, Li F (2020) New insights into the removal of antimony from water using an iron-based metal-organic framework: Adsorption behaviors and mechanisms. Colloid Surf A 602 https://doi.org/10.1016/j.colsurfa.2020.125054
13. Costa ALG, Sarabia A, Zazo JA, Casas JA (2020) UV-assisted catalytic wet peroxide oxidation and adsorption as efficient process for arsenic removal in groundwater. Catal Today:1–7. doi: https://doi.org/10.1016/j.cattod.2020.03.054
14. Dastbaz A, Sabet JK, Moosavian MA (2019) Sonochemical synthesis of novel decorated graphene nanosheets with amine functional Cu-terephthalate MOF for hydrogen adsorption: Effect of ultrasound and graphene content. Int J Hydrog Energy 44(48):26444–26458. https://doi.org/10.1016/j.ijhydene.2019.08.116
15. Delgado CN, Martínez JG, Méndez JRR (2019) Modified activated carbon with interconnected fibrils of iron-oxyhydroxides using Mn^{2+} as morphology regulator, for a superior arsenic removal from water. J Environ Sci 76:403–414. https://doi.org/10.1016/j.jes.2018.06.002
16. Dhaka S, Kumar R, Deep A, Kurade MB, Ji SW, Jeon BH (2019) Metal–organic frameworks (MOFs) for the removal of emerging contaminants from aquatic environments. Coord Chem Rev 380:330–352. https://doi.org/10.1016/j.ccr.2018.10.003
17. Ding M, Cai X, Jiang HL (2019) Improving MOF stability: approaches and applications. Chem Sci 10:10209–10230. https://doi.org/10.1039/c9sc03916c
18. Dong R, Zhang Z, Tranca DC, Zhou S, Wang M, Adler P, Liao Z, Liu F, Shi SY, W, Zhang Z, Zschech E, Mannsfeld SCB, Felser C, Feng X, (2018) A coronene-based semiconducting two-dimensional metal-organic framework with ferromagnetic behavior. Nat Commun 9(1):1–9. https://doi.org/10.1038/s41467-018-05141-4

19. Du J, Zou G (2016) A novel microporous zinc(II) metal-organic framework with highly selectivity adsorption of CO_2 over CH_4. Inorg Chem Commun 69:20–23. https://doi.org/10.1016/j.inoche.2016.04.015
20. Du Y, Fan H, Wang L, Wang J, Wu J, Dai H (2013) α-Fe_2O_3 Nanowires deposited diatomite: highly efficient absorbents for the removal of arsenic. J Mater Chem A 1(26):7729–7737. https://doi.org/10.1039/c3ta11124e
21. Dubey RJ, Bajpai BAK (2016) Chitosan-alginate nanoparticles (CANPs) as potential nanosorbent for removal of Hg (II) ions. Environ Nanotechnol Manag 6:32–44. https://doi.org/10.1016/j.enmm.2016.06.008
22. Dzhardimalieva GI, Baimuratova RK, Knerelman EI, Davydova GI, Kudaibergenov SE, Kharissova OV, Uflyan IE, Zhinzhilo VA, Uflyand IE (2020) Synthesis of copper(II) trimesinate coordination polymer and its use as a sorbent for organic dyes and a precursor for nanostructured material. Polymers 12(5). https://doi.org/10.3390/POLYM12051024
23. Férey G (2008) Hybrid porous solids: past, present, future. Chem Soc Rev 37(1):191–214. https://doi.org/10.1039/b618320b
24. Folens K, Leus K, Nicomel NR, Meledina M, Turner S, Tendeloo GV, Laing GD, Voort PVD (2016) Fe_3O_4@MIL-101–A selective and regenerable adsorbent for the removal of as species from water. Eur J Inorg Chem:4395–4401. https://doi.org/10.1002/ejic.201600160
25. Furukawa H, Cordova KE, Keeffe MO, Yaghi OM (2013) The chemistry and applications of metal-organic frameworks. Science 341(6149). https://doi.org/10.1126/science.1230444
26. Gao Q, Xu J, Bu XH (2019) Recent advances about metal–organic frameworks in the removal of pollutants from wastewater. Coord Chem Rev 378:17–31. https://doi.org/10.1016/j.ccr.2018.03.015
27. Ge J, Guha B, Lippincott L, Cach S, Wei J, Su TL, Meng X (2020) Challenges of arsenic removal from municipal wastewater by coagulation with ferric chloride and alum. Sci Total Environ 725. https://doi.org/10.1016/j.scitotenv.2020.138351
28. Ghanbarian M, Zeinali S, Mostafavi A (2018) A novel MIL-53(Cr-Fe)/Ag/CNT nanocomposite based resistive sensor for sensing of volatile organic compounds. Sensor Actuat B-Chem 267:381–391. https://doi.org/10.1016/j.snb.2018.02.138
29. Giannozzi P, Baroni S, Bonini N, Calandra M, Car R, Cavazzoni C, Ceresoli D, Chiarotti GL, Cococcioni M, Dabo I, Corso AD, Gironcoli SD, Fabris S, Fratesi G, Gebau R (2009) A modular and open-source software project for quantum simulations of materials. J Phys Condens Matter 21(39). https://doi.org/10.1088/0953-8984/21/39/395502
30. Gu ZG, Li DJ, Zheng C, Kang Y, Wöll C, Zhang J (2017) MOF-Templated synthesis of ultrasmall photoluminescent carbon-nanodot arrays for optical applications. Angew Chem Int Ed 56(24):6853–6858. https://doi.org/10.1002/anie.201702162
31. Haldar D, Duarah P, Purkait MK (2020) MOFs for the treatment of arsenic, fluoride and iron contaminated drinking water. Chemosphere 251. https://doi.org/10.1016/j.chemosphere.2020.126388
32. Haque E, Jun JW, Jhung SH (2011) Adsorptive removal of methyl orange and methylene blue from aqueous solution with a metal-organic framework material, iron terephthalate (MOF-235). J Hazard Mater 185(1):507–511. https://doi.org/10.1016/j.jhazmat.2010.09.035
33. Hasanzadeh R, Moghadam, PN, Laleh NB, Sillanpää M (2017). Effective removal of toxic metal ions from aqueous solutions: 2-Bifunctional magnetic nanocomposite base on novel reactive PGMA-MAn copolymer@Fe_3O_4 nanoparticles. J Colloid Interface Sci 490. https://doi.org/10.1016/j.jcis.2016.11.098
34. He X, Deng F, Shen T, Yang L, Chen D, Luo J, Luo X, Min X, Wang F (2019) Exceptional adsorption of arsenic by zirconium metal-organic frameworks: engineering exploration and mechanism insight. J Colloid Interface Sci:223–234. https://doi.org/10.1016/j.jcis.2018.12.065
35. Hua M, Zhang S, Pan B, Zhang W, Zhang LLQ (2012) Heavy metal removal from water/wastewater by nanosized metal oxides: a review. J Hazard Mater 211–212:317–331. https://doi.org/10.1016/j.jhazmat.2011.10.016

36. Huang Z, Zhao M, Wang C, Wang S, Dai L, Zhang L (2020) Preparation of novel Zn(II)-Imidazole framework as an efficient and regenerative adsorbent for Pb, Hg and As Ions removal from water. ACS Appl Mater Interfaces 12(37):41294–41302. https://doi.org/10.1021/acsami.0c10298
37. Hung DQ, Nekrassova O, Compton RG (2004) Analytical methods for inorganic arsenic in water: a review. Talanta 64(2):269–277. https://doi.org/10.1016/j.talanta.2004.01.027
38. Ibrahim D, MD, Froberg B, Wolf A, Rusyniak DE, (2006) Heavy metal poisoning: clinical presentations and pathophysiology. Lab Med 26(1):67–97. https://doi.org/10.1016/j.cll.2006.02.003
39. Issa NB, Ognjanovic VNR, Jovanovic BM, Rajakovic LV (2010) Determination of inorganic arsenic species in natural waters-benefits of separation and preconcentration on ion exchange and hybrid resins. Anal Chim Acta 673(2):185–193. https://doi.org/10.1016/j.aca.2010.05.027
40. Jadhav SV, Bringas E, Yadav GD, Rathod VK, Ortiz I, Marathe KV (2015) Arsenic and fluoride contaminated groundwaters: A review of current technologies for contaminants removal. J Environ Manage 162:306–325. https://doi.org/10.1016/j.jenvman.2015.07.020
41. Jang M, Chen W, Cannon FS (2008) Preloading hydrous ferric oxide into granular activated carbon for arsenic removal. Environ Sci Technol 42(9):3369–3374. https://doi.org/10.1021/es7025399
42. Jian M, Liu B, Zhang G, Liu R, Zhang X (2015) Adsorptive removal of arsenic from aqueous solution by zeolitic imidazolate framework-8 (ZIF-8) nanoparticles. Colloid Surface A 465:67–76. https://doi.org/10.1016/j.colsurfa.2014.10.023
43. Jiao L, Seow JYR, Skinner WS, Wang ZU, Jiang HL (2019) Metal–organic frameworks: structures and functional applications. Mater Today Commun 27:43–68. https://doi.org/10.1016/j.mattod.2018.10.038
44. Kadhom M, Deng B (2018) Metal-organic frameworks (MOFs) in water filtration membranes for desalination and other applications. Appl Mater Today 11:219–230. https://doi.org/10.1016/j.apmt.2018.02.008
45. Kanel SR, Choi H (2017) Removal of arsenic from groundwater by industrial byproducts and its comparison with zero-valent iron. J Hazard Toxic Radioact Waste 21(3):1–7. https://doi.org/10.1061/(ASCE)HZ.2153-5515.0000349
46. Kang X, Fu G, Song X, Huo G, Si F, Deng X, Fu XZ, Luo JL (2019) Microwave-assisted hydrothermal synthesis of MOFs-derived bimetallic CuCo-N/C electrocatalyst for efficient oxygen reduction reaction. J Alloys Compd 795:462–470. https://doi.org/10.1016/j.jallcom.2019.04.325
47. Kang Z, Fan L, Sun D (2017) Recent advances and challenges of metal-organic framework membranes for gas separation. J Mater Chem A 5(21):10073–10091. https://doi.org/10.1039/c7ta01142c
48. Kartinen EO, Martin CJ (1995) An overview of arsenic removal processes. Desalination 103(1–2):79–88. https://doi.org/10.1016/0011-9164(95)00089-5
49. Khan NA, Hasan Z, Jhung SH (2013) Adsorptive removal of hazardous materials using metal-organic frameworks (MOFs): a review. J Hazard Mater 244–245:444–456. https://doi.org/10.1016/j.jhazmat.2012.11.011
50. Kinuthia KG, Veronica N, Beti D, Lugalia R, Wangila A, Kamau L (2020) Levels of heavy metals in wastewater and soil samples from open drainage channels in Nairobi, Kenya: community health implication. Sci Rep 10(1):1–13. https://doi.org/10.1038/s41598-020-65359-5
51. Kohn W, Sham LJ (1965) Self consistent equations including exchange and correlation effects. Phys Rev 140. https://doi.org/10.1103/PhysRev.140.A1133
52. Lagashetty A, Havanoor V, Basavaraja S, Balaji SD, Venkataraman A (2007) Microwave-assisted route for synthesis of nanosized metal oxides. Sci Technol Adv 8(6):484–493. https://doi.org/10.1016/j.stam.2007.07.001
53. Landers J, Gor GY, Neimark AV (2013) Density functional theory methods for characterization of porous materials. Colloid Surface A:3–32. https://doi.org/10.1016/j.colsurfa.2013.01.007

54. Lázaro IA, Forgan RS (2019) Application of zirconium MOFs in drug delivery and biomedicine. Coord Chem Rev 380:230–259. https://doi.org/10.1016/j.ccr.2018.09.009
55. Lázaro IA, Wells CJR, Forgan RS (2020) Multivariate modulation of the Zr MOF UiO-66 for defect-controlled combination anticancer drug delivery. Angew Chem Int Ed 59(13):5211–5217. https://doi.org/10.1002/anie.201915848
56. Lekić BM, Marković DD, Ognjanović NR, Dukić AR, Rajaković LV (2013) Arsenic removal from water using industrial By-products. J Chem:121024. https://doi.org/10.1155/2013/121024
57. Li J, Wu YN, Li Z, Zhu M, Li F (2014) Characteristics of arsenate removal from water by metal-organic frameworks (MOFs) Water Sci Technol 70(8):1391–1397. https://doi.org/10.2166/wst.2014.390
58. Li Y, Wang X, Xu D, Chung JD, Kaviany M, Huang B (2015) H_2O adsorption/desorption in MOF-74: Ab initio molecular dynamics and experiments. J Phys Chem C 119(23):13021–13031. https://doi.org/10.1021/acs.jpcc.5b02069
59. Li Z, Ma S, Chen C, Qu G, Jin W, Zhao Y (2020) Efficient capture of arsenate from alkaline smelting wastewater by acetate modulated yttrium based metal-organic frameworks. Chem Eng Trans 397. https://doi.org/10.1016/j.cej.2020.125292
60. Lin JYS (2016) Molecular sieves for gas separation. Science 353(6295):8–10. https://doi.org/10.1126/science.aag2267
61. Lin S, Zhao Y, Yun YS (2018) Highly effective removal of nonsteroidal anti-inflammatory pharmaceuticals from water by Zr(IV)-based metal-organic framework: adsorption performance and mechanisms. ACS Appl Mater Interfaces 10(33):28076–28085. https://doi.org/10.1021/acsami.8b08596
62. Litter MI, Ingallinella AM, Olmos V, Savio M, Difeo G, Botto TEMF, Taylor S, Frangie S, Herkovits J, Schalamuk I, González MJ, Berardozzi E, Einschlag FSG, Bhattacharya P, Ahmad A (2019) Arsenic in Argentina: Technologies for arsenic removal from groundwater sources, investment costs and waste management practices. Sci Total Environ 690:778–789. https://doi.org/10.1016/j.scitotenv.2019.06.358
63. Lustig WP, Mukherjee S, Rudd ND, Desai AV, Li J, Ghosh SK (2017) Metal-organic frameworks: functional luminescent and photonic materials for sensing applications. Chem Soc Rev 46(11):3242–3285. https://doi.org/10.1039/c6cs00930a
64. Lv D, Chen Y, Li Y, Shi R, Wu H, Sun X, Xiao J, Xi H, Xia Q, Li Z (2017) Efficient mechanochemical synthesis of MOF-5 for linear alkanes adsorption. J Chem Eng Data 62(7):2030–2036. https://doi.org/10.1021/acs.jced.7b00049
65. Lv Z, Fanb Q, Xiec Y, Chena Z, Alsaedid A, Hayat T, Wanga X, Chen C (2019) MOFs-derived magnetic chestnut shell-like hollow sphere NiO/Ni@C composites and their removal performance for arsenic(V). J Chem Eng 362:413–421. https://doi.org/10.1016/j.cej.2019.01.046
66. Ma S, Zhou HC (2010) Gas storage in porous metal-organic frameworks for clean energy applications. Chem Comm 46(1):44–53. https://doi.org/10.1039/b916295j
67. Mohan D, Pittman CU (2007) Arsenic removal from water/wastewater using adsorbents-a critical review. J Hazard Mater 142(1–2):1–53. https://doi.org/10.1016/j.jhazmat.2007.01.006
68. Mondal P, Bhowmick S, Chatterjee D, Figoli A, Bruggen BVD (2013) Remediation of inorganic arsenic in groundwater for safe water supply: a critical assessment of technological solutions. Chemosphere 92(2):157–170. https://doi.org/10.1016/j.chemosphere.2013.01.097
69. Nasir AM, Nordin NAH, Goh PS, Ismail AF (2018) Application of two-dimensional leaf-shaped zeolitic imidazolate framework (2D ZIF-L) as arsenite adsorbent: kinetic, isotherm and mechanism. J Mol Liq 250:269–277. https://doi.org/10.1016/j.molliq.2017.12.005
70. Nicomel NR, Leus K, Folens K, Voort PVD, Laing GD (2015) Technologies for arsenic removal from water: current status and future perspectives. Int J Environ Res 13(1):1–24. https://doi.org/10.3390/ijerph13010062
71. Pan Y, Wang J, Guo X, Liu X, Tang X, Zhang H (2018) A new three-dimensional zinc-based metal-organic framework as a fluorescent sensor for detection of cadmium ion and nitrobenzene. J Colloid Interface Sci 513:418–426. https://doi.org/10.1016/j.jcis.2017.11.034

72. Pandi K, Lee DW, Choi J (2020) Facile synthesis of zirconium-organic frameworks@biomass-derived porous graphitic nanocomposites: arsenic adsorption performance and mechanism. J Mol Liq 314. https://doi.org/10.1016/j.molliq.2020.113552
73. Pang D, Wang CC, Wang P, Liu W, Fu H, Zhao C (2020) Superior removal of inorganic and organic arsenic pollutants from water with MIL-88A(Fe) decorated on cotton fibers. Chemosphere 254.
74. Pang D, Wang P, Fu H, Zhao C, Wang CC (2020) Highly efficient removal of As(V) using metal–organic framework BUC-17. SN Applied Sciences 2(2). https://doi.org/10.1007/s42452-020-1981-3
75. Perdew JP, Burke K, Ernzerhof M (1996) Generalized gradient approximation made simple. Phys Rev Lett 77(18):3865–3868. https://doi.org/10.1103/PhysRevLett.77.3865
76. Pirzadeh K, Ghoreyshi AA, Rahimnejad M, Mohammadi M (2018) Electrochemical synthesis, characterization and application of a microstructure $Cu_3(BTC)_2$ metal organic framework for CO_2 and CH_4 separation. Korean J Chem Eng 35(4):974–983. https://doi.org/10.1007/s11814-017-0340-6
77. Ploychompoo S, Chen J, Luo H, Liang Q (2020) Science Direct Fast and efficient aqueous arsenic removal by functionalized MIL-100 (Fe)/ rGO/d-MnO_2 ternary composites: adsorption performance and mechanism. J Environ Sci 91:1–13.https://doi.org/10.1016/j.jes.2019.12.014
78. Pous N, Casentini B, Rossetti S, Fazib S, Puiga S, Aulenta F (2015) Anaerobic arsenite oxidation with an electrode serving as the sole electron acceptor: a novel approach to the bioremediation of arsenic-polluted groundwater. J Hazard Mater 283:617–622. https://doi.org/10.1016/j.jhazmat.2014.10.014
79. Prabhu SM, Imamura S, Sasaki K (2019) Mono-Di- and Tricarboxylic Acid facilitated Lanthanum-based organic frameworks: insights into the structural stability and mechanistic approach for superior adsorption of arsenate from water. ACS Sustain Chem Eng 7(7):6917–6928. https://doi.org/10.1021/acssuschemeng.8b06489
80. Rajakovic L, Ognjanovic VR (2018) Arsenic in water: determination and removal. J Anal Toxicol. https://doi.org/10.5772/intechopen.75531
81. Rajaković LV, Marković DD, Ognjanović VNR, Antanasijević DZ (2012) Review: the approaches for estimation of limit of detection for ICP-MS trace analysis of arsenic. Talanta 102:79–87. https://doi.org/10.1016/j.talanta.2012.08.016
82. Ramanayaka S, Vithanage M, Sarmah A, An T, Kim KH, Ok YS (2019) Performance of metal-organic frameworks for the adsorptive removal of potentially toxic elements in a water system: a critical review RSC Adv 9(59):34359–34376. https://doi.org/10.1039/c9ra06879a
83. Rao MM, Rao GPC, Seshaiah K, Choudary NV, Wang MC (2008) Activated carbon from Ceiba pentandra hulls, an agricultural waste, as an adsorbent in the removal of lead and zinc from aqueous solutions. J Waste Manag 28(5):849–858. https://doi.org/10.1016/j.wasman.2007.01.017
84. Rasheed T, Hassan AA, Bilal M, Hussain RK (2020) Metal-organic frameworks based adsorbents: a review from removal perspective of various environmental contaminants from wastewater. Chemosphere 9(20).
85. Rebah FB, Mnif W, Siddeeg SM (2018) Microbial flocculants as an alternative to synthetic polymers for wastewater treatment: a review. Symmetry 10(11):1–19. https://doi.org/10.3390/sym10110556
86. Remya VR, Kurian M (2019) Synthesis and catalytic applications of metal–organic frameworks: a review on recent literature. Int Nano Lett 9(1):17–29. https://doi.org/10.1007/s40089-018-0255-1
87. Rodríguez C, Briano S, Leiva E (2020) Increased adsorption of heavy metal Ions in. Molecules 25:3106. https://doi.org/10.3390/molecules25143106a
88. Ruixia L, Jinlong G, Hongxiao T (2002) Adsorption of fluoride, phosphate, and arsenate ions on a new type of ion exchange fiber. J Colloid Interface Sci 248(2):268–274. https://doi.org/10.1006/jcis.2002.8260

89. Sandoval VJL, Castillo DIM, Villarreal IAA, Avila HER, Ponce HAG (2020) Valorization of agri-food industry wastes to prepare adsorbents for heavy metal removal from water. J Environ Chem Eng 8(5) https://doi.org/10.1016/j.jece.2020.104067
90. Shaji E, Santosh M, Sarath KV, Prakash P, Deepchand V, Divya BV (2020) Arsenic contamination of groundwater: a global synopsis with focus on the Indian peninsula. GSF 20:1674–9871. https://doi.org/10.1016/j.gsf.2020.08.015
91. Shankar S, Shanker U, Shikha (2014) Arsenic contamination of groundwater: a review of sources, prevalence, health risks, and strategies for mitigation. Sci World J. https://doi.org/10.1155/2014/304524
92. Shao P, Ding L, Luo J, Luo Y, You D, Zhan Q, Luo X (2019) Lattice-defect-enhanced adsorption of Arsenic on Zirconia Nanospheres: a combined experimental and theoretical study. ACS Appl Mater Interfaces 11(33):29736–29745. https://doi.org/10.1021/acsami.9b06041
93. Shih MC (2005) An overview of arsenic removal by pressure-driven membrane processes. Desalination 172(1):85–97. https://doi.org/10.1016/j.desal.2004.07.031
94. Singh R, Singh S, Parihar P, Singh VP, Prasad SM (2015) Arsenic contamination, consequences and remediation techniques: a review. Ecotoxicol Environ Saf 112:247–270. https://doi.org/10.1016/j.ecoenv.2014.10.009
95. Solis KLB, Kwon YH, Kim MH, An HR, Jeon C Hong Y (2020) Metal organic framework UiO-66 and activated carbon composite sorbent for the concurrent adsorption of cationic and anionic metals. Chemosphere 238. https://doi.org/10.1016/j.chemosphere.2019.124656
96. Sun J, Zhang X, Zhang A, Liao C (2019) Preparation of Fe–Co based MOF-74 and its effective adsorption of arsenic from aqueous solution. J Environ Sci 80:197–207. https://doi.org/10.1016/j.jes.2018.12.013
97. Tahoon MA, Siddeeg SM, Alsaiari NS, Mnif W, Rebah FB (2020) Effective heavy metals removal from water using nanomaterials: a review. Processes 8(6):1–24. https://doi.org/10.3390/PR8060645
98. Tanaka S (2020) Mechanochemical synthesis of MOFs. In Met - Org Fram Biomed Appl. https://doi.org/10.1016/b978-0-12-816984-1.00012-3
99. Tarboush BJA, Chouman A, Jonderian A, Ahmad M, Hmadeh M, Ghoul MA (2018) Metal-organic Framework-74 for Ultratrace Arsenic removal from water: experimental and density functional theory studies. ACS Appl Nano Mater 1(7):3283–3292. https://doi.org/10.1021/acsanm.8b00501
100. Tchalala MR, Bhatt PM, Chappanda KN, Tavares SR, Adil K, Belmabkhout Y, Shkurenko A, Cadiau A, Heymans N, De WG, Maurin G, Salama KN, Eddaoudi M (2019) Fluorinated MOF platform for selective removal and sensing of SO_2 from flue gas and air. Nat Commun 10(1):1–10. https://doi.org/10.1038/s41467-019-09157-2
101. Tian C, Zhao J, Ou X, Wan J, Cai Y, Lin Z, Dang Z, Xing B (2018) Enhanced adsorption of p -Arsanilic acid from water by Amine-modified UiO-67 as examined using extended X-ray absorption fine structure, X-ray Photoelectron Spectroscopy, and density functional theory calculations. Environ Sci Technol 52(6):3466–3475. https://doi.org/10.1021/acs.est.7b05761
102. Tuutijärvi T, Lu J, Sillanpää M, Chen G (2009) As(V) adsorption on maghemite nanoparticles. J Hazard Mater 166(2–3):1415–1420. https://doi.org/10.1016/j.jhazmat.2008.12.069
103. Tyson J (2013) The determination of arsenic compounds: a critical review. ISRN anal chem: 1–24. https://doi.org/10.1155/2013/835371
104. Uddin MDJ, Jeong YK (2020) Review: efficiently performing periodic elements with modern adsorption technologies for arsenic removal. Environ Sci Pollut R 27(32):39888–39912. https://doi.org/10.1007/s11356-020-10323-z
105. Vo TK, Le VN, Quang DT, Song M, Kim D, Kim J (2020) Rapid defect engineering of UiO-67 (Zr) via microwave-assisted continuous-flow synthesis: effects of modulator species and concentration on the toluene adsorption Micropor Mesopor Mat 306 https://doi.org/10.1016/j.micromeso.2020.110405
106. Vu TA, Giang. Le GH, Dao CD, Dang LQ, Nguyen KT, Nguyen QK, Dang PT, Tran HTK, Duong QT, Nguyen TV, Lee GD, (2015) Arsenic removal from aqueous solutions by adsorption using novel MIL-53(Fe) as a highly efficient adsorbent. RSC Adv 5(7):5261–5268. https://doi.org/10.1039/c4ra12326c

107. Wadhawan S, Jain A, Nayyar J, Mehta SK (2020) Role of nanomaterials as adsorbents in heavy metal ion removal from wastewater: a review. J Water Process Eng 33. https://doi.org/10.1016/j.jwpe.2019.101038
108. Wan P, Yuan M, Yu X, Zhang Z, Deng B (2020) Arsenate removal by reactive mixed matrix PVDF hollow fiber membranes with UIO-66 metal organic frameworks. Chem Eng Trans 382. https://doi.org/10.1016/j.cej.2019.122921
109. Wang C, Liu X, Chen JP, Li K (2015) Superior removal of arsenic from water with zirconium metal-organic framework UiO-66. Sci Rep 5:1–10. https://doi.org/10.1038/srep16613
110. Wang C, Luan J, Wu C (2019) Metal-organic frameworks for aquatic arsenic removal. Water Res 158:370–382. https://doi.org/10.1016/j.watres.2019.04.043
111. Wang CC, Ying JY (1999) Sol-gel synthesis and hydrothermal processing of anatase and rutile titania nanocrystals. Chem Mater 11(11):3113–3120. https://doi.org/10.1021/cm990180f
112. Wu C, Cui MQ, Xue SG, Li WC, Huang L, Jiang XX, Qian ZY (2018) Remediation of arsenic-contaminated paddy soil by iron-modified biochar. Environ Sci Pollut Res 25(21):20792–20801. https://doi.org/10.1007/s11356-018-2268-8
113. Wu H, Ma MD, Gai WZ, Yang H, Zhou JG, Cheng Z, Xu P, Deng ZY (2018) Arsenic removal from water by metal-organic framework MIL-88A microrods MIL-88A. ESPR. https://doi.org/10.1007/s11356-018-2751-2
114. Xu Y, Cheng Y, Jia Y, Ye BC (2020) Synthesis of MOF-derived Ni@C materials for the electrochemical detection of histamine Talanta 219. https://doi.org/10.1016/j.talanta.2020.121360
115. Yaghi OM, Li G, Li H (1995) Selective binding and removal of guests in a microporous metal-organic framework. Nature 378:703–706. https://doi.org/10.1038/378703a0
116. Yan H, Lin Y, Wu H, Zhang W, Sun Z, Cheng H, Liu W, Wang C, Li J, Huang X, Yao T, Yang J, Wei S, Lu J (2017) Bottom-up precise synthesis of stable platinum dimers on graphene. Nat Commun 8(1):1–10. https://doi.org/10.1038/s41467-017-01259-z
117. Yang JC, Yin XB (2017) $CoFe_2O_4$@MIL-100(Fe) hybrid magnetic nanoparticles exhibit fast and selective adsorption of arsenic with high adsorption capacity. Sci Rep 7:1–15. https://doi.org/10.1038/srep40955
118. Yang X, Huang G, An C, Chen X, Shen J, Yin J, Song P, Xu Z, Li Y (2021) Removal of arsenic from water through ceramic filter modified by nano-CeO2: a cost-effective approach for remote areas. Sci Total Environ 750. https://doi.org/10.1016/j.scitotenv.2020.141510
119. Yoo Y, Guerrero VV, Jeong HK (2011) Isoreticular metal-organic frameworks and their membranes with enhanced crack resistance and moisture stability by surfactant-assisted drying. Langmuir 27(6):2652–2657. https://doi.org/10.1021/la104775d
120. Yu W, Luo M, Yang Y, Wu H, Huang W, Zeng K, Luo F (2019) Metal-organic framework (MOF) showing both ultrahigh As(V) and As(III) removal from aqueous solution. J Solid State Chem 269(V):264–270. https://doi.org/10.1016/j.jssc.2018.09.042
121. Zeng J, Qi P, Shi J, Pichler T, Wang F, Wang Y, Sui K (2020) Chitosan functionalized iron nanosheet for enhanced removal of As(III) and Sb(III): synergistic effect and mechanism. Chem Eng Trans 382. https://doi.org/10.1016/j.cej.2019.122999
122. Zheng T, Wang J, Wang Q, Meng H, Wang L (2017) Research trends in electrochemical technology for water and wastewater treatment. Appl Water Sci 7(1):13–30. https://doi.org/10.1007/s13201-015-0280-4
123. Zhou HC, Long JR, Yaghi OM (2012) Introduction to metal-organic frameworks. Chem Rev 112(2):673–674. https://doi.org/10.1021/cr300014x
124. Zhu X, Song T, Lv Z, Ji G (2016) High-efficiency and low-cost α-Fe_2O_3 nanoparticles-coated volcanic rock for Cd(II) removal from wastewater. Process SAF Environ 104:373–381. https://doi.org/10.1016/j.psep.2016.09.019

Metal Organic Frameworks (MOFs) as Formidable Candidate for Pharmaceutical Wastewater Treatment

Sadaf Ahmad, Bakar bin Khatab Abbasi, Muhammad Shahid Nazir, and Mohd Azmuddin Abdullah

Abstract In recent years, the pharmaceutical field has significantly achieved magnificent progress owing to the necessities of human health and life; however, it also led to drastic environmental issues. The existence of pharmaceuticals in water bodies, which could cause adverse effects on human beings and environment, rose up distress worldwide. The pharmaceutical components found in water bodies have mainly two origins: manufacturing procedures in pharmaceutical industry and common usage of pharmaceutics. The essence of pharmaceutical wastewater (PWW) is intricate, including large amount of organic matter, high salt, microbial toxicity, and non-biodegradable. In sight of water scarcity means, it is essential to figure out and expand techniques for pharmaceutics derived wastewater in water management. Nevertheless, numerous treatment methods have been established to serve pharmaceutical wastewater including biological treatments, membrane technologies, hybrid technologies, advanced oxidation processes, absorption methods, etc. Recently, metal organic frameworks (MOFs), metallic ions clusters linked with organic bridging linkers, have been utilized in number of uses such as storage, separation, sensing, catalysis, adsorption, and many others. The viability of MOFs toward wastewater treatment (WWT) for various pollutants is fundamentally because of the extreme porosity, discrete pore structure, and superficial modification. This chapter highlights the origin and treatment of pharmaceutical wastewaters via the utilization of MOFs and their hybrid systems. A brief perception of the future work in the field has also been discussed.

Keywords Wastewater · Pharmaceutical · Pollution · Organic pollutant · Antibiotic · Treatment · Biological treatment · Membrane technology · Advanced

S. Ahmad · B. K. Abbasi · M. S. Nazir (✉)
Department of Chemistry, COMSATS University Islamabad, Lahore campus (CUI), Lahore 54000, Pakistan
e-mail: shahid.nazir@cuilahore.edu.pk

M. A. Abdullah
Department of Chemical Engineering, COMSATS University Islamabad, Lahore Campus, 54000 Lahore, Pakistan
e-mail: azmuddin@umt.edu.my

E. Lichtfouse et al. (eds.), *Inorganic-Organic Composites for Water and Wastewater Treatment*, Environmental Footprints and Eco-design of Products and Processes,
https://doi.org/10.1007/978-981-16-5928-7_2

oxidation process · Adsorption · Hybrid · Metal organic framework · Composite · Adsorbent · Catalyst · Magnetic Removal

1 Introduction

1.1 Origin of Pharmaceutical Wastewater (PWW)

It has been reported that various types of wastewaters originating from different industries and from domestic usage have been turned into severe danger to the life of living organisms (plants, animals, and human beings) [25]. Different types of wastewaters are presented in Fig. 1.

In recent times, the contaminated wastewaters generated in the pharmaceutical industries have been transformed into one of the major evolving environmental concerns. Commonly, pharmaceutics are produced with great stability to be operational in animals and humans. These are ordered into eight crews of beta-blockers, anti-inflammatory medications, antibiotics, hormones, antiepileptic, lipid-lowering mediators, anti-depressants, and centered on their healing applications [80].

The existence of pharmaceuticals and personal care products (PPCPs) was earliest recognized in the Europe and United States in surface and wastewater in the 1960s. Alarms about their budding harm were upraised in 1999 with the problem enticing substantial attention when the existence of medicines in stream was associated to feminization of fish existing waters of wastewater treatment plant

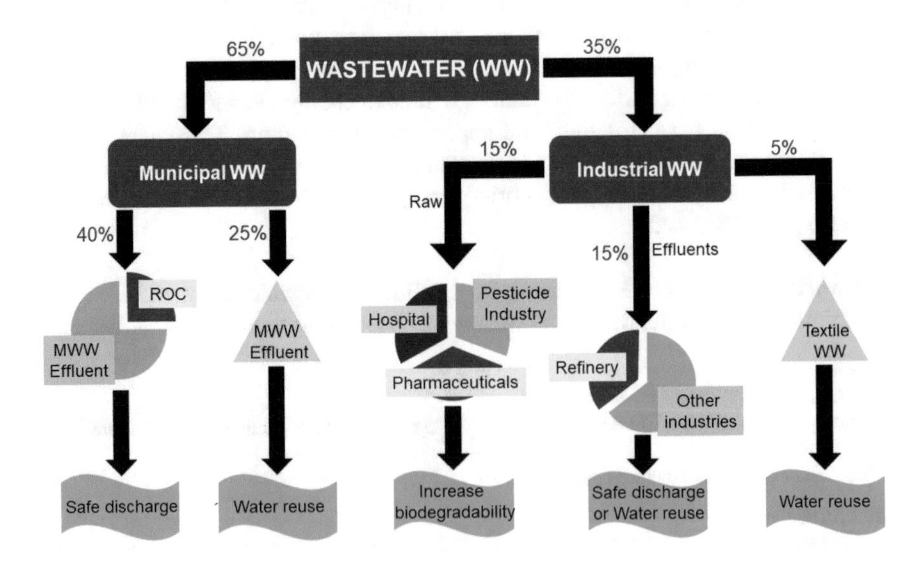

Fig. 1 Various types of wastewater. (Modified after [66])

(WWTP) outlets [17]. Additionally, a public study has raised public awareness that organic contaminants in wastewater, counting PPPCs, are existing in 80% of 139 US watercourses [35].

Owing to the fast expansion of the pharmacological chemical industries and the extensive utilization of health care substances, in recent years, the pharmaceutical manufacturing process has produced large amounts of toxic wastewater [56].

Pharmaceuticals come into the environment from thousands of distributed places. Major causes of contamination comprise pharmaceutical areas, WWTPs, landfills, hospices, and cemeteries [32, 45]. The most examined way for the entrance of pharmaceutical products into the surroundings is through the municipal WWTP. Human releases of unmodified or marginally altered Active pharmaceutical components (APIs) combined with polar particles, for instance glucuronides pass in WWTP where these components could be cut and leave the original API in the atmosphere [26].

1.2 Characteristics of Pharmaceutical Wastewater (PWW)

In contrast to domestic wastewater, PWW characteristically comprises more complex constituents (counting pharmaceutical products and raw resources), greater COD, dusky color, greater toxicity, and deprived biodegradability. PWW, chiefly chemical pharmaceutical wastewater, has now converted into most hard-to-manage and risky categories of wastewater [39]. Pharmaceutical wastewater possesses small amount of biochemical oxygen demand (BOD) and greater concentration of chemical oxygen demand (COD) in pharmaceutical wastewater [8].

The chief features of pharmaceutical wastewater are: (i) greater quantity of organic pollutants, greater amount variation and complex configuration; (ii) greater variance in BOD/COD rate in wastewater; (iii) greater NH_3-N amount and Chroma; and (iv) large suspended solid amount and salinity [43]. Composition of pharmaceutical wastewaters is presented in Table 1.

Table 1 Composition of pharmaceutical wastewater (PWW). Data from [9]

Constituent (mgL^{-1})	Minimum–maximum quantity	Average composition
BOD_5	200–6000	2344
COD	375–32,500	8854
BOD_5/COD (ratio)	0.1–0.6	0.32
TOC	860–4940	2467
TKN	165–770	383
TDS	675–9320	6.9
NH_3-N	148–363	244
SO_4^{2-}	890–1500	1260
Cl^-	760–4200	2820

Pharmaceutical trade industries generate a massive variety of constituents by utilizing inorganic and organic combinations as raw sources thus increase in enormous size of poisonous and multifaceted organic liquescent wastes contained greater intensities of mixed solids. Large and reduced power wastewater creeks initiated from industrial unit frequently comprises various kinds of contaminants, counting organic combinations [13]. Furthermore, numerous kinds of wastewater feature from the pharmaceutical industries vary critically, many categories of pharmaceutical wastewater are resilient to biodegradation, have greater nitrogen in ammonia amounts, color depth and toxicity; and are hard to handle [56].

Furthermore, the negative influence of some pharmaceutically active compounds (PhACs), for example endocrine-disrupting substances (perhaps hormones), sedatives, antidepressants, anesthetics, illicit drugs or recreational substances, on aquatic ecologies has been confirmed in laboratories and naturally. The matter that few of the more tenacious and slowly decomposing components outreach the drinking water source and are captivated by plants over irrigation has worsened this issue. These PhACs, thus, give the impression in the human food chain, surprisingly yet their quantity is low [28].

2 Methods for Pharmaceutical Wastewater Treatment (PWWT)

Just like treatment to other wastewaters, various methods are applied for the treatment of pharmaceutical wastewaters, for instance biological treatment (BT), membrane technologies (MTs), advanced oxidation process (AOPs), adsorption methods (AMs), and hybrid systems (HSs). A summary of different types of treatment processes for pharmaceutical wastewaters is presented in Table 2.

2.1 Biological Treatment (BT)

Traditionally, bio-based management approaches have been developed for the management of PWW. These are sectioned into anaerobic and aerobic procedures. Anaerobic systems comprise anaerobic sludge vessels, anaerobic membrane vessels, and anaerobic sieves. Aerobic uses comprise sequence batch reactors, membrane batch reactors, and activated sludge [18].

The wastewater features show an important part in the assortment of biological methods. Active pharmaceutical ingredients (APIs), solvents, intermediary and raw resources signify biologically refractory constituents, which influence the efficacy of biological management systems [17].

In biological method, microorganisms cut down or transform the complex contaminants into lighter arrangements. In both circumstances, the degraded wastewaters

Table 2 Briefing of diverse treatment means for pharmaceutical wastewater

Sr. number	Treatment method	Summary	References
1	Biological	Microorganisms cut down or transform the complex contaminants/pharmaceuticals into simpler arrangements	[51]
2	Membrane technology	The solutions are sieved/filtered by adsorbing pollutants/pharmaceuticals on the surface of membrane	[18]
3	Advanced oxidation	Chemical procedure breakdowns the chemical construction of the pollutant constituents and splits them into combinations with no dangerous characteristics	[61]
4	Adsorption	Surface protocol contracts mainly with the use of surface forces. Pollutants/pharmaceuticals are adsorbed on the surface of adsorbent	[5]
5	Hybrid systems	Combination of two methods/processes for elimination of pollutants/pharmaceuticals	[22]

and transformed into other product, the subsequent product ensures not to generate secondary contamination. Furthermore, enquiry has been done around the globe for the deprivation of pharmaceutical wastewater by the assistance of fungi, for instance, *Trametes sp.* and *Phanerochaete chrysosporium* [51].

Moreover, the anaerobic ammonium oxidation (anammox) with abundant profitable advantage and decent management result is an auspicious procedure to eliminate N_2 from antibiotic-encompassing wastewater. The anammox procedure is a newfangled wastewater treatment expertise been swiftly industrialized recently. Until 2015, 114 whole-scale anammox sewage-managing plans were available around the globe. The anammox method depends on the distinct physical metabolic appliance of *anammox* bacteria and changes ammonia (NH_4^+) and nitrite (NO_2^-) into N_2 in anaerobic circumstances. Likened to outdated denitrification/nitrification procedures, no necessities of aeration and external carbon origins create anammox procedure to save functioning cost by greater than 60% and decreased energy ultimatum plus greenhouse gas creation by greater than 25% [21].

In recent decades, numerous researchers have explored the efficiency of membrane bioreactor (MBR) in the antibiotic pollution treatment that evidenced that MBR is operational in handling antibiotic pollution [14].

2.2 *Membrane Technologies (MTs)*

Membrane technique is a different physical approach, in which the water is sieved/filtered by adsorbing pollutants on the surface of membrane. Furthermore, variable filtration properties (surface charge, hydrophobicity, and pore size), acquired

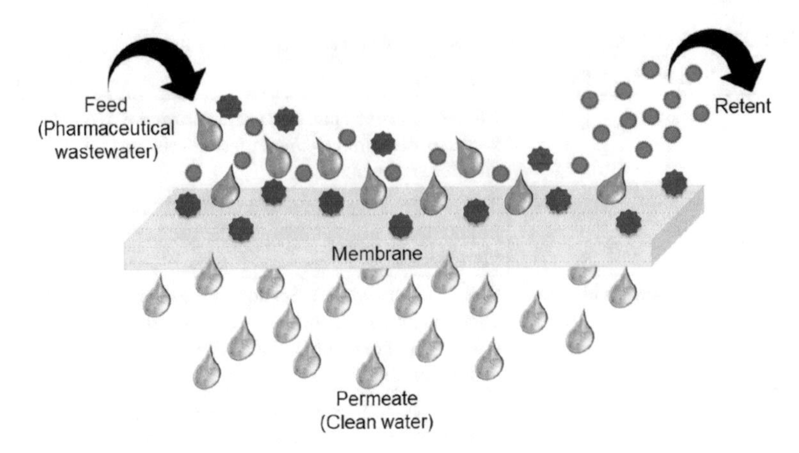

Fig. 2 Sketch of membrane filtration for the exclusion of contaminants from pharmaceutical wastewater. (Modified after [33])

from various membranes, decide the toxins to be detached [18]. General mechanism of membrane filtration for pharmaceutical wastewater is presented in Fig. 2.

Moreover, numerous membrane systems and uses have been investigated for APIs removal at full scale and pilot scale, counting ultra-filtration [10], microfiltration [73], nanofiltration [55], membrane distillation, reverse osmosis [16], membrane bioreactors [40], electrodialysis reversal [23], and arrangements of membranes in sequence [17].

2.3 Advanced Oxidation Processes (AOPs)

Oxidation mechanisms have been utilized predominantly to supplement the traditional schemes and to improve the management of rebellious organic contaminants. Advanced oxidation processes (AOPs) are being considered more for wastewater management because of their great efficacy. AOPs are among the chemical procedures that breakdown the chemical construction of the pollutant constituents and split them into combinations with no dangerous characteristics. In AOPs, decay ensue by creating free radicals, for instance hydro-peroxyl ($HO_2^{\cdot}$), hydroxyl (OH·), sulfate ($SO_4^{\cdot}$), and superoxide ($O_2^{-\cdot}$) that have greater oxidation charge. Furthermore, such radicals are greatly oxidizing with a little life duration that fashioned in the vicinity, correspondingly, rapidly outbreak the chemical composites in the wastewater and end in their oxidation and then disintegration [61].

Successfully, this skill has been employed for the pharmaceutical treatment. A chemical mediator, for instance ozone, hydrogen peroxide, metal oxides, and transition metals, is obligatory for AOPs. Furthermore, an energy basis, for instance UV–Vis radiation, gamma-energy, ultrasound, and electric current is obligatory. AOPs are founded on free radicals creation, specifically radical of hydroxyl and enable the transformation of contaminants to low damaging and greater decomposable combinations. AOPs often comprise ozonation combined with ultraviolet (UV) irradiation and hydrogen peroxide. Moreover, Fenton and TiO_2 photocatalysis have also been utilized. Varied combinations of ozone, Fenton, hydrogen peroxide, and TiO_2 in dark and light have exposed a variety of appropriate treatment approaches depending on the characteristics of the pharmaceuticals and financial deliberations [17].

Advanced oxidation process (AOP) is a better effectual management system for antibiotics than other traditional procedures [50].

2.4 *Adsorption Methods (AMs)*

Adsorption is described as quantity of materials on the solid body's surface. Adsorption is a surface protocol that contracts mainly with the use of surface forces. Moreover, when a solution consisting absorbable solute (adsorbate) approaches into contact with a compact-solid (adsorbent), with extremely porous surface assembly intermolecular forces (liquid–solid) of attraction projects the solute to be concerted at the surface of solid [5].

Adsorption is utilized as the uppermost class treatment technique for the elimination of mixed organic contaminants, for instance antibiotics from PWW [5]. The major characteristics of the adsorbents are robust affinity and greater loading capability for antibiotics elimination [5]. Adsorbents as rice husks [5], spherical biochar [71], nano-rod erdite particles (EPs) [62], Graphene [11], graphene oxide (GO) [7], activated carbon (AC) [74], SiO_2 nanoparticle [49], carbon nanotube (CNT) [3], paper pulp-based adsorbents [54], chitosan, clays, zeolites [37], and low cost: animal waste and agricultural waste [63]. Different adsorbents utilized for wastewater treatment are shown in Fig. 3.

2.5 *Hybrid Systems (HSs)*

It has been witnessed that not any of the procedures could be utilized discretely in wastewater management tenders with decent finances and greater grade of energy competence. Furthermore, the data obligatory for the extensive strategy and implementation are conceivably deficient. Therefore, various hybrid systems have been offered in order to treat wastewaters [22].

In this context, membrane bioreactors and membrane filtrations, for instance nano filtration and reverse osmosis have been integrated for the examination for pharmaceuticals elimination, in recent years [18]. Similarly, anaerobic membrane bioreactor (AnMBR) as a potential substitute for activated sludge procedure has been extensively utilized in management of pharmaceutical wastewater. The blend of conventional anaerobic biological skill and advanced membrane technique expertise holds the benefits of greater organic elimination, less energy ingestion, and biogas retrieval.

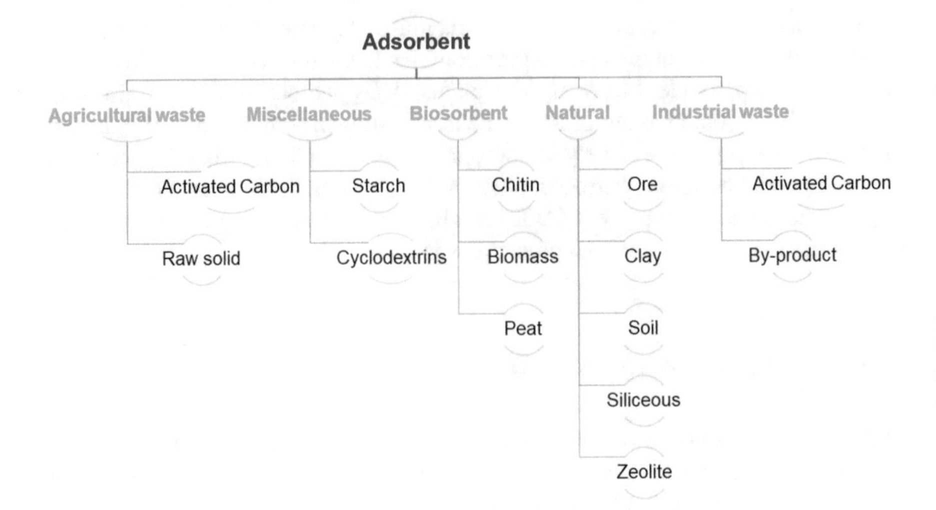

Fig. 3 Different adsorbents utilized for the exclusion of pollutants from wastewaters. (Modified after [65])

In unison, effectual parting of hydraulic retention time (HRT) and the sludge retention time (SRT), less generation of bio-solids and greater permeate value are personified in AnMBR owing to film segment [14].

In a study, the pharmaceutical treatment was evaluated by grouping of photocatalytic oxidation (PcO), electroFenton (EF), and electrocoagulation (EC) procedures. Furthermore, the effect of procedure order and reaction duration, current concentration for EF and EC procedures utilizing iron plate electrodes and lastly photocatalytic action of various catalysts for PcO procedure were inspected in order to remove total organic carbon [8].

AOPs are frequently combined with other methods, for instance adsorption, biodegradation, membrane technique, and electro-coagulation to beat the greater price. AOPs have been utilized in the pretreatment phase to enhance the biodegradability and reduce the price in the improved combination for excessively polluted pharmaceutical wastewaters [59]. Furthermore, hybrid-AOPs serve as greatly effectual and influential management for whole antibiotics demineralization. Recently, a report proposed a novel solution for degradation of antibiotics through three electro catalytic-grounded devices that are electrochemical flotation procedures, electrochemical advanced oxidation, and electrochemical Fenton reagent, though producing peroxide as intermediary. The systems were successful to retrieve targeted antibiotic-Rifampicin from pharmaceutical wastewater [50]. Similarly, pharmaceutical effluent was treated by a hybrid procedure of adsorption and advanced oxidation. Initially, the wastewater was reacted with ozone in the existence of H_2O_2. The effluent comprised of antipsychotic, anti-cancer, anti-biotic, and anti-depressant drugs. Later, the beforehand treated wastewater was transported via a chamber of grained activated carbon (GAC) for additional lessening of the COD [59].

According to the research and scientific approaches, the application of hybrid methodologies has been promising for the effectual real pharmaceutical wastewater treatment. Furthermore, different groupings of traditional wastewater treatment procedures have been applied, for instance chemical and physical approaches (advanced oxidation routes, chemical treatments, adsorption, filtration, and coagulation) besides joined biological courses for the handling of pharmaceutical wastewater. Ultimately, all these systems have different COD elimination capacities, contingent to the kind and organic capacity of the wastewater [13].

3 Metal Organic Frameworks (MOFs) for Pharmaceutical Wastewater Treatment (PWWT)

Recently, vastly ordered structured, porous, and crystal-like metal organic framework (MOF) has been acknowledged as a promising substitute to cover the practical restrictions of traditional porous substances and nano-based substances for ecological uses. Essentially, MOF is a ground-breaking group of hybrid substances comprised of metal ions having definite coordination geometry with organic bridging linkers [36], as shown in Fig. 4.

MOF structures could be designed realistically over superficial regulation on the architecture and pores functionalization. Furthermore, the efficiency of MOFs in many applications (counting catalysis, sensors, energy storage, drug transport system, separation, gas storage, and nonlinear optics) has been largely documented. The technical evidences of MOFs have been reported excellently to deliver well information on their scheme, production, and uses [36].

Additionally, their viability to wastewater treatment (WWT) of numerous contaminants (perhaps heavy metallic ions, volatile organic compounds (VOCs), pesticides, and other dangerous compounds) has been comprehensively assessed [36].

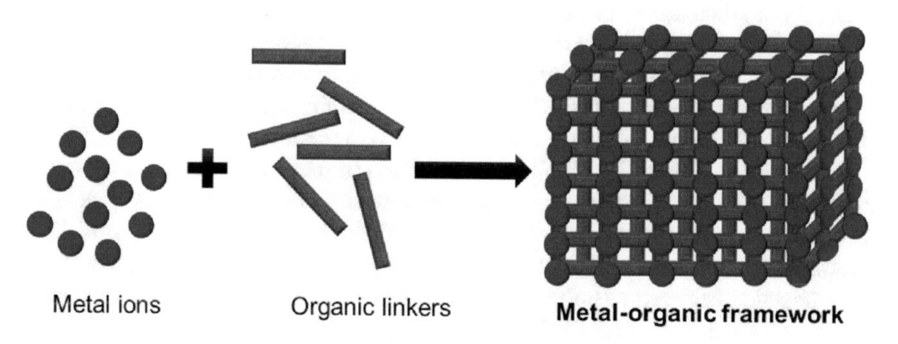

Fig. 4 General representation of assembly of Metal organic framework (MOF). (Modified after [12])

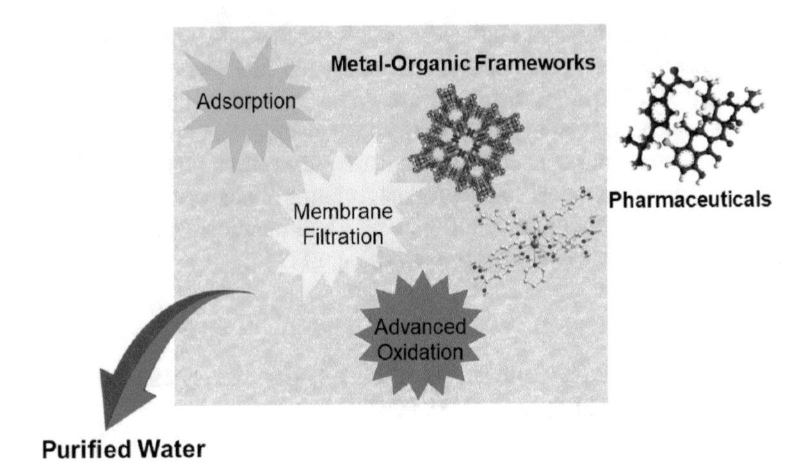

Fig. 5 Exploitation of Metal organic frameworks (MOFs) as adsorbents, in membrane filtration, and in advanced oxidation processes to treat various pharmaceuticals. (Modified after [42])

Subsequently, the matrix, which comprises pharmaceutical pollutants are universal aqueous solution, thus MOFs and their spin-offs water stability, counting primeval MOFs, MOFs-based constituents (metal oxides, porous carbon, metal over porous carbon, and porous over metal oxide carbon), MOFs composites, should be a precedence for the positive applied use [27]. Figure 5 presents application of MOFs for pharmaceutical wastewater treatment.

3.1 Metal Organic Frameworks (MOFs) as Adsorbents

Porous constituents, for instance activated carbons, zeolites, carbon nanotubes, and mesoporous silica, have been utilized to eradicate organic impurities. Conversely, as PPCPs bring developed more miscellaneous, fresh categories of porous constituents with large surface area, ideal aperture dimensions, framework tuning, and control over active parts are prerequisite for the effectual elimination of these contaminants. As a result, MOFs are gifted constituents likened to traditional adsorbents [29].

Furthermore, adsorptive elimination by means of porous crystalline MOFs is an extremely effectual approach to impound pollutants. Miscellaneous interfaces, for instance p–p interactions, electrostatic interactions, hydrogen bonding, acid–base interfaces, connections with metal knots, and hydrophobic interactions, have been recommended between adsorbates and adsorbents. Various connections have been recognized for the effectual PPCPs adsorption [29].

Recent research developments in the adsorptive elimination of PPCP contaminants using MOFs. Numerous intermolecular interactions among MOFs and PPCPs could upsurge the removal efficacy of PPCPs developments in the adsorptive elimination

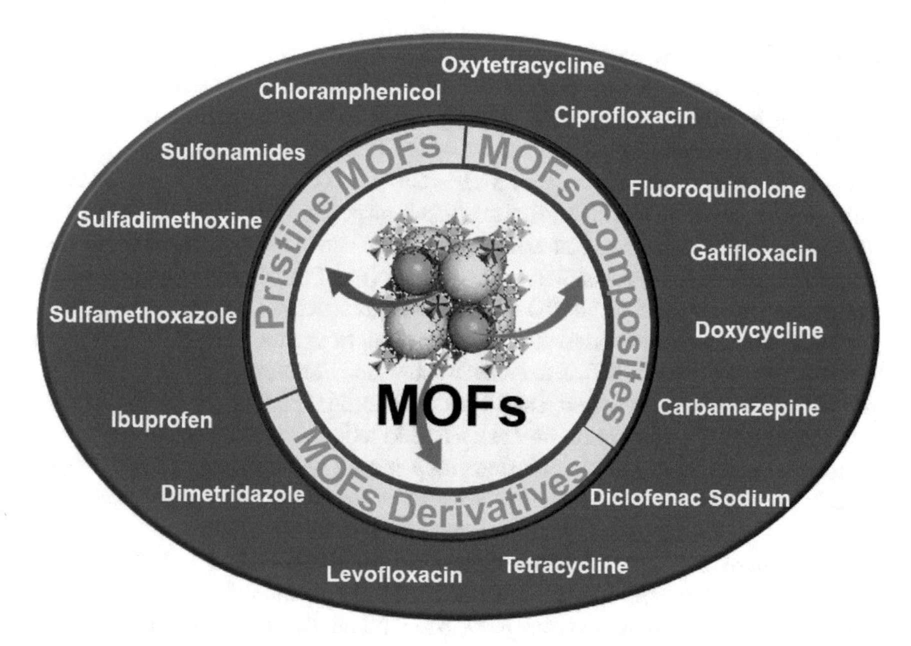

Fig. 6 Application of MOFs as adsorbents for different pharmaceutical eliminations. (Modified after [27])

of PPCP contaminants using MOFs. Conspicuously, multiple intermolecular interactions among MOFs and PPCPs could upsurge the exclusion efficacy of PPCPs [29]. Utilization of MOFs as adsorbents for various pharmaceuticals is shown in Fig. 6.

In a study, for the foremost while, adsorption of the characteristic PPCPs through MOFs, particularly MIL-100-Fe and MIL-101, has been evaluated to comprehend the features of adsorption and opportunity of utilizing MOF-based adsorbents aimed at elimination of PPCPs from polluted water. Successful adsorptive elimination of clofibric acid and naproxen, two characteristic PPCPs, has been planned to utilize MOFs. Furthermore, adsorption procedure might be elucidated with a humble electrostatic interaction among MOFs and PPCPs. Finally, it could be advocated that MOFs possessing greater porous nature and big pore aperture could be impending adsorbents to eliminate injurious PPCPs in polluted water [24]. In alternative report, a new MOF (Basolite A100) showed greater elimination competence to eliminate CBZ (carbamazepine and salicylic acid) and ibuprofen than the market powder activated carbon with kinetics of pseudo second order [30].

Furthermore, less cytotoxic UiO-66 (MOF) and NH_2-UiO-66 (MOF) have been stated as effectual adsorbent mediators. Methotrexate salt (MTX) was designated as the prototypical drug that was adsorbed onto internal pores and networks of MOFs through diffusion means. Finally, the results presented that UiO-66 possessed great adsorbing capability and abundant attraction towards MTX [1]. Similarly, UiO-66 MOFs have been prepared and used for the adsorption-based elimination of diclofenac (DCF) from PWW. UiO-66-NH_2 presented a greater adsorption ability as likened to UiO-66 (MOF), and the upsurge of activation temperature could enhance the adsorption ability of UiO-66-NH_2 for DCF. Furthermore, the adsorption procedure of DCF through UiO-66-NH_2(MOF) ought to be endothermic reaction and conformed the kinetics of pseudo second order. Moreover, the adsorption capabilities were 555 mg/g and 357 mg/g for MOFs-UiO-66-NH_2 and UiO-66, respectively. Ultimately, H-bonding and electrostatic interaction perhaps frolicked a major part for DCF adsorption. Hence, the synthesized UiO-66 formed MOFs evidenced to be effectual DCF elimination adsorbents from wastewater [85]. Furthermore, the comprehensive studies established that the UiO-66 presented great permeability and crystallinity. The MOF has been examined for anti-inflammatory drug (non-steroidal) adsorption, Ketorolac tromethamine (KTC), from the wastewater. Furthermore, it has been established that the supreme adsorption ability of the Uio-66 (729.92 mg g^{-1}) was attained in pH of three. The adsorption process tailed kinetic of pseudo second order with monolayer adsorption. Ultimately, this study served as an effectual approach for the elimination of KTC from wastewater and could endorse the perfect design of progressive MOFs adsorbents for conservational remediation [69].

Furthermore, Zr^{4+}-derived MOFs (imperfect MOF-808 and MOF-UiO-66) have amazing adsorption aptitude to eliminate anti-inflammation drugs from wastewater. Brilliant adsorption routines have been found for MOF-808 and UiO-66, the adsorption abilities are the uppermost in a varied sequence of adsorptive constituents hitherto stated, chiefly for UiO-66. Moreover, it has been explicated that the cationic imperfect-coordinated zirconium in the group has great attraction for the anionic pollutant, i.e., chemical adsorption, pharmaceutical, and that the interface of adsorption amongst benzene ring in pharmaceutical and ligand of MOF is included to improve the adsorption contacts (i.e., π–π interface). Particularly, adsorption of indomethacin, ketoprofen, naproxen, ibuprofen, and furosemide by MOFs; UiO-66 and MOF-808 and π–π interface and the synergetic influence of chemical adsorption are exceptional, providing tremendously great sorption abilities and binding energies [46].

Antibiotic remains in produced wastewater have appealed abundant courtesy for their severe harmfulness to surroundings. In this aspect, a Zr-derived MOF-PCN-777 has been produced as adsorbent to eliminate cephalexin from wastewater. Credited to synergistic influence of the coordination contact and electrostatic interface amongst cephalexin and MOF framework, PCN-777 displayed a great adsorption capability 442.48 mg g^{-1}. Furthermore, the adsorption progression charted pseudo-second-order kinetics. Additionally, this MOF could be recycled through an easy technique. Consequently, PCN-777 might have a gifted potential use in decontamination of wastewater comprising trace antibiotics [81]. Furthermore, great effectual elimination of chloramphenicol (CAP) drug from pharmaceutical wastewater has been observed through MOF-PCN-222. In this regard, PCN-222 revealed a big adsorption aptitude of 370 mg g^{-1} and more significantly, adsorption equilibrium could be rapidly attained at simply 58 s. In addition, approximately 99% of CAP could be detached from wastewater in less quantity (counting the concentrations present in actual water). Additional exploration predicted that hydrogen bonding, electrostatic force, and the distinct pore assembly of PCN-222 all exhibited significant impacts on

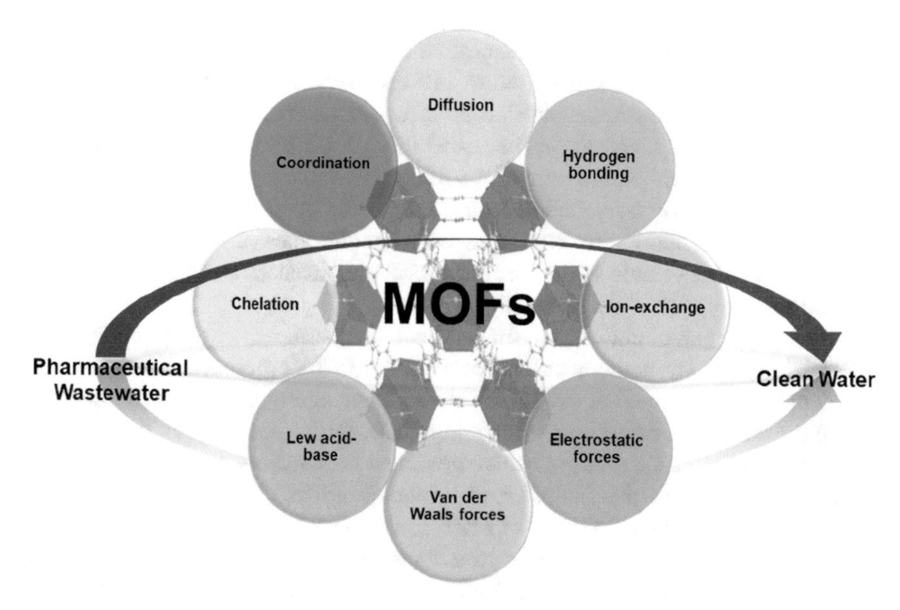

Fig. 7 Different forces/interactions involved in adsorption mechanism of MOFs for pharmaceutical wastewater treatment. (Modified after [48])

the brilliant elimination of CAP [82]. Figure 7 shows the important forces or interactions tangled in the adsorption mechanism of pharmaceuticals present in wastewater on the surface of MOFs.

Therapeutic medications are vital and crucial part for life. Wide-reaching, antibiotics are extensively utilized for prophylactic and therapeutic in veterinary and human medicine. Subsequently, they are obstinately unconfined into the atmosphere as wastewater from hospital, industrial, and domestic wastes. Exploration has demonstrated the release and existence of pharmacological medicines in wastewater management plants, lakes, and rivers. An adsorbent MOF $[Cu(Glu)_2\ (H_2O_2)]\cdot H_2O$ has been manufactured by the action of glutamic acid and Cu^{2+}. The prepared MOF was examined for ciprofloxacin adsorption present in PWWT. The adsorption front data discovered the capacity to adsorb of the MOF $[Cu(Glu)_2(H_2O)]\cdot H_2O$ was found to be 61.35 mg g^{1-} for ciprofloxacin medication. Furthermore, the adsorption of ciprofloxacin tailed a pseudo-second-order kinetics, suggesting the existence of physical sorption. Consequently, the outcomes recommended the usage of MOF $[Cu(Glu)_2(H_2O)]\cdot H_2O$ by way of an adsorbent for ciprofloxacin elimination from the wastewater [53].

Two MOFs, MOF-(525 and 545), contained Zr-oxide groups and porphyrin places in diverse structures have been utilized for adsorption-based elimination of the largely utilized organic pollutant sulfamethoxazole (SMX) from wastewater. Moreover, both MOFs proved to be highly effectual for adsorption of sulfamethoxazole showing supreme adsorption abilities of 690 and 585 mg g^{1-} for MOF-545 and 525, correspondingly. Interestingly, the former rate is the uppermost adsorption capability testified up to now for adsorption of sulfamethoxazole particles on any adsorbent. In addition, the adsorption of sulfamethoxazole led over a collection of typical MOFs with dissimilar physicochemical characteristics and thorough analysis established that great adsorption capability of the porphyrin-based MOFs is attained by hydrogen bonding amongst sulfamethoxazole particle and the nitrogen spots of the porphyrin components in the MOFs, the great surface area, and π–π interaction. Furthermore, the adsorbents were simply renewed by modest washing by acetone and recyclable with more than 95% effectiveness during four-repeated adsorption–desorption sequences [79].

Mounting organic pollutants, for instance endocrine disrupting compounds (EDCs), have been turned into serious concern due to their presence in wastewaters. The detailed arrangement of progressive treatments uniting multi-barrier treatment schemes is desired to remove emerging pollutants, for instance EDCs. Remarkably, numerous studies naked that innovative treatment constituents, for instance MOFs, are more effective for the exclusion of minor quantities of EDCs in wastewater. Furthermore, the technical virtues of MOF constituents have been greatly explored to give logical statistics and consideration of their ultimate strategy and groundwork course. Consequently, the green fabrication of MOFs that are water stable has been established, with numerous remarkable uses, for instance adsorption and separation [2].

Current research and growth exertions have shown evidence of concept elaborations of the utilization of MOFs as water soluble, operational adsorbents in wastewaters. To enhance current development in MOF as adsorbent, frugally practicable MOF adsorption procedures should be industrialized and tried on full scale [38].

3.2 *Metal Organic Frameworks (MOFs) Catalyzed Advanced Oxidation Processes (AOPs)*

Advanced oxidation progressions and adsorption have pursued the utilization of MOFs in water decontamination. Furthermore, numerous efforts have been done on eliminating the pollutants of industries, and an extensive variety of catalytic researches have lately been accomplished on the decontamination of water via MOF-aided AOPs (Fenton-founded, photocatalysis, and $SO_4^{\bullet -}$centered oxidative schemes). Particularly, two recommended main photocatalytic reactions are the metal-oxo bunches excitation in MOFs and charge transfer in ligand-to-metal through adsorbing photons in the presence of UV or/and normal light radiation [68].

Active and water-stable MOFs are significant resources for extenuation of water toxins through catalysis. For instance, a very water-stable cobalt MOF, specifically bio-MOF-11-Co, has been utilized as a catalyst for effective peroxymonsulfate activation in deprivation of sulfachloropyradazine and para-hydroxybenzoic acid. Furthermore, bio-MOF-11-Co displayed prompt deprivation of both sulfachloropyradazine and para-hydroxybenzoic acid and could be recycled many periods deprived of dropping the capacity. Consequently, the accessibility of electron opulent nucleobase adenine armored the kinetics of reaction via electron donation along with atoms of cobalt inside the structure of bio-MOF-11-Co [4].

Recently, iron-derived MOF that is MIL-88-A has been investigated mainly due to its facility to make active persulfate (PS) catalytically in an AOP for the deprivation of naproxen (NPX) in wastewater. Outcomes displayed that (i) MIL-88-A served as a decent PS activator while 65–70% NPX degradation was perceived, (ii) MIL-88-A proved to be reusable for four consecutive rounds with noteworthy elimination degree, (iii) phosphates posed no influence on the NPX degradation but, bicarbonates showed a robust reticence, (iv) activation of PS and degradation of NPX found to be best under acidic circumstances, and (v) PS produced SRs and HRs in the middle and was exposed to be greater oxidant than hydrogen peroxide in the existence of MIL-88-A. Consequently, the system established that deprivation of NPX stood at the MIL-88-A surface creating the heterogeneous catalysis AOP-grounded expertise fruitful for the exclusion of pharmacological composites from wastewater [19]. General representation of MOF catalyzed advanced oxidation process is presented in Fig. 8.

A distinctive iron-grounded MOF (namely MIL-100(Fe)) has been selected as a specimen in the peroxydisulfate (PDS) activation for the elimination of antibiotic contaminants. Remarkably, an effect of auto-acceleration has been detected in the mechanism of activating PDS with MIL-100(Fe) assisted by visible irradiation. Comparatively, the decaying competence of sulfamethoxazole (SMX) attained in the normal light aided activation of PDS by MIL-100(Fe) procedure was improved by 5.6 and 2.1 terms to the procedures with photo-activated PDS unaccompanied and iron-based MIL-100-activated peroxydisulfate single-handedly. Furthermore, the photo-generated electrons in MIL-100(Fe) caused the reduction of fundamental Fe^{3+} to Fe^{2+} that sequentially enhanced the activation of PDS competence in the creation of $O_2^{-\cdot}$ and ·OH radicals for the elimination of sulfamethoxazole [78]. On

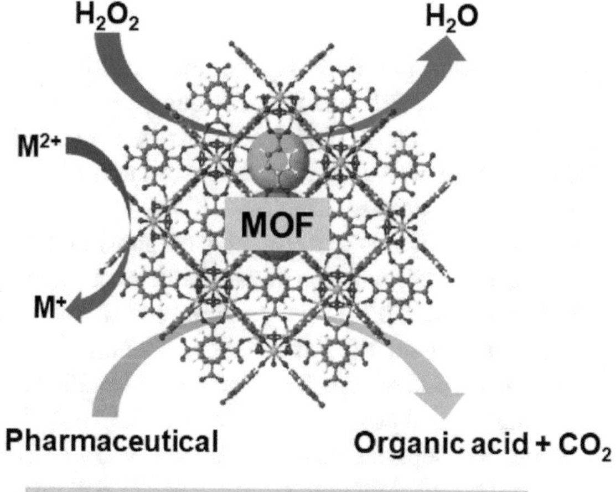

Fig. 8 General representation of Metal organic framework catalyzed advanced oxidation. (Modified after [57])

the other hand, a brilliant Fenton-like catalyst based on CUS-MIL-100(Fe) (MOF) and MIL-100(Fe) (MOF) has been utilized for sulfamethazine (SMT) degradation. Fascinatingly, the outcomes displayed that CUS-MIL-100(Fe) efficiently degraded sulfamethazine having 100% elimination competence in 180 min. Furthermore, CUS-MIL-100(Fe) demonstrated greater catalytic action likened to MIL-100(Fe) for decay of SMT. Ultimately, the improved catalytic action could be credited to the integration of Fe^{2+} and Fe^{3+} CUSs, the huge surface area and construction of mesoporous structures. Finally, CUS-MIL-100(Fe) showed decent constancy and recyclability [70]. Similarly, MIL-100(Fe) has also been utilized as an effective catalyst (photo-Fenton) for eliminating a few PPCPs (ibuprofen, theophylline, in addition to bisphenol A) in UV–Vis. Lately, research has validated the efficiency of MOF heterogeneous catalysts meant for photo-Fenton elimination of PPCPs [44].

Furthermore, a trivalent Fe-tartaric acid MOF (T2-MOF) has been effectively utilized as a catalyst for succinonitrile ozonation. Indeed, T2-MOF possesses clear crystal features and even structure. Thus, T2-MOF unveiled robust catalytic efficiency in succinonitrile ozonation. Furthermore, in the process, the retrieval rate for COD of 100 mg/L succinonitrile extended 73.1% in just 180 min that was 67.3% greater than that found in the procedure deprived of catalyst. Finally, the Fe^{2+} amount at different time intervals confirmed that the homogeneous catalysis happened concurrently with heterogeneous catalysis [75].

Likened to conventional AOPs, methods in which MOFs are taking part display tremendous efficiency in pharmaceuticals degradation. Thus, MOFs show a significant part in the progress in AOP methods, for instance Fenton photo-electro catalysis, ozone oxidation and AOPs comprising sulfate radicals [84].

3.3 Metal Organic Frameworks (MOFs)-Hybrid Systems (HSs)

Among different potential expenditures, MOFs and MOF-derived nanostructured constituents are widely utilized to eliminate environmental pollutants, mainly PPCPs from the wastewaters. The MOFs-derived nano-adsorbents have gained greater interest because of their exclusive construction, physicochemical characteristics, and great adsorptive presentation [64].

MOF-based hybrid systems and composites have been extensively applied in the elimination of pharmaceuticals from wastewater. Different hybrids/composites grounded on MOFs have been prepared and characterized for the successful elimination of pharmaceuticals. For instance, calcium alginate/MOF microsphere composites have been utilized for the levofloxacin (antibiotic) adsorption, from wastewater. Interestingly, great adsorption ability of 86.43 mg g^{1-} has been attained, which was far larger than separate calcium alginate or UiO-66. Furthermore, the recyclability experiment established greater than 70% adsorption of levofloxacin subsequently for five consecutive series [47]. Likewise, in another research, composite of chitosan-sodium alginate derived aluminum-based MOF has been arrayed for the alleviation of bisphenol A present in the wastewater. Tentative outcomes displayed greater adsorptive presentation per Al-MOF/SA complements. Furthermore, principal processes tangled in adsorption were cation–π interface, π–π stacking, and hydrogen bonding. Finally, as-prepared blobs were renewed and reused with the preservation of greater than 95% performance for adsorption next five constant bunch sequences [47].

A decent spongy metal organic framework composite has been produced by joining MIL-101(Cr) to graphene oxide (GnO). In addition, MIL-101/GnO composite' porosity was amplified by putting together graphene oxide with the MIL–101 equipped to a definite ratio. Furthermore, the MIL-101/GnO amalgams were applied for adsorption of anti-inflammatory compounds, for instance ketoprofen (KTP) and naproxen (NAP) from wastewater. Interestingly, MIL-101/GnO composites showed greatly enhanced adsorption headed for both KTP and NAP comparative to virgin MIL-101 and market-activated carbon. Furthermore, MIL-101/GnO(3%) possessed 1.4 and 2.1 times than those of original MIL-101 and market activated carbon, respectively. In addition, the enriched adsorption efficiency for NAP had been attributed to hydrogen bonding owing to the attendance of numerous functional clusters in the composites. Moreover, MIL-101/GnO could be renewed deprived of severe descent by ethanol wash and could be reprocessed for consecutive adsorption. Consequently, the MIL-101/GnO composite had been recommended as an effectual adsorbent to exclude anti-inflammatory compounds from wastewater [67].

With the intention of investigating the Excellency of MOF-composites for the elimination of tetracycline present in pharmaceutical wastewater, classic MOFs (ZIF-67, ZIF-8, Fe-BTC, and HKUST-1) had been incorporated into economic matrix of chitosan. Initially, composites blobs of metal oxide or metal hydroxide/chitosan as of the metallic salt-based chitosan blend were employed as MOF predecessor that provided even and firm MOF filling into chitosan. Among these, ZIF-8-based chitosan showed decent adsorption efficiency for tetracycline antibiotic. Furthermore, the adsorption tailed pseudo-second-order kinetics and the supreme adsorption capability could touch 495.04 mg g^{1-}, which was greater than many of the MOF-derived or regular polymer-derived adsorbents for tetracycline. Additionally, the exclusion competence of ZIF-8-derived chitosan hybrid for tetracycline was found to be greater than 90% even after 10 times adsorption–desorption rounds. Ultimately, the adsorption procedure had π–π stacking, hydrogen bonding, and electrostatic interface [81]. Generally, in view of the less reusability of fine adsorbents, an effectual technique was required to construct fresh three-dimensional (3D) adsorbents. Hence, by considering the benefit of the iron 3D framework and commendable adsorption efficiency of the Fe-grounded MOF (Fe-MOF), a sequence of Fe-founded MOFs (MIL-53(Fe), MIL-101(Fe), and MIL-100(Fe)) against the iron framework had been investigated for the elimination of tetracycline (TC). Unsurprisingly, the Fe/MILs exhibited the greater adsorption capability of powdered MILs(Fe), whereas tremendous reprocessing presentation was successfully achieved. Furthermore, iron-based MIL-100(Fe) unveiled the uppermost efficiency for tetracycline (TC), greater than 95% tetracycline might be detached at fifth cycle of adsorption. Moreover, protocol designated that the hydrogen bonding and π–π connections flanked by TC, and iron-based MIL-100(Fe) served an indispensable part in the mechanism of adsorption. In addition, the adsorbent iron-derived MIL-100(Fe) was found to be water stable with minor iron leakage in solution. Interestingly, even in real wastewater (pharmaceutical wastewater, river water, and tap water), great efficiency was exhibited by the adsorbent Fe@MIL-100(Fe). Ultimately, this research delivered a newfangled technique to make adsorbents of 3D MILs(Fe) for wastewater treatment [39].

Magnetic nanocomposites based on MOFs have grown interest due to their simplistic construction, amendment, expedient retrieval, and eco-friendly. Furthermore, these also possess notable landscapes, for instance enormous surface area, minor size, dispersion, and distinct adsorptive aptitude. Prominently, the solid–liquid retrieval could be effortlessly comprehended utilizing an exterior magnetic source. Subsequently, these nanocomposites based on MOFs could be regularly engaged as magnetic adsorbents to efficiently detach and supplement an extensive collection of minor components in conservational samples. Furthermore, the adsorption's specificity and electivity of projected materials could be achieved by the amendment of such MOF-magnetic nanoparticles [64]. General scheme of magnetic exclusion of pharmaceuticals from wastewater is shown in Fig. 9.

A magnetic MOF Fe_3O_4@(Iron-(benzene-1,3,5-tricarboxylic acid) has been made, evaluated as a magnetic adsorbent to serve in dispersive solid-phase extraction (DSPE) of numerous extensively utilized blood phospholipid controllers (namely, clofibrate, clofibric acid, bezafibrate, gemfibrozil, plus fenofibrate) present in wastewaters. Furthermore, the highest absorption abilities were found to be 620.3 mg g^{1-} for clofibric acid, 197 mg g^{1-} for bezafibrate, 537.6 mg g^{1-} for clofibrate, 223.2 mg g^{1-} for fenofibrate, and 288.7 mg g^{1-} for gemfibrozil. Endorsements of the improved magnetic DSPE technique for investigation in sharp wastewater

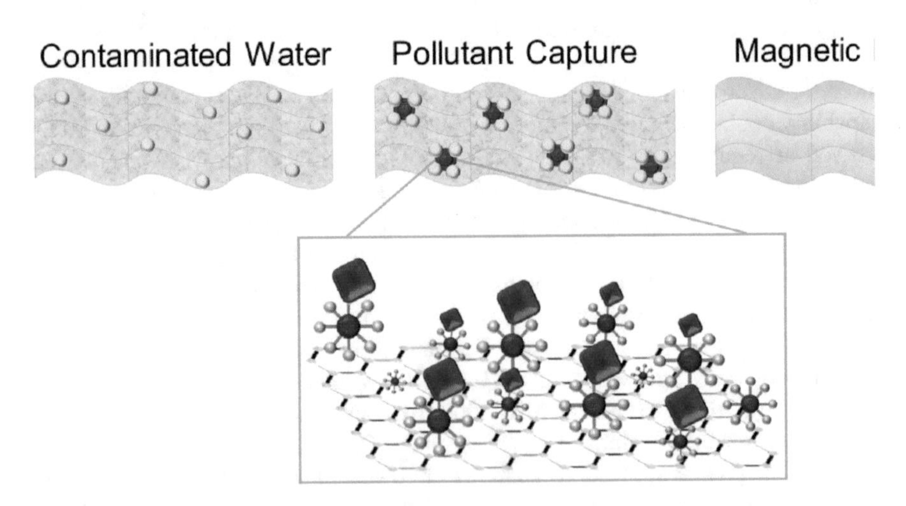

Fig. 9 General representation of magnetic removal of pollutants from wastewaters. (Modified after [72])

provided relative retrieval values equal to 70% (clofibrate) and between the array of 80% and 100% for fenofibrate, gemfibrozil, clofibric acid, and bezafibrate [60]. In another approach, a magnetic copper-derived MOF Fe_3O_4/HKUST-1 had been produced and utilized as an actual and reusable adsorbent for the elimination of antibiotics: ciprofloxacin (CIP) and norfloxacin (NOR) present in wastewater. Furthermore, Fe_3O_4/HKUST-1 possessed great adsorption degree, and it was observed that NOR and CIP could be detached in 30 min. Additionally, the greatest adsorption abilities of the aggregates toward NOR and CIP touched as tall as 513 mg/g and 538 mg/g, respectively, far greater as compared with most of the stated adsorbents for these two antibiotics. Furthermore, the magnetization saturation worth of composite was found to be 44 emu/g that was enough for the parting of the adsorbent-Fe_3O_4@HKUST-1 from wastewater by the use of an outside magnetic arena. Furthermore, the prepared magnetic hybrid displayed a decent recyclability with the capacity of adsorption lessening only a little after recycle for 10 cycles. Consequently, such outcomes designated that the Fe_3O_4@HKUST-1 magnetic composites might be an auspicious adsorbent toward refinement of antibiotics from wastewater because of its great adsorption proficiency, debauched kinetics, calm separation from wastewater/water, and outstanding reusability [77]. Moreover, MIL-53(Fe) MOF along with its magnetic hybrid MIL-53(Fe)/Fe_3O_4 has been applied for the elimination of doxycycline from wastewaters. It was found that under adjusted circumstances, the adsorption ability of 322 mg g^{-1} was attained. Furthermore, the utilized adsorbent was effortlessly parted from the water by putting on outside magnetic arena. The renewed adsorbent reserved most of its original capability after six renewal periods [52].

Recently, MOF-derived hybrid systems have gained extensive attention due to their double benefits. For example, MOFs have been united with ultrafiltration (UF) to form hybrid systems (MOF-UF) for the treatment of selected pharmaceutically active compounds (PhACs), counting α-ethinyl estradiol and ibuprofen. Owing to the great porosity of MOFs, the hybrid systems exhibited enhanced capacity for eliminating pollutants and decreasing entangling in UF-adsorbent hybrid modules. Furthermore, the typical retaining rate of PhACs in (53.2%) MOF-based UF was improved comparative to the (36.7%) UF only. Ultimately, the PhACs had been successfully adsorbed onto the MOF surface because of their robust porous features. Moreover, in comparison of MOF-based UF and activated carbon-based UF (AC-UF) hybrid schemes, the typical retaining rates of PhACs for the PAC-UF came out to be lesser than MOF-UF [34].

In contrast, MOFs-derived nanofiltration (NF) films have also been fabricated for the effectual elimination of organic pollutants from wastewater. For instance, MOF-UiO-66 derived widespread, efficient, and malleable nanofiltration (NF) membrane had been investigated antibiotics parting. Initially, the substratum was built via two periods; (i) graphene oxide (GO) sheets doping in solution of membrane casting comprising polyacrylonitrile (PAN) thus founding 2D-3D linking pores by process of phase inversion (GO@PAN) and (ii) dipping of the substrate of GO@PAN in solution of dopamine for self-polymerization to chain macromolecules that acquired good attuned and malleable substrate (PGP). Furthermore, the synthesized UiO-66 was coated onto the substrate of PGP through vacuum-aided filtration. Owing

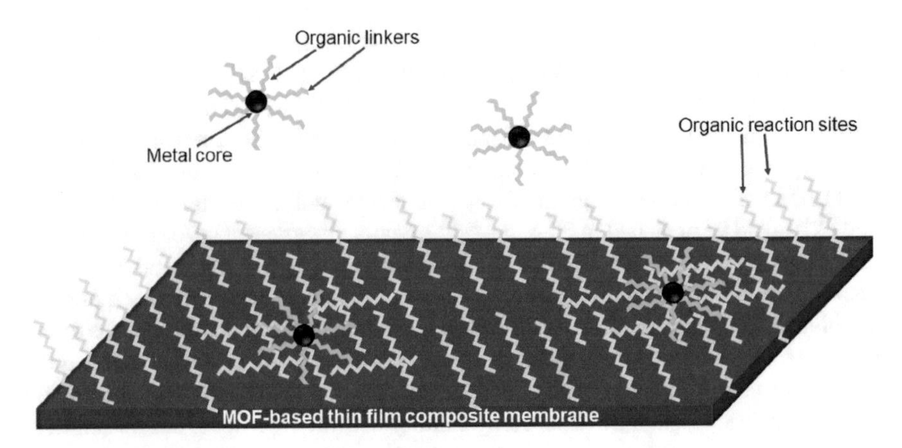

Fig. 10 General representation of metal organic framework-based thin film composite membrane. (Modified after [31])

to stability toward water, negative charge, and permeable structure of UiO-66, an excellent filtration presentation of the thin film composite (TFC) membrane of UiO-66/PGP was attained. Furthermore, the refusal rates for antibiotics (Oxytetracycline, Ciprofloxacin, and Tetracycline hydrochloride) were found to be greater than 94%. Consequently, the equipped UiO-66/PGP TFC membrane proved to be capable of wastewater handling, purification, and separation, in numerous industrial and pharmacological arenas [20]. General representation of MOF-based TFC membrane is shown in Fig. 10.

As nanofiltration can serve as an expedient means to eliminate pharmaceuticals in wastewater. The efficiency of the supremely utilized TFC layers, characteristically having a reedy polyamide-polymer (PA) cover, could be enhanced utilizing thin film nanocomposite (TFN) films gained after the institution of filler inside the PA layer. For instance, to regulate the filler incorporation, two types of MOF-based PA bilayer TFC (BTFC) films, HKUST-1 (MOF)-based PA and ZIF-93-based PA have been manufactured. Initially, the interfacial production was employed for the fabrication of a MOF film, and then, synthesis of PA film through interfacial polymerization. Furthermore, prepared BTFC films had been evaluated in the nanofiltration of naproxen and diclofenac present in wastewater attaining a supreme rejection of $\geq$ 98% while HKUST-1 was applied. Consequently, the enhancements were linked to the thickness of PA layer, membrane hydrophilicity, membrane roughness, and MOF porosity [58].

Recently, MOF-based membranes have been utilized in extensive applications because of their decent stability and greatly adaptable pore structure. Particularly, both the great permeability and selectivity MOF have occasioned in the increasing uses of MOF-based membranes toward membrane-dependent liquid segment parting, specifically for management of wastewater and water renewal [39].

4 Conclusion and Future Perspective

Earth waters are polluted by pharmaceutically active compounds (PhACs) that are considered as prevalent pollutants. There is mounting proof that lingering pharmaceuticals could consequence in adverse effects on non-targeted beings at low concentration intensities. Hence, the operative management of chemical-derived pharmaceutical wastewater is of pronounced importance to resolve water ecological contamination.

Indeed, the distinguished residual drug amount in wastewater is inferior to other contaminants, but the detrimental influences on aquatic creatures, environment, and human life could be very imperative. Thus, employing appropriate treatment schemes for pharmaceutical wastewater has lately grown interest by researchers/scientists. In fact, there are numerous treatment procedures for industrial wastewater, for instance chemical, biological, and physical procedures. The small quantity of BOD and large COD in pharmaceutical wastewater pose an argument for biological treatment procedures, as resistant chemical constituents in wastewater might extremely limit the microbes' activities. Hence, physical and chemical procedures apart from biological methods are required to manage this type of wastewater.

Among the effective chemical and physical processes utilized in industrial wastewater management: are efficiently eliminating stubborn pollutants from wastewaters assisted with metal-organic frameworks (MOFs). Research reports have proved that MOFs could eliminate the pharmaceutical pollutants from wastewater using adsorption or/and as a catalyst in advanced oxidation processes. Furthermore, the hybrid systems of MOFs have widened their potential to serve as a formidable candidate in wastewater treatment generated by pharmaceutical industries. These hybrid systems result in more efficiency as compared with individual MOFs.

Nevertheless, the elimination of PPCPS by means of MOFs is unfortunately in its prompt phases. Hence, more research and attention is required in order to achieve excellence in the utilization of MOF-based systems for the treatment of pharmaceutical wastewaters, as these wastewaters have become one of the alarming concerns of the environment.

References

1. Aghajanzadeh M, Zamani M, Molavi H, Khieri Manjili H, Shojaei DH, A, (2018) Preparation of Metal-Organic Frameworks UiO-66 for Adsorptive Removal of Methotrexate from Aqueous Solution. J Inorg Organomet Polym Mater 28:177–186. https://doi.org/10.1007/s10904-017-0709-3
2. Aris AZ, Mohd Hir ZA, Razak MR (2020) Metal-organic frameworks (MOFs) for the adsorptive removal of selected endocrine disrupting compounds (EDCs) from aqueous solution: A review. Appl Mater Today 21. https://doi.org/10.1016/j.apmt.2020.100796

3. Awfa D, Ateia M, Yoshimura FM, C, (2019) Novel Magnetic Carbon Nanotube-TiO2 Composites for Solar Light Photocatalytic Degradation of Pharmaceuticals in the Presence of Natural Organic Matter. Journal of Water Process Engineering 31. https://doi.org/10.1016/j.jwpe.2019.100836
4. Azhar MR, Vijay P, Tadé MO, Wang SH, S, (2018) Submicron sized water-stable metal organic framework (bio-MOF-11) for catalytic degradation of pharmaceuticals and personal care products. Chemosphere 196:105–114. https://doi.org/10.1016/j.chemosphere.2017.12.164
5. Balarak D, Bandani F, Ahmed SZ, N, (2020) Adsorption Properties of Thermally Treated Rice Husk for Removal of Sulfamethazine Antibiotic from Pharmaceutical Wastewater. Journal of Pharmaceutical Research International 84–92. https://doi.org/10.9734/jpri/2020/v32i830475
6. Balarak D, Chandrika KAD, K, (2020) Antibiotics Removal from Aqueous Solution and Pharmaceutical Wastewater by Adsorption Process: A Review. International Journal of Pharmaceutical Investigation 10:106–111. https://doi.org/10.5530/ijpi.2020.2.19
7. Balasubramani K, Naushad SN, M, (2020) Effective adsorption of antidiabetic pharmaceutical (metformin) from aqueous medium using graphene oxide nanoparticles: Equilibrium and statistical modelling. J Mol Liq 301. https://doi.org/10.1016/j.molliq.2019.112426
8. Başaran Dindaş G, Çalışkan Y, Çelebi EE, Tekbaş M, Yatmaz BN, HC, (2020) Treatment of pharmaceutical wastewater by combination of electrocoagulation, electro-fenton and photocatalytic oxidation processes. J Environ Chem Eng 8. https://doi.org/10.1016/j.jece.2020.103777
9. Bhandari M, Bhushan S, Rana M, Raychaudhuri S, Simsek H Prajapati SK (2020). Algae-and bacteria-driven technologies for pharmaceutical remediation in wastewater: 373–408.
10. Cao V, Yunessnia lehi A, Bojaran M Fattahi M, (2020) Treatment of Lasalocid A, Salinomycin and Semduramicin as ionophore antibiotics in pharmaceutical wastewater by PAMAM-coated membranes. Environ Technol Innov 20. https://doi.org/10.1016/j.eti.2020.101103
11. Carmalin Sophia A, Lima EC, Rajan AN, S, (2016) Application of graphene based materials for adsorption of pharmaceutical traces from water and wastewater- a review. Desalin Water Treat 57:27573–27586. https://doi.org/10.1080/19443994.2016.1172989
12. Carrasco S (2018) Metal-Organic Frameworks for the Development of Biosensors: A Current Overview. Biosensors 8. https://doi.org/10.3390/bios8040092
13. Changotra R, Rajput H, Guin JP, Dhir KSA, A, (2020) Techno-economical evaluation of coupling ionizing radiation and biological treatment process for the remediation of real pharmaceutical wastewater. J Clean Prod 242. https://doi.org/10.1016/j.jclepro.2019.118544
14. Chen L, Cheng P, Ye L, Chen H, Zhu XuX, L, (2020) Biological performance and fouling mitigation in the biochar-amended anaerobic membrane bioreactor (AnMBR) treating pharmaceutical wastewater. Biores Technol 302. https://doi.org/10.1016/j.biortech.2020.122805
15. Chen Z, Min H, Hu D, Wang H, Zhao Y, Cui Y, Zou X, Wu P, Ge H, Luo K, Liu ZL, W, (2020) Performance of a novel multiple draft tubes airlift loop membrane bioreactor to treat ampicillin pharmaceutical wastewater under different temperatures. Chem Eng J 380. https://doi.org/10.1016/j.cej.2019.122521
16. Couto CF, Santos AV, Amaral MCS, Lange LC, de Andrade LH, Fernandes FAFS, BS, (2020) Assessing potential of nanofiltration, reverse osmosis and membrane distillation drinking water treatment for pharmaceutically active compounds (PhACs) removal. Journal of Water Process Engineering 33. https://doi.org/10.1016/j.jwpe.2019.101029
17. Deegan AM, Shaik B, Nolan K, Urell K, Oelgemöller M, Morrissey TJ, A, (2011) Treatment options for wastewater effluents from pharmaceutical companies. Int J Environ Sci Technol 8:649–666. https://doi.org/10.1007/BF03326250
18. Dhangar K, Kumar M (2020) Tricks and tracks in removal of emerging contaminants from the wastewater through hybrid treatment systems: A review. Sci Total Environ 738. https://doi.org/10.1016/j.scitotenv.2020.140320
19. El Asmar R, Baalbaki A, Abou Khalil Z, Naim S, Ghauch BA, A, (2021) Iron-based metal organic framework MIL-88-A for the degradation of naproxen in water through persulfate activation. Chem Eng J 405. https://doi.org/10.1016/j.cej.2020.126701

20. Fang S-Y, Zhang P, Gong J-L, Tang L, Zeng G-M, Song B, Cao W-C, Ye LJ, J, (2020) Construction of highly water-stable metal-organic framework UiO-66 thin-film composite membrane for dyes and antibiotics separation. Chem Eng J 385. https://doi.org/10.1016/j.cej.2019.123400
21. Fu J, Zhang Q, Huang B, Jin FN, R, (2020) A review on anammox process for the treatment of antibiotic-containing wastewater: Linking effects with corresponding mechanisms. Front Environ Sci Eng 15:17. https://doi.org/10.1007/s11783-020-1309-y
22. Gogate PR, Pandit AB (2004) A review of imperative technologies for wastewater treatment II: hybrid methods. Adv Environ Res 8:553–597. https://doi.org/10.1016/S1093-0191(03)00031-5
23. Guedes-Alonso R, Montesdeoca-Esponda S, Pacheco-Juárez J, Sosa-Ferrera Z Santana-Rodríguez JJ (2020) A Survey of the Presence of Pharmaceutical Residues in Wastewaters. Evaluation of Their Removal using Conventional and Natural Treatment Procedures. Molecules 25:doi:https://doi.org/10.3390/molecules25071639
24. Hasan Z, Jhung JJ, SH, (2012) Adsorptive removal of naproxen and clofibric acid from water using metal-organic frameworks. J Hazard Mater 209–210:151–157. https://doi.org/10.1016/j.jhazmat.2012.01.005
25. Hasanpour M, Hatami M (2020) Photocatalytic performance of aerogels for organic dyes removal from wastewaters: Review study. J Mol Liq 309. https://doi.org/10.1016/j.molliq.2020.113094
26. Heberer T (2002) Occurrence, fate, and removal of pharmaceutical residues in the aquatic environment: a review of recent research data. Toxicol Lett 131:5–17. https://doi.org/10.1016/S0378-4274(02)00041-3
27. Huang L, Shuai SR, Q, (2021) Adsorptive removal of pharmaceuticals from water using metal-organic frameworks: A review. J Environ Manage 277. https://doi.org/10.1016/j.jenvman.2020.111389
28. Jakab G, Szalai Z, Michalkó G, Ringer M, Filep T, Szabó L, Maász G, Pirger Z, Ferincz Á, Staszny Á, Kondor DP, AC, (2020) Thermal baths as sources of pharmaceutical and illicit drug contamination. Environ Sci Pollut Res 27:399–410. https://doi.org/10.1007/s11356-019-06633-6
29. Jin E, Lee S, Kang E, Choe KY, W, (2020) Metal-organic frameworks as advanced adsorbents for pharmaceutical and personal care products. Coord Chem Rev 425. https://doi.org/10.1016/j.ccr.2020.213526
30. Jun B-M, Kim S, Heo J, Her N, Jang M, Yoon PCM, Y, (2019) Enhanced sonocatalytic degradation of carbamazepine and salicylic acid using a metal-organic framework. Ultrason Sonochem 56:174–182. https://doi.org/10.1016/j.ultsonch.2019.04.019
31. Kadhom M, Deng HuW, B, (2017) Thin Film Nanocomposite Membrane Filled with Metal-Organic Frameworks UiO-66 and MIL-125 Nanoparticles for Water Desalination. Membranes 7. https://doi.org/10.3390/membranes7020031
32. Khetan SK, Collins TJ (2007) Human Pharmaceuticals in the Aquatic Environment: A Challenge to Green Chemistry. Chem Rev 107:2319–2364. https://doi.org/10.1021/cr020441w
33. Khulbe KC, Matsuura T (2018) Removal of heavy metals and pollutants by membrane adsorption techniques. Appl Water Sci 8:19. https://doi.org/10.1007/s13201-018-0661-6
34. Kim S, Muñoz-Senmache JC, Jun B-M, Park CM, Jang A, Yu M, Yoon H-M, Y, (2020) A metal organic framework-ultrafiltration hybrid system for removing selected pharmaceuticals and natural organic matter. Chem Eng J 382. https://doi.org/10.1016/j.cej.2019.122920
35. Kolpin DW, Furlong ET, Meyer MT, Thurman EM, Zaugg SD, Buxton BLB, HT, (2002) Pharmaceuticals, Hormones, and Other Organic Wastewater Contaminants in U.S. Streams, 1999–2000: A National Reconnaissance. Environ Sci Technol 36:1202–1211. https://doi.org/10.1021/es011055j
36. Kumar P, Bansal V, Kwon K-H, EE, (2018) Metal-organic frameworks (MOFs) as futuristic options for wastewater treatment. J Ind Eng Chem 62:130–145. https://doi.org/10.1016/j.jiec.2017.12.051
37. Kyzas GZ, Fu J, Lazaridis NK, Matis BDN, KA, (2015) New approaches on the removal of pharmaceuticals from wastewaters with adsorbent materials. J Mol Liq 209:87–93. https://doi.org/10.1016/j.molliq.2015.05.025

38. Lee Y-J, Chang Y-J, Hsu L-J, J-P, (2018) Water stable metal-organic framework as adsorbent from aqueous solution: A mini-review. J Taiwan Inst Chem Eng 93:176–183. https://doi.org/10.1016/j.jtice.2018.06.035
39. Li C, Mei Y, Qi G, Xu W, Shen ZY, Y, (2020) Degradation characteristics of four major pollutants in chemical pharmaceutical wastewater by Fenton process. J Environ Chem Eng 104564. https://doi.org/10.1016/j.jece.2020.104564
40. Li J, Wang H, Yuan X, Chew ZJ, JW, (2020) Metal-organic framework membranes for wastewater treatment and water regeneration. Coord Chem Rev 404. https://doi.org/10.1016/j.ccr.2019.213116
41. Li W, Cao J, Xiong W, Yang Z, Sun S, Xu JM, Z, (2020) In-situ growing of metal-organic frameworks on three-dimensional iron network as an efficient adsorbent for antibiotics removal. Chem Eng J 392. https://doi.org/10.1016/j.cej.2020.124844
42. Li X, Wang B, Cao Y, Zhao S, Wang H, Feng X, Ma ZJ, X, (2019) Water Contaminant Elimination Based on Metal-Organic Frameworks and Perspective on Their Industrial Applications. ACS Sustainable Chemistry & Engineering 7:4548–4563. https://doi.org/10.1021/acssuschemeng.8b05751
43. Li Z, Yang P (2018) Review on Physicochemical, Chemical, and Biological Processes for Pharmaceutical Wastewater. IOP Conference Series: Earth and Environmental Science 113. https://doi.org/10.1088/1755-1315/113/1/012185
44. Liang R, Luo S, Jing F, Shen L, Wu QN, L, (2015) A simple strategy for fabrication of Pd@MIL-100(Fe) nanocomposite as a visible-light-driven photocatalyst for the treatment of pharmaceuticals and personal care products (PPCPs). Appl Catal B 176–177:240–248. https://doi.org/10.1016/j.apcatb.2015.04.009
45. Lillenberg M, Yurchenko S, Kipper K, Herodes K, Pihl V, Lõhmus R, Ivask M, Kuu A, Kutti S, Nei LSV, L, (2010) Presence of fluoroquinolones and sulfonamides in urban sewage sludge and their degradation as a result of composting. Int J Environ Sci Technol 7:307–312. https://doi.org/10.1007/BF03326140
46. Lin S, Yun ZY, Y-S, (2018) Highly Effective Removal of Nonsteroidal Anti-inflammatory Pharmaceuticals from Water by Zr(IV)-Based Metal-Organic Framework: Adsorption Performance and Mechanisms. ACS Appl Mater Interfaces 10:28076–28085. https://doi.org/10.1021/acsami.8b08596
47. Luo Z, Chen H, Wu S, Cheng YC, J, (2019) Enhanced removal of bisphenol A from aqueous solution by aluminum-based MOF/sodium alginate-chitosan composite beads. Chemosphere 237. https://doi.org/10.1016/j.chemosphere.2019.124493
48. Manousi N, Giannakoudakis DA, Zachariadis RE, GA, (2019) Extraction of Metal Ions with Metal-Organic Frameworks. Molecules 24. https://doi.org/10.3390/molecules24244605
49. Mostafapour FK, Baniasadi BD, M, (2018) Removal of ciprofloxacin from of pharmaceutical wastewater by adsorption on SiO2 nanoparticle. Journal of Pharmaceutical Research International 1–9. https://doi.org/10.9734/jpri/2018/v25i630127
50. Mukimin A, Zen VH, N, (2020) Hybrid advanced oxidation process (HAOP) as highly efficient and powerful treatment for complete demineralization of antibiotics. Sep Purif Technol 241. https://doi.org/10.1016/j.seppur.2020.116728
51. Murshid S, Dhakshinamoorthy GP (2020) Application of an immobilized microbial consortium for the treatment of pharmaceutical wastewater: Batch-wise and Continuous studies. Chin J Chem Eng. https://doi.org/10.1016/j.cjche.2020.04.008
52. Naeimi S, Faghihian H (2017) Application of novel metal organic framework, MIL-53(Fe) and its magnetic hybrid: For removal of pharmaceutical pollutant, doxycycline from aqueous solutions. Environ Toxicol Pharmacol 53:121–132. https://doi.org/10.1016/j.etap.2017.05.007
53. Olawale MD, Tella AC, Olatunji OJA, JS, (2020) Synthesis, characterization and crystal structure of a copper-glutamate metal organic framework (MOF) and its adsorptive removal of ciprofloxacin drug from aqueous solution. New J Chem 44:3961–3969. https://doi.org/10.1039/D0NJ00515K

54. Oliveira G, Calisto V, Santos SM, Esteves OM, VI, (2018) Paper pulp-based adsorbents for the removal of pharmaceuticals from wastewater: A novel approach towards diversification. Sci Total Environ 631–632:1018–1028. https://doi.org/10.1016/j.scitotenv.2018.03.072
55. Oulebsir A, Chaabane T, Tounsi H, Omine K, Sivasankar V, Darchen FA, A, (2020) Treatment of artificial pharmaceutical wastewater containing amoxicillin by a sequential electrocoagulation with calcium salt followed by nanofiltration. J Environ Chem Eng 8. https://doi.org/10.1016/j.jece.2020.104597
56. Ouyang E, Liu Y, Wang OJ, X, (2019) Effects of different wastewater characteristics and treatment techniques on the bacterial community structure in three pharmaceutical wastewater treatment systems. Environ Technol 40:329–341. https://doi.org/10.1080/09593330.2017.1393010
57. Pan Y, Jiang S, Xiong W, Liu D, Li M, He B, Luo FX, D, (2020) Supported CuO catalysts on metal-organic framework (Cu-UiO-66) for efficient catalytic wet peroxide oxidation of 4-chlorophenol in wastewater. Microporous Mesoporous Mater 291. https://doi.org/10.1016/j.micromeso.2019.109703
58. Paseta L, Antorán D, Téllez CJ, C, (2019) 110th Anniversary: Polyamide/Metal–Organic Framework Bilayered Thin Film Composite Membranes for the Removal of Pharmaceutical Compounds from Water. Ind Eng Chem Res 58:4222–4230. https://doi.org/10.1021/acs.iecr.8b06017
59. Patel S, Mondal S, Majumder SK, Ghosh DP, P, (2020) Treatment of a Pharmaceutical Industrial Effluent by a Hybrid Process of Advanced Oxidation and Adsorption. ACS Omega 5:32305–32317. https://doi.org/10.1021/acsomega.0c04139
60. Peña-Méndez EM, Mawale RM, Conde-González JE, Socas-Rodríguez B, Ruiz-Pérez HJ, C, (2020) Metal organic framework composite, nano-Fe3O4@Fe-(benzene-1,3,5-tricarboxylic acid), for solid phase extraction of blood lipid regulators from water. Talanta 207. https://doi.org/10.1016/j.talanta.2019.120275
61. Pirsaheb M, Janjani HH, H, (2020) Reclamation of hospital secondary treatment effluent by sulfate radicals based–advanced oxidation processes (SR-AOPs) for removal of antibiotics. Microchem J 153. https://doi.org/10.1016/j.microc.2019.104430
62. Qu Z, Dong G, Zhu S, Yu Y, Huo M, Liu XuK, M, (2020) Recycling of groundwater treatment sludge to prepare nano-rod erdite particles for tetracycline adsorption. J Clean Prod 257. https://doi.org/10.1016/j.jclepro.2020.120462
63. Quesada HB, Baptista ATA, Cusioli LF, Seibert D, de Oliveira BC, Bergamasco R (2019) Surface water pollution by pharmaceuticals and an alternative of removal by low-cost adsorbents: A review. Chemosphere 222:766–780. https://doi.org/10.1016/j.chemosphere.2019.02.009
64. Rasheed T, Bilal M, Hassan AA, Nabeel F, Bharagava RN, Romanholo Ferreira LF, Iqbal THN, HMN, (2020) Environmental threatening concern and efficient removal of pharmaceutically active compounds using metal-organic frameworks as adsorbents. Environ Res 185. https://doi.org/10.1016/j.envres.2020.109436
65. Razi M, Hamdan HM, R, (2017) Factor Affecting Textile Dye Removal Using Adsorbent From Activated Carbon: A Review. MATEC Web of Conferences 103:06015. https://doi.org/10.1051/matecconf/201710306015
66. Rueda Márquez JJ, Levchuk I, Sillanpää MM, M, (2020) Toxicity Reduction of Industrial and Municipal Wastewater by Advanced Oxidation Processes (Photo-Fenton, UVC/H2O2, Electro-Fenton and Galvanic Fenton): A Review. Catalysts 10:612. https://doi.org/10.3390/catal10060612
67. Sarker M, Jhung SJY, SH, (2018) Adsorptive removal of anti-inflammatory drugs from water using graphene oxide/metal-organic framework composites. Chem Eng J 335:74–81. https://doi.org/10.1016/j.cej.2017.10.138
68. Sharma VK, Feng M (2019) Water depollution using metal-organic frameworks-catalyzed advanced oxidation processes: A review. J Hazard Mater 372:3–16. https://doi.org/10.1016/j.jhazmat.2017.09.043

69. Singh S, Sharma S, Umar A, Jha M, Kansal MSK, SK, (2018) Nanocuboidal-shaped zirconium based metal organic framework for the enhanced adsorptive removal of nonsteroidal anti-inflammatory drug, ketorolac tromethamine, from aqueous phase. New J Chem 42:1921–1930. https://doi.org/10.1039/C7NJ03851H
70. Tang J, Wang J (2018) Metal Organic Framework with Coordinatively Unsaturated Sites as Efficient Fenton-like Catalyst for Enhanced Degradation of Sulfamethazine. Environ Sci Technol 52:5367–5377. https://doi.org/10.1021/acs.est.8b00092
71. Tran HN, Tomul F, Thi Hoang Ha N, Nguyen DT, Lima EC, Le GT, Chang C-T, Woo MV, SH, (2020) Innovative spherical biochar for pharmaceutical removal from water: Insight into adsorption mechanism. J Hazard Mater 394. https://doi.org/10.1016/j.jhazmat.2020.122255
72. Ventura K, Arrieta RA, Marcos-Hernández M, Jabbari V, Powell CD, Turley R, Lounsbury AW, Zimmerman JB, Gardea-Torresdey J, Villagrán WMS, D, (2020) Superparamagnetic MOF@GO Ni and Co based hybrid nanocomposites as efficient water pollutant adsorbents. Sci Total Environ 738. https://doi.org/10.1016/j.scitotenv.2020.139213
73. Viegas RMC, Mesquita E, Rosa CM, MJ, (2020) Pilot Studies and Cost Analysis of Hybrid Powdered Activated Carbon/Ceramic Microfiltration for Controlling Pharmaceutical Compounds and Organic Matter in Water Reclamation. Water 12. https://doi.org/10.3390/w12010033
74. Wang K, Liu S, He ZQ, Y, (2009) Pharmaceutical wastewater treatment by internal micro-electrolysis–coagulation, biological treatment and activated carbon adsorption. Environ Technol 30:1469–1474. https://doi.org/10.1080/09593330903229164
75. Wang S, Ma X, Liu Y, Yi X, Li DuG, J, (2020) Fate of antibiotics, antibiotic-resistant bacteria, and cell-free antibiotic-resistant genes in full-scale membrane bioreactor wastewater treatment plants. Biores Technol 302. https://doi.org/10.1016/j.biortech.2020.122825
76. Wang S, Zhu G, Yu Z, Li C, Cao WD, X, (2020) Trivalent iron-tartaric acid metal-organic framework for catalytic ozonation of succinonitrile. Water Sci Technol 81:2311–2321. https://doi.org/10.2166/wst.2020.272
77. Wu G, Ma J, Li S, Guan J, Jiang B, Wang L, Li J, Chen WX, L, (2018) Magnetic copper-based metal organic framework as an effective and recyclable adsorbent for removal of two fluoroquinolone antibiotics from aqueous solutions. J Colloid Interface Sci 528:360–371. https://doi.org/10.1016/j.jcis.2018.05.105
78. Yin R, Chen Y, He S, Li W, Zeng L, Zhu GW, M, (2020) In situ photoreduction of structural Fe(III) in a metal–organic framework for peroxydisulfate activation and efficient removal of antibiotics in real wastewater. J Hazard Mater 388. https://doi.org/10.1016/j.jhazmat.2019.121996
79. Yu K, Ahmed I, Won D-I, Ahn LWI, W-S, (2020) Highly efficient adsorptive removal of sulfamethoxazole from aqueous solutions by porphyrinic MOF-525 and MOF-545. Chemosphere 250. https://doi.org/10.1016/j.chemosphere.2020.126133
80. Zakeritabar SF, Jahanshahi M, Akhtari PM, J, (2020) Photocatalytic study of nanocomposite membrane modified by CeF3 catalyst for pharmaceutical wastewater treatment. J Environ Health Sci Eng 18:1151–1161. https://doi.org/10.1007/s40201-020-00534-4
81. Zhao R, Ma T, Zhao S, Rong H, Zhu TY, G, (2020) Uniform and stable immobilization of metal-organic frameworks into chitosan matrix for enhanced tetracycline removal from water. Chem Eng J 382. https://doi.org/10.1016/j.cej.2019.122893
82. Zhao X, Zhao H, Dai W, Wei Y, Wang Y, Zhang Y, Zhi L, Gao HH, Z, (2018) A metal-organic framework with large 1-D channels and rich OH sites for high-efficiency chloramphenicol removal from water. J Colloid Interface Sci 526:28–34. https://doi.org/10.1016/j.jcis.2018.04.095

83. Zhao Y, Zhao H, Zhao X, Liu QuY, D, (2020) Synergistic effect of electrostatic and coordination interactions for adsorption removal of cephalexin from water using a zirconium-based metal-organic framework. J Colloid Interface Sci 580:256–263. https://doi.org/10.1016/j.jcis.2020.07.013
84. Zhu G, Wang S, Yu Z, Zhang L, Wang D, Sun PB, W, (2019) Application of Fe-MOFs in advanced oxidation processes. Res Chem Intermed 45. https://doi.org/10.1007/s11164-019-03820-5
85. Zhuang S, Wang CR, J, (2019) Adsorption of diclofenac from aqueous solution using UiO-66-type metal-organic frameworks. Chem Eng J 359:354–362. https://doi.org/10.1016/j.cej.2018.11.150

Performance of Metal-Based Nanoparticles and Nanocomposites for Water Decontamination

M. K. Mohammad Ziaul Hyder and Sajjad Husain Mir

Abstract Water comprises an integral component of human life and its accessibility is essential for all life in the entire planet. Due to climate changes and other man-made activities, the world is facing shortage of drinking water. There are a number of pollutants present in the water such as gases, chemicals and heavy metals. Therefore, it is imperative to decontaminate water for a healthy planet. There are numerous problems and challenges of wastewater treatment. For better ecological and health issues some measures are required to take in advance to avert possible evil or to secure good results. Metal-based nanomaterials have found exceptional use in the decontamination purpose due to their nature which arises from nanosize, such as better adsorption and catalytic activity. Metal-based nanomaterials can productively remove different contaminants from water and they have been effectively applied in decontamination of water. Due to having larger surface area and having ability to work at low concentration these metal-based nanomaterials are very efficient in wastewater treatment. Nanoengineered nanoparticles impart a promising and effective treatment method to wastewater and thus can be adapted simply. Modern techniques for treatment of wastewater must be cost-effective and accessible for commercial use. In this chapter, we outline the role of metal-based nanoparticles and nanocomposites applied in water decontamination. Moreover, we discuss the advantages, disadvantages, shortcomings and future prospects associated with these nanomaterials.

M. K. M. Z. Hyder (✉)
Department of Chemistry, Chittagong University of Engineering & Technology, Chattogram, Bangladesh
e-mail: ziaulhyder@cuet.ac.bd

S. H. Mir
School of Chemistry, Trinity College Dublin, Dublin 2, Ireland

E. Lichtfouse et al. (eds.), *Inorganic-Organic Composites for Water and Wastewater Treatment*, Environmental Footprints and Eco-design of Products and Processes,
https://doi.org/10.1007/978-981-16-5928-7_3

1 Introduction

Water is the basic component required for living being on earth. Safe and clean water is vital for direct uses and improved and good health of people. Besides water for drinking and other domestic uses, the second significant application of water is irrigation. The productivity of agriculture largely depends on water and thus indirectly affects food security of the population. There are other precious uses of water apart from human direct consumption and food production. Water for washing, cooking, sanitation and cleaning are essential preconditions for hygiene and health. Hydropower generation and watering of livestock are other prolific applications. Collectively, these principle applications of water infer that the abundance of water in superior quality and smooth supply has a great influence on social development. The change in climate conditions and increasing pollution are making water even scarce, particularly in developing countries. Currently, FAO (Food and Agriculture Organization) published that by 2025, fifty percent of the world will be living in water-scarce area creating more demand to use wastewater directly or indirectly (FAO, United Nations 2020). Table 1 presents types of water contaminants with origins and their impacts on health and environment.

Water sources are diminishing gradually because of overuse and misuse. Most of the normal sources of freshwater such as lakes, rivers, canals, reservoirs and rainwater have been found to be polluted with many types of precarious and poisonous materials or organic waste from different industries, household waste or originated from the

Table 1 Types of water contaminants with origins and their impacts on human health and environment

Water Contaminants	Origin of Contaminants	Impacts of contaminants
Sewage and contaminated water [197]	Domestic wastewater	Diarrhea, cholera, typhoid, etc.
Macroscopic pollutants [93]	Marine debris	Environmental pollution
Organic pollutants (Wang et al. 2019)	Fungicides, detergents, insecticides	Endangering aquatic life, dysgenic
Radioactive contaminants [24]	Different isotopic elements	Carcinogenic, tooth decay, damages bones and skin
Industrial contaminants [131]	Municipal contaminants	Induce air and water Pollution
Pathogens [265]	Germs	Diarrhea, cholera, typhoid, etc.
Suspended solids and sediments [206]	Land demolition, mining, land cultivation, etc.	Endangering aquatic life such as fishes, insects and affecting fish spawning
Inorganic contaminants [234]	Inorganic salts, Heavy metals, Mineral acids, Trace elements	Damage to flora and fauna in aquatic, public health issues
Agricultural contaminants (Tang et al. 2016)	Chemicals used in farming	Freshwater pollution

various point [139, 165]. These contaminants are detrimental to the people and other living beings and devastate the environment with permanent impacts [87, 204, 248].

Present wastewater control frameworks have generally been effectively examined though there are plenty of impediments [42]. For example, there is wide interest in developing advanced technologies to relieve toxicity and to ensure a secure living environment for humans. In this condition, several methods have been utilized such as chemical precipitation [112], electro-dialysis [83], reverse osmosis, ion exchange [58], adsorption [103, 104, 119], solvent extraction [289] and ultrafiltration [6]. The abovementioned technologies are expansive, inadequate and require a large amount of chemicals. These conventional methods of water treatments are no longer productive to eliminate many of the contaminants found in water with a view to attain water quality benchmarks. They regularly depend on a centralized framework, which the distribution and discharge processes are not sustainable for present day's requirements [191]. Upon this issue, nanotechnology could be used as an improved method to treat wastewater due to the size of nanomaterials which have the bigger surface area, high reactivity, fast kinetics; specificity to contaminants and, another advantage is the cost of nanomaterials that are going to decrease [290]. It is assessed that approximately 663 million people don't have access to potable water, mainly in developing countries (World Health Organization 2017). So, it becomes important to ensure basic water treatment to these people, where water treatment often is not available. The removal of pollutants from contaminated water is essential to avoid harm to public health and to the ecosystems [213]. On the aforementioned problems, the present review pointed on the utilization of metal-based nanomaterials to upgrade the standard of water with respect to the removal of metals, pathogens, salinity, oil and discuss the antimicrobial activity and the possible risks that these nanomaterials can affect the environment. Nanofiltration, nanoadsorbent, nano photocatalyst, disinfection and sensing with nanomaterials are the main techniques to treat wastewater by nanotechnology. This chapter emphasizes the method in wastewater control system with metal-based nanoparticle and attempts to point out the modern technology, outlooks, advantages and disadvantages of this emerging field.

The developing field of nanotechnology offers potential advancements to decontaminate water with cost-effective, improved working capability in removing contaminants and recycling capacity. Over the years, nanoparticles are effectively used in several fields such as in medicinal science, photocatalysis, etc. Presently, as the world confronting vital challenges of safe and clean water, scientists discovered that nanomaterial is a way superior choice wastewater treatment since it has some basic characteristics with greater surface area, nanosize, better reactivity [279], tough mechanical criteria, good porousness, hydrophilicity, dispersity. Some organic and inorganic contaminants, heavy metal like Hg, Cr, Pb, etc,. and numerous detrimental pathogens are presented to be effectively removed by utilizing distinctive nanomaterials [64, 153, 265]. Currently, the wastewater decontamination processes are

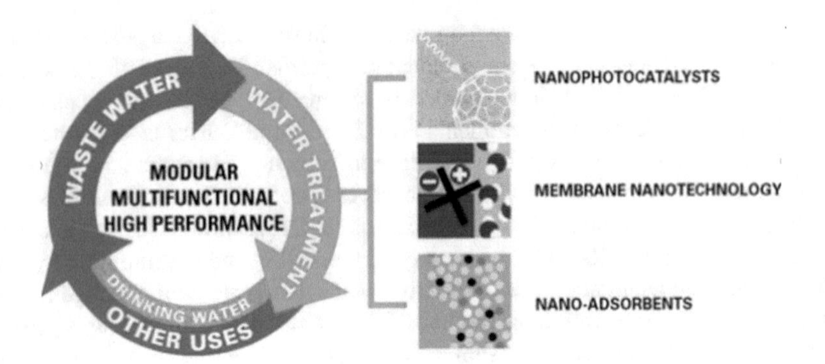

Fig. 1 Different categories of nanomaterials that are used for safe and sustainable water supply (Reproduce with license from American Chemical Society, Copyright (2013))

progressing with the advancement occurring in nanomaterials such as nanoadsorbent, nano photocatalysts and some imprinted polymers. Moreover, we have recently developed range of hybrid nanomaterials based on polymers-metal complexes, which have potential applications in water treatment and pollutant removal [10, 115, 154–160, 219]. In brief, the investigation on using nanomaterial in water treatment is regarded to measure positive viewpoint [23, 160, 221]. Figure 1 presents several types of nanomaterials applied in safe and sustainable water supply [191].

Nanomaterials are generally classified into different groups related to their surface and physical characteristics. Nanomaterials include metallic nanoparticles (Au & Ag nanoparticles), metal oxide nanoparticles (ZnO nanoparticles, Al_2O_3 nanoparticles, CeO_2 nanoparticles and TiO_2 nanoparticles), magnetic nanocomposites, nanocomposite with organic and inorganic supports, carbon nanoadsorbents, polymer nanoadsorbent. These nanomaterials are utilized as nanoadsorbent, nanomembrane, nanocatalyst, disinfectant and nanosensor for waste water treatment. Thus, we have outlined the importance of metal-based nanomaterials for treating wastewater to subdue the crises of fresh water problems in this chapter. A prospective and substantial method with easy accessibility can be obtained by using metal-based nanomaterials but a few flaws still require advanced consideration which is specifically outlined in this chapter. Besides, we also point out the limits, benefits, drawbacks and future prospects with relation to the metal-based nanoparticles. Moreover, merits and demerits of the metal-based nanoparticles with their other diverse uses in treatment of wastewater are shortly explained that can be beneficial to scientists for designing new strategy.

2 Categories of Metal-Based Nanoparticles in Water Treatment

2.1 Nanoadsorbents

One of the important technologies to separate contaminants from water is the notable adsorption method. Nanosorbents exhibit high and efficient adsorption capability with extensive uses in decontamination and purification of wastewater. Here, the nanoparticles absorb the contaminants in the water which can be separated from water after reaching the equilibrium. The method of adsorption of contaminants from wastewater is generally considered as a better process over other methods. Adsorption technology of wastewater control systems is usually a better technique over conventional methods. Due to its inexpensiveness, good performance, easy operation, it has high ability to remediate wide variety of pollutants from water [123]. Nanosorbents possess great properties such as high sorption ability which gives the nanosorbents more capability and more effectiveness for treating wastewater.

Nanoadsorbents have extraordinary ability for unique, more effective and quick decontamination procedures with a view to separation of inorganic and organic contaminants. Scientists are doing a great deal of work to prepare nanosorbents in a bigger amount at commercial level as they are exceptionally uncommon in commercial form. The field of nanotechnology is progressing by doing extensive research in this area to resolve the issues in removing contaminant metals from water with a view to find better nanoparticle combinations. In this method, titanium oxide, iron oxide and aluminum oxide-based nanomaterials have shown promising characteristics with cost-effectiveness and high adsorption property. Besides, nanoadsorbents have high porousness and larger active surface area which make them capable of removing different sizes of pollutants without discharging any toxic elements.

Adsorption method can be employed to extract the metal contaminants from contaminated water of various sources. Pb, Hg, Cr, Cd, Co, Zn, As, Cu, Ni, etc., are the kind of major metal contaminants responsible for water pollution [105]. Current investigations reveal that the nanoparticles are highly efficient for competent removal of abovementioned metal contaminants from wastewater. Nowadays, the nanomaterials of metals and metal oxides are widely utilized in decontamination of water through adsorption process. Nanoadsorbents made from metal nanoparticles are less expensive nanomaterials showing effective sorption quality. They are frequently utilized for the treatment of water containing heavy metals. Among the nanoadsorbents fabricated from metal nanoparticles, TiO_2, Fe_2O_3, MnO_2, MgO_2 and Al_2O_3 are well investigated and were observed to remove the heavy metals from wastewater very effectively. Metal oxide nanomaterials are regarded to be more capable than the normal adsorbent due to their larger active surface area. There are different points of interests related with metal oxide-based nanoadsorbents. The simplicity of synthesis, lesser toxicity, higher active surface area for contact and chemical stability impart the more distinctive properties to these metal oxides-based nanoadsorbents and make them more lucrative than other adsorbents [199].

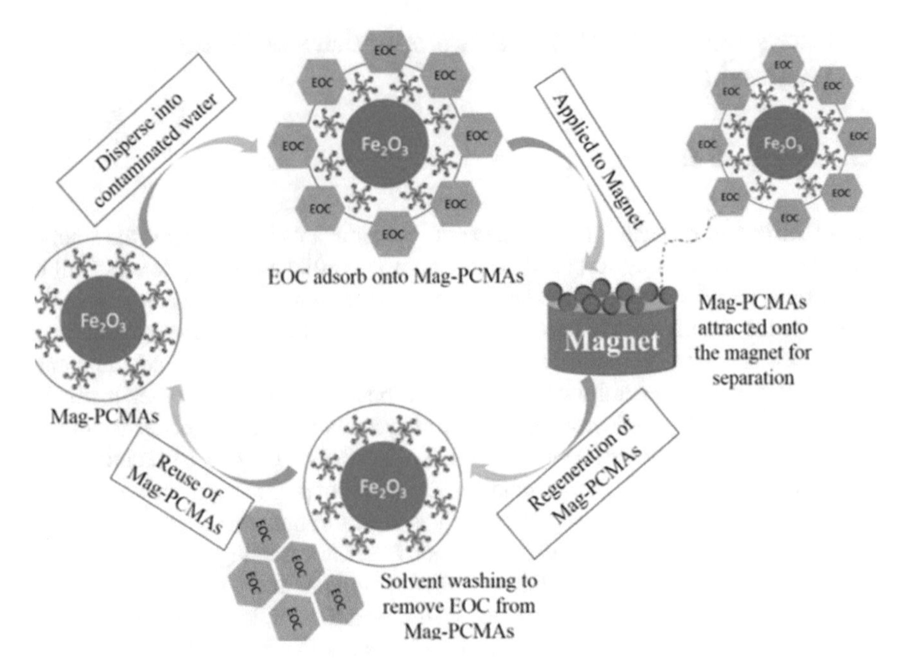

Fig. 2 Magnatic nanoparticle permanently confined micelle arrays adsorbents for complex emerging organic contaminants from wastewater. Reproduce with license from American Chemical Society, Copyright (2013)

The metal-based nanoadsorbents are prominent presently. Huang et al. showed the significant interaction between magnetic nanoparticle permanently confined micelle arrays (Mag-PCMAs) adsorbent and complex emerging organic contaminants (EOCs) which made magnetic nanoparticles efficient to be applied in complex chemical environments, like wastewater treatment (Fig. 2) [100]. The nanocomposite of various materials such as silver/carbon, silver/polyalanine, carbon/titanium oxide, etc., possessing tremendous significance with a view to remove the impact of poisonous properties in the treatment process of wastewater. Metal/metal oxide nanosorbents bear significant adsorption positions and due to their large active surface area they are the efficient materials for the contaminants by adsorbing methods. Likely, organic contaminants and heavy metals from wastewater are effectively removed by the polymeric nanoadsorbents [72]. For instance, with the help of dendrimer-ultrafiltration systems, copper ions can be reduced [226]. They are easily recovered at wide ranges of pH and exhibit biodegradability and non-toxicity. Moreover, dyes and other organic contaminants can be removed with the efficiency of 99% [182]. Zeolite is another vital nanoadsorbent where different nanomaterials such as copper and silver ions could be implanted [76].

The magnetic nanosorbents have a significant and unique capacity in water treatment to eliminate different organic contaminants from water. Nanoadsorbent for magnetic separation with particular affinity to contaminants was synthesized through ligands coating with magnetic nanoparticles [183]. Individual or combined metals can be utilized for the effluent treatment depending on the nature of pollutants. Iron oxides can be simply prepared and modified as the availability of iron is high on earth. The super magnetism and large surface to volume ratio of iron oxide give it the rank of a very good adsorbent with lesser toxicity, chemical inertness. These distinguished criteria of nanoadsorbents create a very fine option for the treatment of wastewater. Magnetic nanosorbents are also conducive in treating wastewater and are tested as very promising tools especially for organic contaminants elimination. Various procedures like cleaning agents, magnetic forces, ion exchangers are applied to eliminate nanoparticles from the system to prevent unwanted perniciousness. Recovered nanosorbents could be a better option for commercialization due to their cost-effectiveness. The surface interaction of magnetic nanoparticles and their aggregation can be restrained by using non-ionic, amphoteric, cationic or anionic surfactants [84]. Different forms of iron oxides that have been studied greatly having the capability to act as nanoadsorbents include maghemite (γ-Fe_2O_3), hematite (α-Fe_2O_3) and magnetite (Fe_2O_3), goethite (α-FeOOH), Iron oxide (Fe_xO_y), etc. [17, 39]. The magnetic properties of Fe_2O_3 nano adsorbent cause the separation process to be very simple from the dilute or even from viscous solutions. The removal of heavy metals like Cr^{6+} and Pb^{2+} were carried out efficiently where the protonation or deprotonation of magnetite surface hydroxyl group followed by water loss causes Cr^{6+} and Pb^{2+} adsorption. Likewise, different types of nano-structured wastewater metal adsorbents have been noted with various characteristics such as ZrO_2, TiO_2, CuO, MgO, etc. [109, 241].

2.2 *Nanomembranes*

Membrane technology is one of the most substantial developed techniques in the water treatment process. There is a broad range of new membrane materials applied to process water for reuse. For example, ceramic and polymeric membranes are familiar in the water treatment process. Presently, the applications of membrane technology are rising due to development of this method that has made them more accessible, flexible and effective. Accordingly, the water treatment industry is witnessing a flourish worldwide for all those factors. The membrane-based on nanofilter is the relatively modern technology in the treatment of wastewater. Nanomembrane removes the ions through ultrafiltration electrical effects following the reverse osmosis ion interaction mechanism as well as combination of ions. Novel properties of nanomembrane make it capable of selectively removing pollutants from the system. The improvement of nanomembrane innovation in recent years makes it for multiple use such as in pharmaceutical industry, demineralization in the dairy industry, bleaching in the textile industry, metal recovery from wastewater. Nanofiltration membrane is one of the suitable methods to treat organic and inorganic contaminants in surface water. Nanofiltration is more credible in treating surface water due to its low pressure activity as

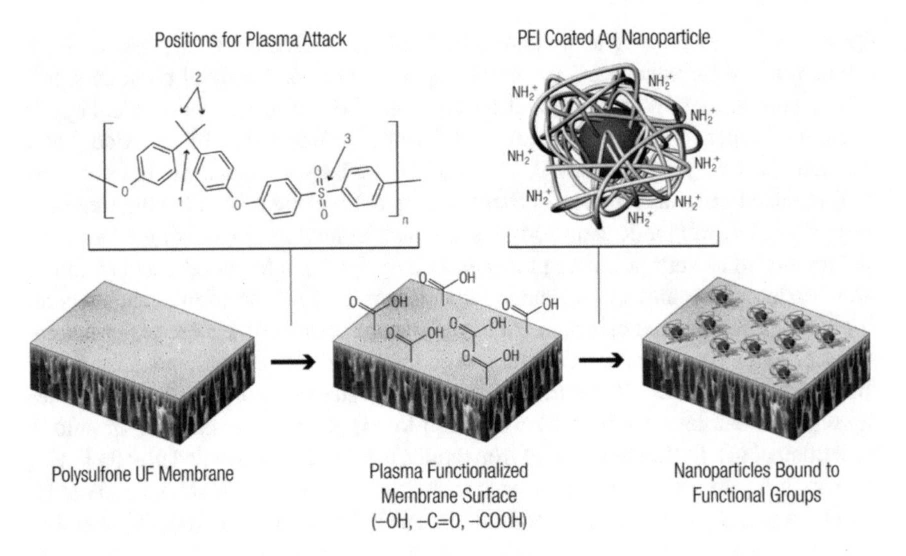

Fig. 3 Silver nanoparticles encapsulated ultrafiltration membranes for water treatment. (Reproduce with license from American Chemical Society, Copyright (2011))

surface water has low osmotic pressure [31]. Application of reverse osmosis methods is a normal procedure to make the water drinkable by filtering process.

Nanometal-based membranes are utilized to eliminate industrial contaminants from wastewater. The merits of nanometal membrane-based wastewater treatment are its easy operation, greater efficacy and low space demands. Moreover, by employing proper chemicals and nanoparticles the filtration capability can be improved [297]. Nanomembranes can be prepared with diversified characteristics of antimicrobial, anti-fouling, improved permeability, photocatalytic activity and selectivity on the basis of applications [168]. Ultrafiltration membrane shows evaluative treatment process in improved wastewater treatment. Figure 3 shows the ultrafiltration membrane where silver nanoparticles are encased in positively charged polyethyleneimine which provides an effective way of water treatment [150].

Multilayered inorganic—organic hybrid membranes using metallic molybdenum disulfide (MoS_2) as two-dimensional transition metal dichalcogenide nanosheets and one-dimensional silk nanofibrils were utilized for water purification [295]. Because of its possessing of negatively charged layers and interaction sites, the hybrid film could adsorb metal ions and dyes from water (Fig. 4). The separation performance can be tuned by changing the component ratios of these two nanomaterials. During filtration, due to the reducing effect of the MoS_2 nanosheets, precious metal ions are reduced to their nanoparticle form without any further thermal or chemical treatments. In addition to the one-step removal and recovery of metal ions, the hybrid membranes exhibit excellent potential for the determination and removal of different dyes from water.

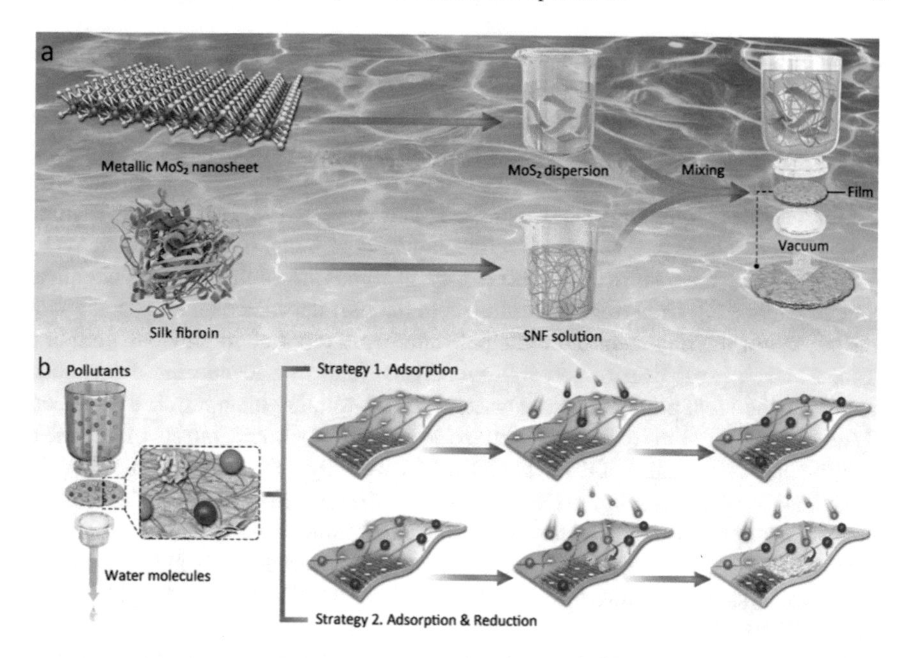

Fig. 4 Fabrication process of the mixed-dimensional MoS_2 silk nanofibrils hybrid membrane and the water purification mechanisms. **a** Scheme of the fabrication steps of the hybrid membrane. **b** Mechanism for the removal of contaminants from water. (Reproduced with license from American Chemical Society, Copyright (2020))

Nanofibers are one of the first species of membrane filters. They are porous and have high surface area with high interlinkage and can be prepared by a simple electrospun method. Nanofibers prepared by electrospinning with the combination of high surface area nanomaterials have shown efficiency in adsorption of pollutants. For instance, trace amounts of arsenite can be effectively removed by chitosan electrospun nanofiber which is manufactured by crosslinking ammonia vapor with the mixtures of Fe^{3+} and chitosan, poly (ethylene oxide) [152].

2.3 *Nanophotocatalyst*

Photocatalytic technologies have drawn most concentration for water pollution management. Photocatalysis is observed as a more efficient method for the purification of water purification which subdues the environmental pollution. The basic fundamental of photocatalyst is that the catalyst oxidizes the pollutants in water by utilizing radiation from sunlight. Metal oxide-based nanophotocatalysts are the vital candidate for the rectification of environmental pollution through recent application of it in water decontamination. In this method, electron-hole pairs are produced by irradiation of nanophotocatalyst. These photoelectrons create holes by jumping from the valence band to the conduction band.

A good photocatalyst absorbs visible or near ultraviolet more efficiently. For the prevention of the recombination of electron-hole pairs adequate electron vacant states are required. Nanophotocatalysts should be biologically inactive and nonpoisonous due to their ongoing extensive uses in microbiological and agricultural sectors. Nano photocatalysts prepared from semiconductors of metal oxide like TiO_2, WO_3, Zn_2SnO_4, ZnO have shown high efficiency in removing biological and chemical contaminants [21]. Comprehensive studies in the past decades on uses of nanophotocatalysts for the treatment of municipal water were reported in previous literature [145, 214]. A great deal of research work was done to alleviate the detrimental impact of chemical pollutants from wastewater by utilizing nanoparticles photocatalysts such as titanium oxide and zinc oxide [92]. Nanophotocatalysts of activated carbon-supported nano-FeOOH (FeOOH/AC) were synthesized with the help of air oxidation of ferrous hydroxide suspension method [288]. FeOOH/AC heterogeneous nanophotocatalyst owns remarkable adsorption capacity and the oxidation of amaranth happens through the homogeneous and heterogeneous in bulk solution and on catalyst/solution interface, respectively, because of releasing of iron from the nano-FeOOH (Fig. 5).

Ternary oxide zinc stannate is drawing concentration from researchers as viable photocatalyst [97, 195]. Due to improved photocatalytic activity and non-toxicity, TiO_2 nanophotocatalysts are observed to show the effective and prominent activities for photodegradation dyes from contaminated water. The demerits of TiO_2 as

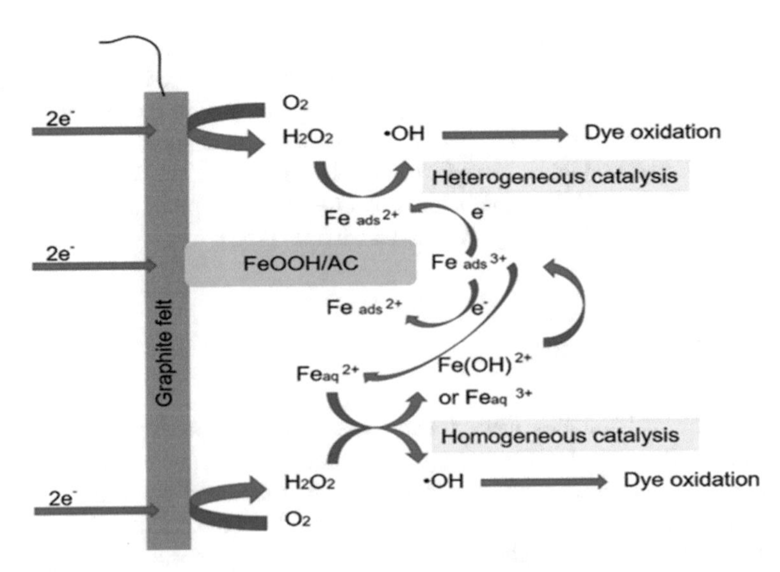

Fig. 5 Effective adsorptibility and Fenton oxidation with the combination heterogeneous/homogeneous process of amaranth utilizing supported nano-FeOOH. Reproduce with license from American Chemical Society, Copyright (2012)

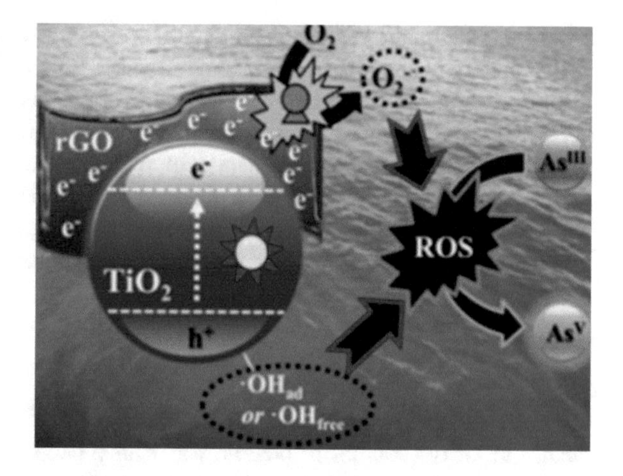

Fig. 6 Reduced Graphene Oxide/TiO_2 for the Effective Photocatalytic Oxidation of Arsenite. Reproduce with license from [163]. American Chemical Society, Copyright (2014)

a photocatalysts is low quantum efficiency in visible regions, decreasing photocatalytic ability on account of wide band gap and quick recombination of charge carriers. Several studies have been noted depicting the antimicrobial activity of TiO_2 nanophotocatalysts against various waterborne pathogens including protozoans and bacteria [86]. Photocatalytic activity of TiO_2 has also been investigated utilizing the TiO_2 mediated photodegradation of Rhodamine B dye and bromoethane [111]. Cyanobacteria have an immense toxic effect on human health. Lawton et al. showed the mechanism of TiO_2 mediated mineralization of cyanobacterial hepatotoxin [124]. The reduced graphene oxide hybridized with TiO_2 was prepared as a cost-effective catalyst compared with Pt/TiO_2 and found to show improved activity for the photocatalytic oxidation of As(III) [163]. The photocatalytic activity and arsenic oxidation mechanism observed with reduced graphene oxide implanted TiO_2 are almost the same activities shown by Pt/TiO_2 (Fig. 6). The nanocomposite of reduced graphene oxide/TiO_2 can be considered as a useful environmental photocatalyst for pretreating the water polluted with As(III).

ZnO nanoparticles also displayed photocatalytic activity by creating hole-electron pair [27]. ZnO-NPs have prominent photocatalytic activity for elimination of different organic contaminants having their high binding energy and broad band energy, powerful oxidation capability and high active surface area [256]. Silver nanoparticles are noted to show antimicrobial activity against waterborne pathogens [71]. Mpenyana-Monyatsi fabricated Ag NPs coated filters and showed the efficiency of the elimination of microbes from water with 100% effectivity [166]. Photocatalytic degradation based on the heterogeneous semiconductors is one of the safest simple and cost-effective techniques for dyes and organic compounds removal from water polluted by industries and residences [68, 118, 152, 169]. Various metal oxides NPs like ZnO, CuO and TiO_2, etc., are being used for photodegradation of organic dyes [232, 233, 239]. This process assists to remove contaminants such as pathogens, organic dyes and micro pollutants, etc. [192, 273]. For instance, a hetero structured nanocomposite $BiVO_4/CH_3COO$ was synthesized for the degradation of the organic contaminations from water by photocatalytic activity [293]. TiO_2 effectively eliminates toxic chemical tartrazine from water utilizing its photocatalytic activity [82]. Polyaniline/ZnO nanocomposites show improved degrading capability toward colored dye through producing enough electrons at the conduction band of zinc oxide semiconductor [215]. Different nanocomposites of zinc oxide or the compounds of zinc oxide with materials have been observed to degrade the contaminants in wastewater very effectively [198]. In a similar way, filtration technology could be developed by integrating photocatalytic characteristics of a photocatalyst[138].

2.4 Disinfection

Nearly all of the sources of potable water have been observed to be polluted with various poisonous materials and pathogenic microorganisms. The World Health Organization (WHO) reported that approximately 12 million people die every year from waterborne illness. 90% of all diseases resulting from impure water were found in developing countries. The global disease infecting people with the use of impure water is nearly 4 billion. The responsible microorganism in water which causes diseases to people is known as pathogen. Different technologies are applied to treat the pathogen in water. Deactivation of pathogens is generally known as disinfection. Presently, the disinfection of drinking water is carried out by chemical or physical method. Various techniques are used to disinfect water such as UV treatment, chlorination and ozonation. The well-accepted method of disinfection by chlorination has some limitations. The excess chlorine beyond the permissible level is toxic and may be responsible for bladder or colorectal cancer. Chemical treatment of water by antibacterial disinfectants like triclosan and triclocarban may cause hormone-disrupting effects. In the presence of natural organic matter, ozone can form non-halogenated organic disinfection by-products such as aldehydes, ketone, carboxylic acids. The effect of UV treatment is temporary and water can be infected by pathogens if the water is stored for a long time. These traditional water disinfection methods have definite limitations to apply at large scale.

Nanotechnology in water purification shows huge potential to decontaminate water [20, 22, 22]. This is a viable way to remove the pathogens from wastewater. In the present condition, nanomaterials can be utilized to eliminate microbes more efficiently. The nanomaterials accommodate different processes to kill the organisms. These nanomaterials may connect the organisms specifically through hindering the electron transfer to transmembrane, destroying cell enclosure or by producing reactive oxygen species they can damage cell walls [129]. Various nanoparticles with antimicrobial inherent were reported with action against organisms [129, 246]. A biomass-based renewable film with good mechanical strength and porous network structure was facilely fabricated via Fe (III) crosslinking induced with collagen fibers and gallic acid-protected silver nanoparticles self-assembly (Fig. 7) [132]. This film exhibited both excellent bacterial anti-adhesive and bactericidal activities, which effectively prevented biofouling during the filtration process, due to the anionic

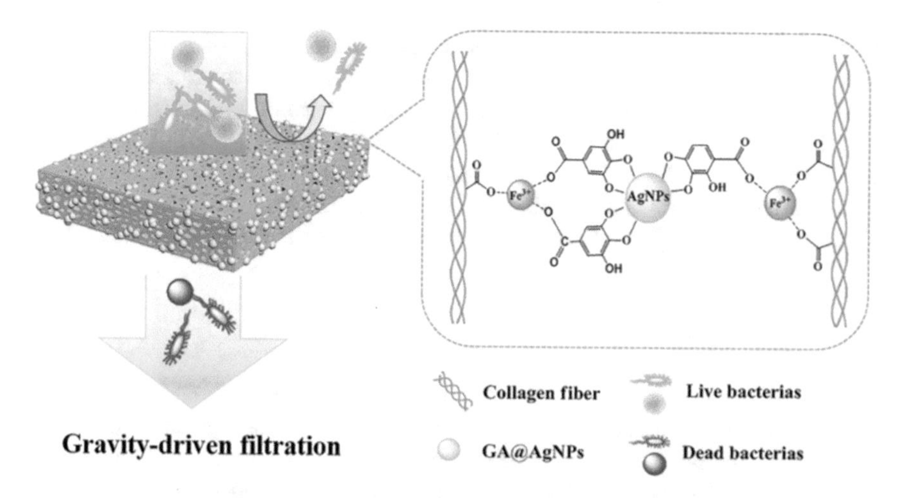

Fig. 7 Point-of-use water disinfection by a cost-effective porous renewable film incorporated with silver nanoparticle. Reproduce with license from American Chemical Society, Copyright (2020)

gallic acid-protected silver nanoparticles. As bactericidal filter driven by gravity, 1 L natural water sample was treated by the film in 20 min, and the water quality is in full compliance with the drinking water guidelines of WHO, demonstrating the potential application of the proposed filter in point-of-use water disinfection.

TiO_2 produces hydroxyl free radicals and forms peroxide with photocatalytic activity which is responsible for antimicrobial properties of its [129]. TiO_2 with the incorporation of other nanomaterials displays enhanced antimicrobial photocatalytic properties [47, 110]. The nanoparticles of zinc oxide exhibit notable antimicrobial properties against waterborne pathogens, and hence they are utilized to purify the wastewater [51]. Salem et al. made a comparison of the antimicrobial properties of Ag nanoparticles and Zn nanoparticles toward *V. cholerae* and enterotoxin *E. coli* [56]. Iron nanoparticles also exhibit antimicrobial activities by eliminating *Entamoeba histolytica* cysts from water [231]. For centuries, silver has been considered a well-known antibacterial material. The release of silver ions efficiently destroys the cell envelope and retards the DNA replication [191]. Nanofiltration techniques are another method to remove the microbes by filtration [218].

As waterborne disease causes serious health effects to humans, the disinfection technology is drawing more attention recently. Titanium dioxide with its environmental friendliness behavior was exhibited to be prepared as antimicrobial agents in more recent studies. The investigations showed that TiO_2 improves the capability of disinfection through the deactivation organisms such as Escherichia coli, Staphylococcus Aureus, etc. Nano-WO_3 synthesized by sol–gel method displayed enhanced capability for the disinfection of E.coli in water [79]. Copper displays high antimicrobial activity with attractive cost and low toxicity. Moreover, it has been reported that Cu_2O showed more activity toward bacteria than silver and CuO [251].

Deng et al. reported that copper graphene sponge can be used for water purification more efficiently through inactivation of bacteria [48]. Bactericidal activity of gold nanoparticles is scarcely reported for gram-negative or gram-positive bacteria [13, 106]. Contrarily, gold NPs display fungicidal activity [7, 106]. Platinum NPs can destroy cell walls and can release cytosolic proteins bacteria and fungi [16]. Palladium NPs show better antimicrobial activity toward gram-positive bacteria than gram-negative bacteria, and exhibit size-dependent antimicrobial properties [2].

2.5 Sensing

The detection of pathogens is essential because of their precarious impact on human health. The traditional sensing methods are steady and incapable of monitoring the existence of harmful viruses and pathogens such as helicobacter, legionella, norwalk viruses, echoviruses, hepatitis A. Most of these microbes are biological operant in the rise of contamination in drinking water. Water sterilization process depends on pathogen recognition. There is great progress in research to develop nanomaterial-enabled nanosensors. Present studies are concentrating on the improvement of three principal parts of nanosensor: (i) nanomaterials (ii) recognition materials and iii) signal transduction mechanism. The recognition materials selectively interact with pathogens. Rapid feedback and selectivity are obtained by using nanomaterials. Nanomaterials intensify the detection speed and sensing capability to perform multiple target identification with their novel optical, electrochemical and magnetic characteristics. Nanosensors may be used for the detection of biomolecules cells.

A numerous research has been done on the appropriate design and application of nanosensors [19, 35, 60, 61, 73, 172, 271]. These nanosensors can be utilized in the central distribution system, at the location of point-of-use or in the water treatment plant. The monitoring of sensing may be online to determine the quality of water during flow through or may be offline by collecting water samples at different points. Nanosensors are more capable than traditional water quality sensors. Nanosensor rapidly and reversibly measures the analyte whereas nanoprobe selectively determines pathogens with great sensing capability in an irreversible way. [207].

A wide variety of nanosensors was reported to show the capability of identification of pathogens, toxin and pH in water [35, 66, 207, 266, 268]. A percolation method was reported to inactivate pathogens through silver nanoparticles containing paper sheets. Here, on blotting paper sheets of cellulose fibers, silver nanoparticles were accumulated [46] (Fig. 8). The silver nanoparticles sheets showed remarkable antimicrobial properties toward enterococcus faecalis and Escherichia coli with high reducing ability. This outcome of deactivation of pathogens through silver nanoparticle sheets is encouraging enough to utilize it in emergency water treatment.

A direct intrinsic signal from the analyte can be acquired by nanosensor or by employing high quality recognition elements that are bound to analyte.

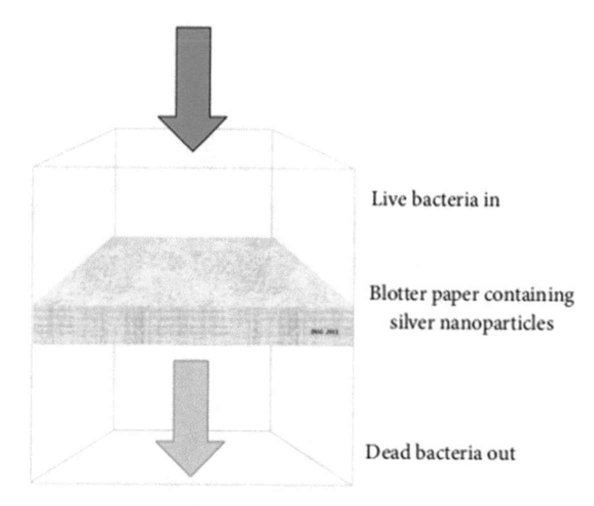

Fig. 8 Blotting paper implanted with Ag nanoparticles for point-of-use treatment of water. Reproduced with license from American Chemical Society, Copyright (2011)

Figure 9 depicts the sensing of environmental analytes by nanosensor architectures. A simple approach to effective detection of bacteria S. aureus through surface-enhanced Raman-scattering with the synthesized gold-coated magnetic nanoparticles core/shell nanocomposites [272].

Ng et al. reported the recent development of fluorescent nanosensors such as metal nanoparticles [174]. Strong electromagnetic field is generated on the nanoparticle surface when silver or gold nanosensors are excited by light [278]. Magnetic nanomaterials are capable of identifying magnetically isolated analytes as they are highly responsive to external magnetic fields [122]. The detection of influenza A and Mycobacterium was carried out through changing the electrical resistance of magnetic nanoparticle-labeled analytes by magnetoresistance sensors [121]. Quantum dots are promising as fluorescent nanosensors which have larger band gaps and narrow fluorescent spectra have been detected through one excitation light source [268]. The Internet of things can be connected to the system of nanosensors used in distribution systems to ensure quality, stability and degrability of nanosensors [151].

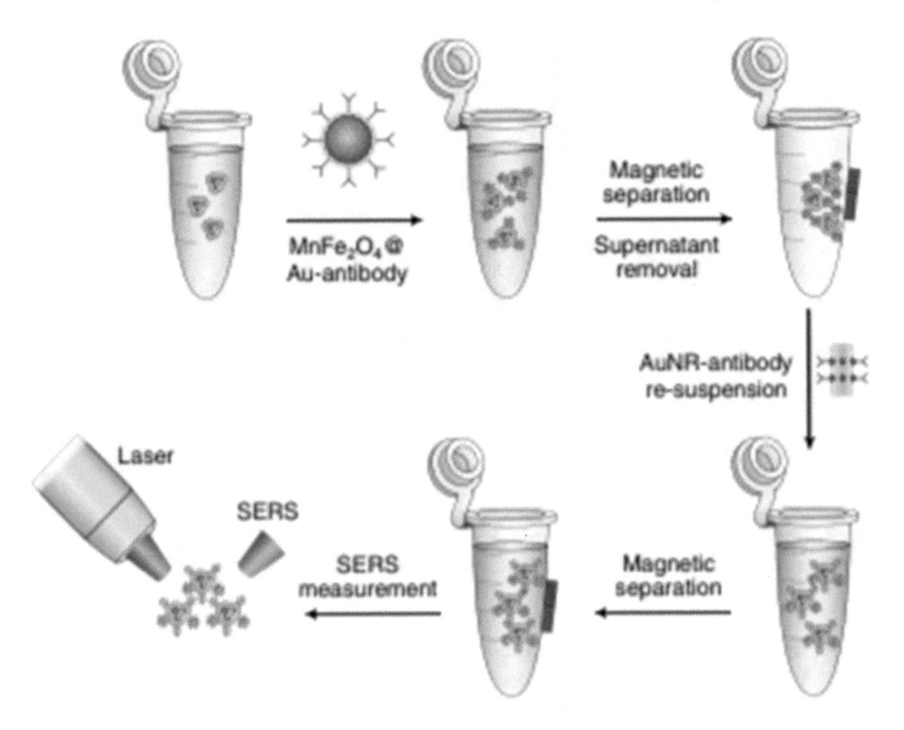

Fig. 9 Schematic illustration of the operating procedures for bacteria detection via a surface-enhanced Raman-Scattering method. Reprinted from with license from American Chemical Society, Copyright (2016)

3 Metal and Metal Oxide Nanoparticles Used in Water Treatment

3.1 Metal Nanoparticles

Nanometals can be used in water purification with high efficiency. There are various forms of nanometals utilized in wastewater treatment for instance nanostructures, cationic forms and inert or active substances supported form. Silver nanoparticles have been reported to be applied to adsorb Cr(II) and Pb(II) as suspended free nanoparticles in the system [12]. Copper nanoparticles were used as antibiofouling, antioxidant and antibacterial agents for wastewater treatment [36]. Those nanoparticles showed efficiency in inactivation of pathogens, inhibition of lipid oxidation and biofilm formation and scavenging free radicals. Citrate-supported silver nanoparticles were used for degradation of organic pesticides chlorpyrifos [30] (Fig. 10). Octahedral palladium nanoparticles were used for reduction of bromate in municipal water treatment [276]. The supported nano metals have various advantages. Support helps to prevent aggregation of nanoparticles and separate nanoparticles from water after treatment which may be responsible for self toxicity [30, 276]. For synthesis

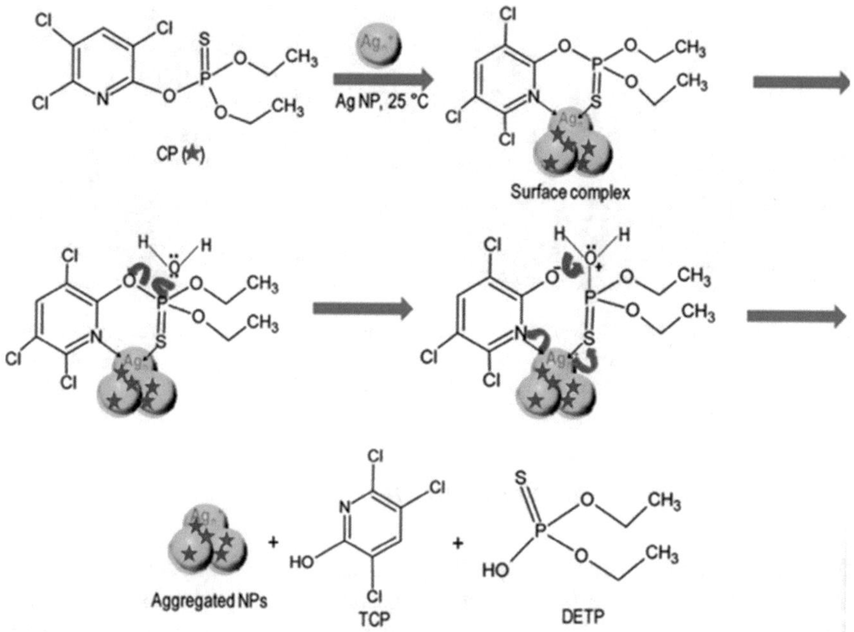

Fig. 10 Representation of degradation of chlorpyrifos on silver nanoparticles. Reprinted with license from American Chemical Society, Copyright (2012)

and stabilization of nano metals, different chemicals are utilized. For instance, to get the reductive and stabilized form of silver nanoparticles, chitosan and polyethylene glycol are used in synthesis [269]. The easiest approach to eliminate harmful contaminants from water is various physicochemical processes such as adsorption, filtration or cuagulation. For instance, silver and iron nanoparticles effectively remove Pb(II), Cr(II) and Cr(VI) ions from aqueous solutions by the physicochemical technique [148].

Ag nanoparticles can effectively remove Hg(III) from aqueous solution [63, 164]. Ag nanoparticles display improved activity due to their decreasing reduction potential with the decrease of particle size [188]. Au nanoparticles with aluminum support could be applied to remove Hg(II) effectively from wastewater. Jiménez et al. reported citrate-coated Au nanoparticles for treating Hg(II) in water [178]. Here, Hg (II) was converted to Hg (0) by weak citrated ions reducing agent without application of $NaBH_4$. The concentration of Hg(II) was reported to decrease from 65 to 5 ppb (Fig. 11). Other noble metal nanoparticles like palladium and ruthenium nanoparticles exhibit effective antimicrobial agents for gram-positive bacteria and display size-dependent antimicrobial activity [2].

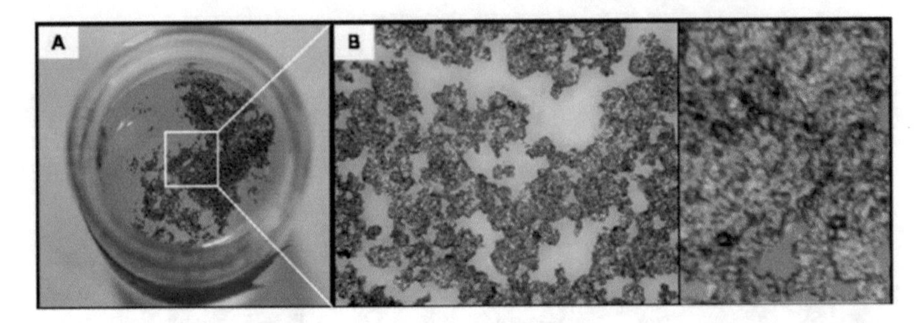

Fig. 11 Hg(II) removal from river water by citrate coated Au nanoparticles. **a** Precipitation image after treatment of Hg(II) with Au nanoparticles (efficiency 40%) **b** Zoomed image of precipitation taken with optical microscope. Reprinted from with license from American Chemical Society, Copyright (2012)

3.2 Zero-Valent Metal Nanoparticles

Wastewater treatment process is greatly advanced through using zero-valent metal nanoparticles. They were found to show excellent antimicrobial ability, degradation ability as well as high removal ability of heavy metal from wastewater. Zero-valent iron was well studied for the elimination of heavy metals and for deactivating pathogens from wastewater. Zero-valent iron (nZVI) nanoparticles consist of Fe (0) and Fe_2O_3 coating [177]. It is applied widely to treat heavy metals like Cr (VI), Hg (II), Cu (II), Ni (II), etc. [133, 222]. Principally, Fe (0) produces the reduction ability whereas the Fe_2O_3 coating creates the active position to attract heavy metals through electrostatic attraction. Moreover, the shape of nZVI could be easily manageable and huge reactive sites could be created on the surface of nZVI [43]. The high reducing ability and high active surface area impart the nZVI higher performance for the removal of heavy metals from contaminated water [98]. Furthermore, nZVI has been displayed to have a promising bactericidal effect and toxicity toward pathogens [49, 125].

The high efficiency and versatility of nZVI have made it perfect technology for practical utilization in wastewater treatment. Nano zero-valent iron can also be applied for improving the quality of groundwater contaminated with perchlorates and chlorinated hydrocarbons. nZVI is more reactive than conventional iron because of its high active surface. On the other hand, the lifetime of nZVI is very low due to its high reactivity characteristic. As a result, more research on surface modification of nZVI is necessary to make it stable [15, 94]. Zhang et al. deposited synthesized nZVI particles on the surface of biomass activated carbon and applied to remove 98% methyl orange from water [287]. nZVI has been efficiently used to treat the wastewater and groundwater with arsenic [173], chlorinated hydrocarbons [53, 247], heavy metals [190, 292], nitroaromatic [285], phenol [220], heavy metals [190, 292], nitrate [102], dyes [229] and phenol [220].

Application of nZVI for the treatment of wastewater has some drawbacks because of its instability, quick aggregating and problematic separation process. To resolve these disadvantages, nZVI could be supported by zeolite, bentonite, resin, etc. Deposition of nZVI nanoparticles on supporting materials for the elimination of contaminants makes the procedure easy and also enhances the reduction ability. The reactivity of ZVI could be improved by depositing a thin film of any other metals like Ni, Pt, or Pd on iron as principal metal which could efficiently remove chlorinated hydrocarbons from wastewater. For instance, Xu et al. synthesized novel Ni–Fe bimetal for effective removal of 4-chlorophenol with enhanced catalytic hydrogenation [280]. Another Pd/Fe bimetallic system shows very effective removal of tetrabromobisphenol A, 2,4-dichlorophenol and polychlorinated biphenyls and displays better dechlorination that than normal nZVI [98]. In addition, deposition of Pd on nZVI decreases the release of toxic intermediate on nZVI's surface [40]. The translocations and transformations of contaminants such as arsenic species at and within the nZVI particle are distinctly depicted in Fig. 12 [283].

Despite a lot of research on decontamination of wastewater by nZVI, zero-valent zinc (nZVZ) has been found as an alternative. nZVZ nanoparticles were shown to degrade dioxin excellently [29]. The reducing ability of Zn is higher than Fe. Thus it is clear that the power of contaminant degradation of nZVZ particles will be higher than nZVI particles. It is reported that degradation of CCl_4 happened more quickly by nZVZ compared to nZVI [261]. Moreover, an investigation was done for the comparison of degradation ability toward halogenated hydrocarbons in water with nZVI, nZVZ, nano zero-valent aluminum (nZVAl), nano zero-valent nickel (nZVN) nanoparticles. The study showed that only nZVZ was capable of degrading octachlorodibenzo-p-dioxin effectively into less chlorine concentrated materials

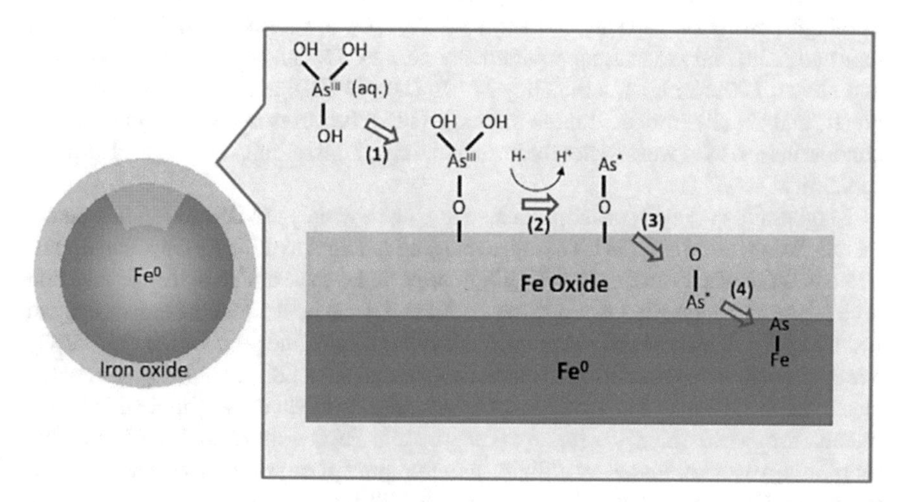

Fig. 12 The translocations and transformations of contaminants such as arsenic species at and within the nZVI particle. Reprinted with permission from American Chemical Society, Copyright (2012)

[29]. Though nZVZ efficiently degrades halogenated hydrocarbons, treatment of other contaminants with nZVZ was not reported a lot yet [261].

3.3 *Iron Oxides Nanoparticles*

The use of iron oxide nanoparticles in wastewater treatment is remarkably increasing. There has been rising attention on the application of iron oxide-based nanoparticles for the removal of heavy metals and remediation of wastewater in recent years [11, 212]. Due to the higher abundance of iron on earth and simple synthesis method of iron oxide-based nanoparticles, extensive research has been done on it. The challenge of using nanoparticles in water treatment is their recovery and separation from treated water. But, with the help of external magnetic fields, most of the iron nanoparticles can be separated. As a result, iron oxide nanoparticles could be efficiently employed to remove heavy metal from contaminated water and could thus be separated successfully from the systems [127, 175, 257]. Goethite (α-FeOOH) is studied a lot where it is manifested that they are competitive adsorbent of heavy metals owing to their cost-effectiveness, good adsorption capability and environmental friendliness [149]. Goethite was reported to be synthesized from ferrous and ferric salts to remove uranium from water [250].

Nanoscale α-FeOOH shows photocatalytic activity and good adsorption quality toward heavy metals [39]. For the present, nanoscale α-FeOOH has shown high adsorption capability toward heavy metals [70, 128]. The most stable and corrosion resistance form of iron oxide is hematite (α-Fe_2O_3) [255]. Hematite nanoparticles have been shown very effective to adsorb heavy metals such as Cr (VI) [3, 7, 50]. The high adsorption capacity of nanoscale α-Fe_2O_3 toward heavy metals has been reported [228]. Very recent, superparamagnetic α-Fe_2O_3 nanoparticles were prepared and shown 100% removal efficiency of Mg (II), Al (III), and Mn (II) and 80% of Ni (II) and Zn (II) from acid mine drainage [113]. It proves α-Fe_2O as an excellent nanoparticle to treat wastewater for its low toxicity, high stability and high adsorption capability.

Maghemite (γ-Fe_2O_3) nanoparticles have been widely studied to remove heavy metals from wastewater [59]. There are many advantages to utilize γ-Fe_2O_3 nanoparticles in wastewater treatment. γ-Fe_2O_3 nanoparticles have a high active surface and high adsorption capacity toward heavy metal and it can be separated from the system just by applying an external magnetic field. Furthermore, the preparation of γ-Fe_2O_3 nanoparticles is easy and they behave environmentally [263]. γ-Fe_2O_3 nanoparticles of particle size 14 nm synthesized by single-step method were applied to heavy metals from wastewater [9]. Superparamagnetic γ-Fe_2O_3 nanoparticles with tunable morphology were prepared by utilizing a flame spray pyrolysis approach and applied to remove Cu(II) and Pb(II) from wastewater [200]. Magnetite-based nanoparticles are extensively applied as nanoadsorbent because of their simple preparation, easy use, cost-effectiveness, friendly behavior to the environment and easy separation from systems [146, 223, 277]. Fe_3O_4 nanoparticles are generally altered on surface by –SH [179], $-NH_2$ [258], –COOH [227]. Pan et al. studied adsorption of Cr(IV) on engineered iron oxide nanoparticles [180] (Fig. 13). Damino activated Fe_3O_4 nanoparticles were prepared through utilizing one-pot synthesis method and applied to test the adsorption capacity toward Cr (VI) and Ni (II) [176].

Core–shell structure of Fe_3O_4 nanoparticles have been prepared by utilizing various coating materials such as sodium dodecyl sulfate [4], tannic acid [18], silica [141], oleate [143], p-nitro aniline [140], polyethylene glycol [210], chitosan[194], etc., and used for the treatment of heavy metals in wastewater. For instance, a core–shell structure magnetite NPs was prepared by spraying the polymer of organo disulfifide polymer onto the $-NH_2$ activated Fe_3O_4 nanoparticles and exhibited efficient adsorption capacity toward heavy metals in a high concentration solution [99]. Figure 14 represents the core–shell structure of amphiphilic polyisopreneblock-

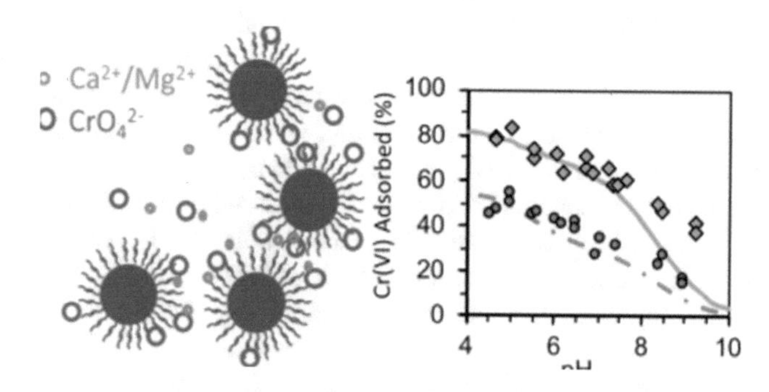

Fig. 13 Cr(VI) Adsorption on engineered iron oxide nanoparticles. Reprinted with license from American Chemical Society, Copyright (2019)

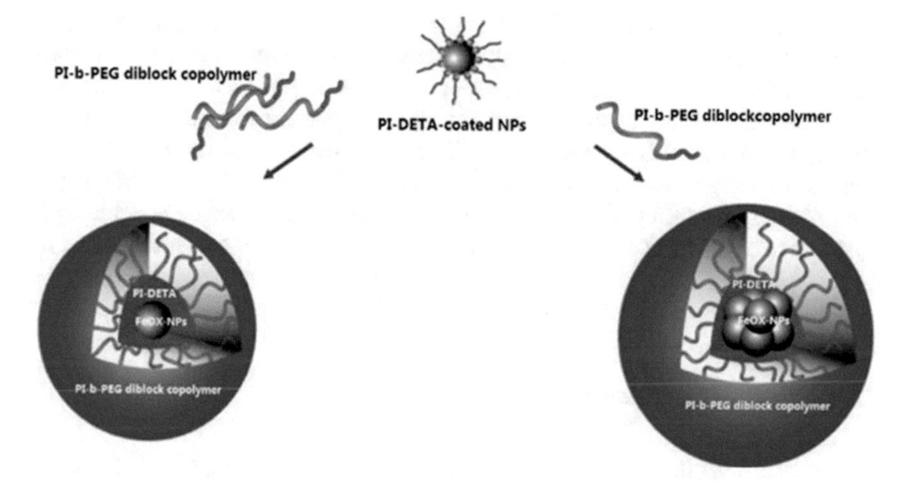

Fig. 14 Encapsulation of single or multiple nanoparticles by polyisopreneblock-poly(ethylene glycol) diblock copolymer. Reprinted with license from American Chemical Society, Copyright (2014)

poly(ethylene glycol) (PI-b-PEG) copolymer which encaged the Fe_3O_4 nanoparticles [217]. These core–shell structure magnetite nanoparticles were found to remove heavy metal with high efficiency and were easily separated from wastewater.

3.4 Titanium Oxide Nanoparticles

Titanium dioxide (TiO_2) nanoparticles with high chemical stability, lower toxicity and low cost are employed as competitive materials in disinfection and decontamination of wastewater. Thus, TiO_2 nanoparticles have drawn more concentration among researchers because of their extensive properties [117, 170, 216, 259]. TiO_2 nanoparticles do not change for a long time during degradation of pathogens and organic compounds. TiO_2 nanoparticles were widely investigated on degradation of organic contaminants with high effectiveness [14]. It was also represented with the effective removal of heavy metals from contaminated water [235]. Nanowires with diameter of 30–50 nm were synthesized from TiO_2 and were applied to eliminate Cu (II), Pb (II), Fe (III), Zn (II) and Cd (II) from contaminated water with high efficiency [286]. Iron-doped TiO_2 nanoparticles were prepared and utilized to remove arsenic with higher effectiveness than pure TiO_2 nanoparticles [171]. TiO_2 NPs coating with starch- were synthesized to eliminate 90% of Ni (II), Cd (II), Pb (II), Co (II) and Cu (II) from tap-water [25]. Microwave-synthesized TiO_2-chitosan nanoparticles were synthesized and were used for the removal of heavy metals applying the microwave-enforced sorption approach. This approach was observed as environmentally friendly and fast removal efficiency. TiO_2 nanoparticles displayed promising adsorption capacity toward organic and inorganic contaminants [252]. TiO_2, the semiconductor photocatalyst exhibits a variety in the case of mineralization or decontamination of harmful substance in water [253]. It is evident that TiO_2 nanoparticles in anatase phase possess strong catalytic activities for having high active surface and redox properties. Magnetic TiO_2 nanoparticles were prepared for the treatment of wastewater and this nanowire could easily be separated from the system with external magnetic fields showing suitability to commercial applications [147]. The demerits of TiO_2 nanoparticles are complex production processes and difficulty in removal from the system after use [137]. It is generally not easy to separate TiO_2 nanoparticles NPs when it is used to treat a slurry suspension of contaminated water [54].

3.5 Other Metal Oxide Nanoparticles

ZnO nanoparticles have come out as a promising material in decontamination of water because of their distinctive characteristics with large band gap in the near-UV electromagnetic spectrum spectral, powerful oxidation capability, enhanced photocatalytic ability [38, 201]. Moreover, having the almost identical band energy gap, the ZnO nanoparticles show similar photocatalytic activity as displayed by TiO_2 nanoparticles. Besides, ZnO nanoparticles are advantageous in the case of cost-effectiveness compared to TiO_2 nanoparticles [45]. ZnO nanoparticles posses the higher adsorption capability of light from the electromagnetic spectrum in a wide range in comparison with some other metal oxides nanoparticles [26]. Rapid reunification of photogenerated charges causes low photocatalytic efficiency of ZnO nanoparticles [78]. Photocatalytic efficiency of ZnO nanoparticles could be enhanced by doping metal. Different kinds of dopants mainly of metals such as inner transition elements dopants, codopants, anionic or cationic dopants, etc., were utilized for improving the photocatalytic efficiency of ZnO nanoparticles [126]. ZnO nanoparticles could be employed as a good nanoadsorbent for its non-toxicity, well antimicrobial activity, chemical, thermal and mechanical stability and overall efficient adsorption quality. ZnO nanoparticles exhibit higher adsorption efficiency toward heavy metals than TiO_2 nanoparticles [193]. ZnO nanoparticles were reported to show enhanced sorption capacity toward inorganic and organic contaminants [41]. ZnO nanoparticles have significant photocatalytic potential for exclusion of various organic compounds and contaminants due to their wide band gap energy, i.e., 3.37 eV, high exciton binding energy, i.e., 60 meV, strong oxidation ability and larger surface to volume ratio [240].

Manganese oxides (MnO_2) nanoparticles have been reported to show good sorption performance toward metal ions [167]. It has also been noted that MnO_2 nanoparticles and hydrous manganese oxide showed good removal efficiency of heavy metals from wastewater [134]. MnO_2/gelatin was prepared to remove Cd (II) and Pb (II) from wastewater through adsorption [274]. Guo et al. reported the effective removal of arsenite from water with synthesized paper-like, free-standing membrane of Mn_3O_4/CeO_2 hybrid nanotubes (Fig. 15) [81]. MnO_2 nanoparticles were noted to adsorb Tl (I) in wastewater [101]. MnO_2 nanoparticles were reported to show high capability to remove Cu(II), Hg(II), Pb(II), U, Cd(II), etc., from wastewater [1, 116, 130]. HMO is reported to exhibit advantageous characteristic in adsorption of heavy metals because of its porosity, ample active sites and high surface area [62].

Recently, hydrous manganese oxide-biochar nanocomposites were synthesized by implanting the hydrous manganese oxide nanoparticles into the biochar [270]. This composite material was applied to remove Pb (II) and Cd (II) in a broad pH range with high efficiency. Hence, hydrous manganese oxide-biochar could be a thriving candidate for the removal of heavy metals from contaminated water.

Aluminum oxides (Al_2O_3)-based nanoparticles are extensively utilized as adsorbent for removal of heavy metals. The major advantages of Al_2O_3 nanoparticles are low preparation cost and efficient decontamination capability [75, 187]. γ-Al_2O_3 nanoparticles were prepared through a sol–gel process and showed the removal capacity of 97% for Pb (II) and 87% for Cd (II) [254]. The effect of phosphate, humic acid and citrate on Al_2O_3 nanoparticles' adsorption behavior toward Cd (II) and Zn (II) has also been investigated and phosphate and humic acid were observed to show improved adsorption capacity toward Cd (II) and Zn (II) while citrate could reduce the capacity of adsorption toward Zn (II) [244]. Beside the abovementioned heavy metals, Al_2O_3 nanoparticles exhibit efficient removal capabilities toward Hg (II), As (III), Cu (II), Ni (II), Cr (VI), etc. [144, 181, 230, 275].

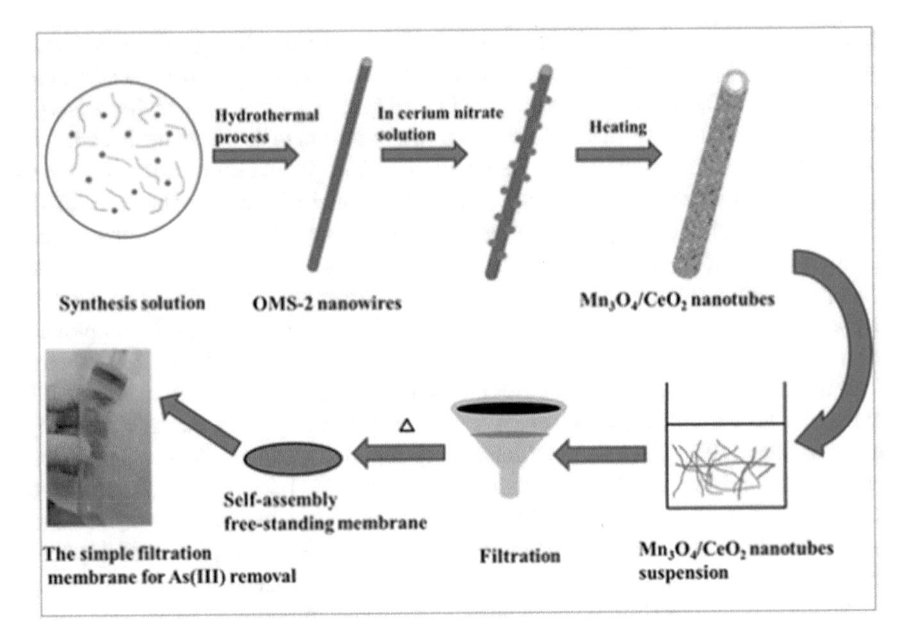

Fig. 15 Removal of arsenite from water by paper-like membrane of Mn_3O_4/CeO_2 Hybrid Nanotubes. Reprinted with license from American Chemical Society, Copyright (2015)

Magnesium oxide (MgO) nanoparticles are promising sorption materials for the removal of heavy metals due to their abundance, non-toxicity, environmental friendliness and overall cost-effectiveness. It was reported that MgO nanoparticles effectively remove Pb (II), Cd (II) and *Escherichia coli* from wastewater [34]. Furthermore, MgO nanoparticles showed extraordinary antibacterial properties toward gram-positive and gram-negative bacteria [245]. In another investigation, mesoporous MgO nanosheets were synthesized and were displayed to be excellently removed 1684.25 $mg{\cdot}g^{-1}$ Ni (II) from aqueous solution [67]. MgO nanoparticles were synthesized through the incineration process and were found to remove 96% Cu(II) from 10 ppm aqueous copper solution with high adsorption capability.

Cerium oxide (CeO_2) nanoparticles are non-toxic substances which have been utilized as photocatalysis and sensing [264], water treatment, etc. [203]. CeO_2 nanoparticles exhibit superior performance in heavy metal removals due to their active surface area, stability, selectivity and dispersion behavior. The sorption criteria of CeO_2 nanoparticles were investigated for the removal of Cr (VI) from aqueous solution [202]. The maximum adsorption capacity for Cr (VI) was reported as 121.95 $mg{\cdot}g^{-1}$. CeO_2 nanoparticles were reported to be prepared and were applied to remove As (V) and As (III) from aqueous solution [162]. The adsorption efficiency toward these two ions were observed as 36.8 and 71.9 $mg{\cdot}g^{-1}$, respectively,

Zirconium oxides (ZrO) nanoparticles are excellent metallic oxide adsorbent for the treatment of wastewater containing heavy metals. The merit of ZrO nanoparticles is the abundance of functional hydroxyl groups and high active surface areas.

Furthermore, ZrO nanoparticles have the chemical stability and show extraordinary sorption capability toward Pb (II), Zn (II) and Cd (II) [108]. A novel e ZrO_2/B_2O_3 nanocomposites were reported to be synthesized and were used to remove Cu (II), Co (II) and Cd (II) [282]. The removal efficiency for Cu (II), Co (II) and Cd (II) were found as 46.5, 32.2 and 109.9 $mg \cdot g^{-1}$, respectively. Polystyrene-supported $Zr(OH)_4$ nanoparticles were fabricated and were applied to remove Cd (II) from aqueous solution in varying pH [291]. The experimental outcome manifested that Cd (II) could be removed effectively in a wide pH range.

4 Nanocomposite in Water Treatment

Applications of nanoparticles in wastewater treatment have some issues regarding aggregation, intensive pressure drop during flow process, difficulties in separation from systems [95]. Though the types of metal nanoparticles discussed above have their own merits, they have often some problems in practical applications. For instance, nZVI aggregate and oxidized rapidly. TiO_2 nanoparticles and ZnO nanoparticles absorb electromagnetic spectrum only in the UV region because of their wide band gap. Carbon nanotube has difficulty in uniform suspension in various solvents and nZVI are easily oxidizable [89]. In order to overcome these problems, a general approach is adopted by synthesizing hybrid nanocomposites for wastewater treatment. For these reasons, the preparation of different nanocomposites has been gaining much attention to the researchers. Qian et al. briefly review the nanocomposite used in water treatment [189]. Figure 16 presents the nanoconfinement mediated water treatment by nanocomposite.

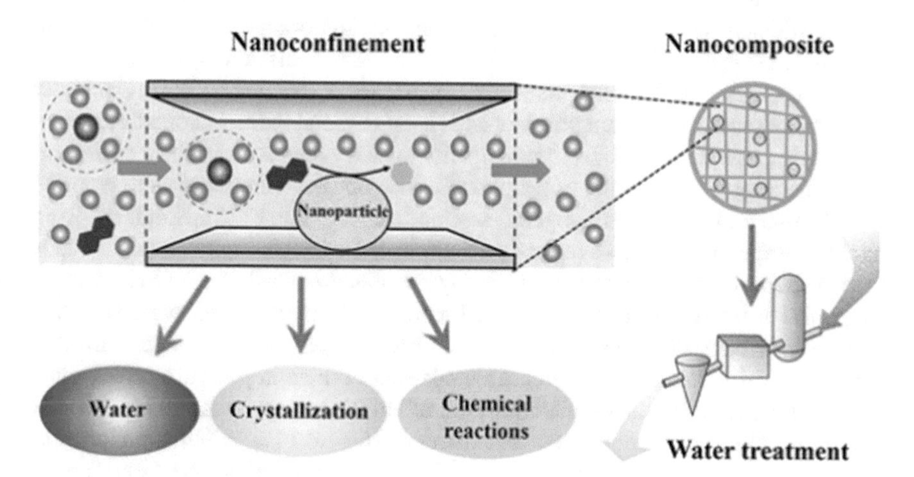

Fig. 16 Nanoconfinement mediated water treatment with nanocomposite. Reprinted with license from American Chemical Society, Copyright (2020)

Accordingly, many investigations have been done to fabricate useful nanocomposite for wastewater treatment throughout the world. For instance, a novel nanocomposite material was synthesized using nano zero-valent iron and carbon nanotubes where chemical deposition of nano zero-valent iron was done on the surface of carbon nanotubes. This nanocomposite adsorbent showed efficient capability to remove nitrate from water and it can easily be separated from the system by external magnetic fields [85]. Nanofiltration membranes of thin film nanocomposite have been synthesized through in situ implantation of TiO_2 nanoparticles on a polyimide support where TiO_2 nanoparticles were functionalized with both amine and chloride compounds to improve its compatibility. Nanofiltration membranes thus prepared displayed effective dye degradation and methanol flux [185]. Perfect nanocomposites for practical uses should be reactive as nanomaterials and continuous [260]. The more important thing is that treatment of wastewater requires non-toxic, cost-effective and log-time stable nanocomposites. To find suitable nanocomposites, further research in this field is still under way. In this section, various types of nanocomposite synthesized and applied for water treatment have been extensively discussed.

4.1 Nanocomposites with Inorganic Support

Nanocomposites are materials of multiple substances where one of the materials must be nanostructured. The combination of materials during preparation of nanocomposites offers suitable characteristics to it for the practical application in water treatment. Combination of TiO_2 and SiO_2 for the preparation of nanocomposites offers advantages of both materials by adsorbing virus on SiO_2 and showing enhanced antimicrobial activity with TiO_2 [107]. Ag_2S@Ag nanocomposite was fabricated which displayed enhanced sorption capability toward methyl orange and methyl blue in contaminated water [211]. In a review, Yin and Deng discussed about different nanocomposites with polymer-matrix for wastewater treatment [284]. Nanofiber membranes synthesized from polymer and metal or metal oxide nanoparticles were reported to show improved adsorption quality to heavy metals and enhanced antimicrobial activity. For example, Polyaniline/FeO composite nanofibers were reported for effective removal of carcinogenic arsenic from the water [28]. Similarly, from drinking water, the arsenic was effectively eliminated using bio-nanocomposite beads fabricated from chitosan goethite [91]. Many investigations on the use of hybrid nanomaterials for the removal of heavy metal from contaminated water were reported. As nanoadsorbent, the discarded parts of Zn-Mn dry batteries have been utilized to remove As, Cd and Pb [262]. Selenium nanoparticles containing polyurethane sponge have been reported for the efficient removal of Hg (II) from very rapidly because of the better affinity of selenium toward mercury [8]. Novel Fe_3O_4@diaminophenol-formaldehyde core–shell ferromagnetic nanorods for the elimination of Pb(II) from water was noted [267]. The nanorods displayed magnificent recovery time (25 s) due to the ferromagnetic properties with a high saturation magnetization value of the nanorod and hence possess better reusability among reported materials. So, the

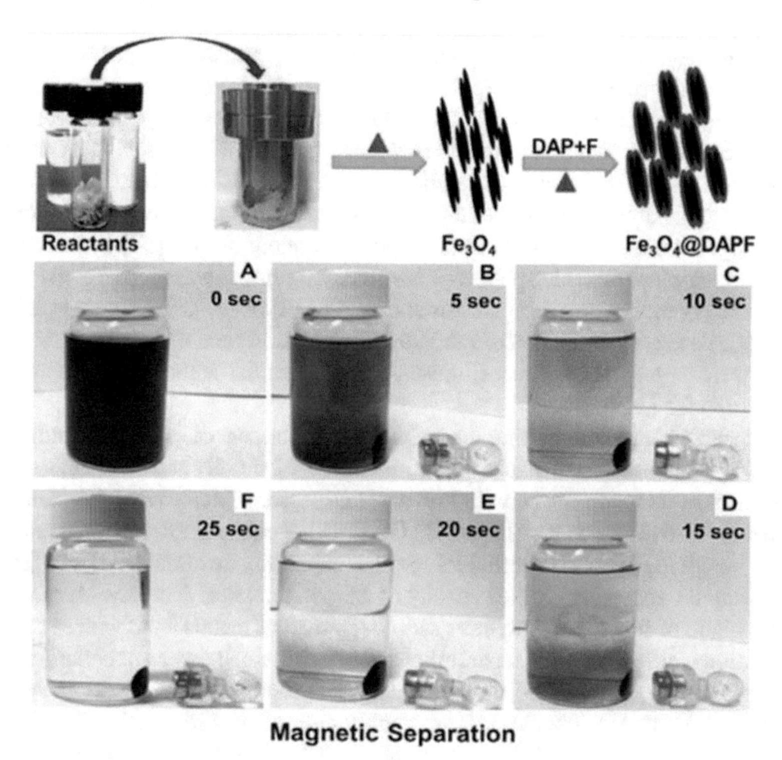

Fig. 17 Polymer composite of Fe_3O_4@diaminophenol-formaldehyde core–shell ferromagnetic nanorods based on core–shell ferromagnetic nanorod for the rapid removal of Pb(II). Reprinted with license from American Chemical Society, Copyright (2015)

Fe_3O_4@diaminophenol-formaldehyde core–shell ferromagnetic nanorods can act as good recyclable adsorbent alternatives to commonly utilized adsorbing materials for the fast removal of heavy metals from aqueous solutions (Fig. 17). Bentonite is excellent competitive material for the treatment of concentrated heavy metal contamination [52]. nZVI were found to be used with bentonite and applied for the removal of heavy metals [5].

4.2 Nanocomposites with Organic Supports

Organic polymer has numerous excellent properties with extraordinary mechanical strength, simple regeneration, easy degradability and modifiable functional group which enable it a promising candidate for being the host of nanocomposites [296]. Generally, there are two types of polymer-supported nanocomposite namely biopolymer-supported nanocomposites and synthetic organic polymer-supported nanocomposites [136]. The common example of the synthetic organic polymer used to support materials for preparing nanocomposites is polyaniline, polystyrene, etc. [196]. For instance, polypyrrole-polyaniline/Fe_3O_4 magnetic nanocomposites were synthesized and were reported to remove 100% of Pb (II) from 20 ppm aqueous solution [5]. Beside the synthetic organic polymers, natural polymers such as chitosan, cellulose, alginate, etc., were also applied as host materials for nanocomposites. The most abundant natural polymer cellulose has ample coordination sites for which make it amazing materials for adsorbent and support for nanoadsorbent [34]. Nanocellulose -Ag nanoparticles embedded pebbles-based nanocomposite was prepared and used to remove heavy metals, microorganisms and dyes from wastewater. Complete removal of Pb (II), 98% removal of Cr (III) and 99% disinfection capability toward microbial agents were displayed by Nanocellulose-Ag nanoparticles embedded pebbles-based nanocomposite [249].

Chitosan is another starting material for fabrication of promising adsorbent for metal contaminants because of the presence of amino and hydroxyl groups. ZnO/chitosan nanocomposite with low cost and lesser toxicity were fabricated and applied to remove Pb (II), Cd (II) and Cu (II) from aqueous solution [208]. The experimental result manifested the efficient sorption capability towards Cd (II), Pb (II) and Cu (II) and the recurring usable capacity of nanocomposites. A review on nanocomposites blend of functional polymers for the removal of metals from water with their preparation method, toxicity, separability and interactivity between nanoparticles and polymer were reported [135]. In another investigation, nanocomposite of hydrous Zr(IV) oxide was fabricated with the combination of a cation exchange resin and hydrous Zr (IV) oxide [96]. The investigated result showed remarkable adsorption capacity of nanocomposite of hydrous Zr(IV) oxide toward Cd (II) and Pb (II) in a column adsorption process. The cyclic column method displayed that the nanocomposites could be applied to practical acid mine wastewater time and again without loss of any capacity.

4.3 Magnetic Nanocomposites

Magnetic nanocomposites are promising candidates for the removal and degradation of contaminants from the polluted systems. The extensive studies of the toxicity of magnetic nanomaterials within or outside of an entire living organism have already been carried out. Hence, the abundant information on the toxicity of magnetic nanoparticles assists improved use of magnetic nanocomposites with less toxicity for treatment of contaminated water. However, there are a limited number of available magnetic nanoparticles such as Fe_2O_3, Fe_3O_4, nZVI, Co_3O_4 and NiO nanoparticles, etc. These are not enough for fabrication of magnetic nanocomposites to apply in the decontamination of water. There are also some issues to use magnetic nanocomposite for commercial purposes. First of all, the magnetic nanocomposite should be cost-effective to the practical application in the environment field. Otherwise, it will not be sustainable for common application for water treatment. Second, the properties of magnetic nanocomposite are essentially needed to improve for avoiding aggregation. The aggregation of the magnetic NPs and composite materials will hinder reusable capacity of the nanocomposite in the practical environmental remediation field. Finally, hazardness to the environment with application of magnetic nanocomposites in wastewater treatment should be minimized.

Studies on the toxicity of magnetic nanocomposites are just at the beginning stage. More research about the toxicity of the magnetic nanocomposites is necessary for the improvement of technology based on magnetic nanocomposite for water purification. Magnetic nanocomposites could be fabricated through surface modification of magnetic nanoparticles by different functional groups, combining magnetic nanoparticles with other organic or inorganic compounds like polyethylenimine, polyrhodanine, humic acid, MnO_2, etc. [116, 136, 161, 238]. Magnetic nanocomposites were synthesized through spraying the magnetic nanoparticles on graphene oxide or carbon nanotubes [57]. A core–shell Fe_3O_4@SiO_2 novel magnetic nanocomposite was synthesized and showed high removal ability toward Pb (II) and methylene blue. [99]. Fe@MgO nanocomposite was synthesized with the combination of nZVI MgO [74]. The advantage of strong magnetism of nZVI and efficient adsorption capability of MgO made it superior material for the effective removal of Pb (II) and methyl orange from wastewater. It is proven that magnetic nanocomposites have a high ability to remove heavy metal and to degrade the organic and inorganic pollutant from contaminated water with some limitations.

5 Conclusion and Perspective

Clean water is the key requirement to human health. The world is confronting critical challenges to meet the increasing demands of clean water as the sources of freshwater are declining due to climate change, population growth, increasing food production, increasing competition for fresh water resources in some areas, etc. Moreover, fresh water is polluted by agricultural contaminants, industrial contaminants, sewage contaminants, radioactive contaminants, microbes, organic and inorganic pollutants. There are several traditional ways for the treatment of wastewater. But nanomaterials have a number of important physicochemical properties that enable them especially attractive as a decontaminator wastewater. Nanomaterials can be modified by different functional materials to enhance their attraction toward contaminants. It is proved that they have the higher ability to remove organic and inorganic pollutants, toxic metal and radionuclides from aqueous solutions. Nanomaterials also give outstanding possibilities for the improvement of water purification systems more efficiently due to their high active surface areas and their size-dependent catalytic, optical and electronic characteristics. Nanomaterials are also being applied as active antimicrobial agents to treat pathogens containing water. Nanomaterials are widely applied to remove heavy metals from wastewater or aqueous solution of metal ions due to their excellent adsorption capabilities toward heavy metals.

In this chapter, metal-based nanomaterials are used in decontamination of wastewater which are fabricated from metal and metal oxide nanoparticles such as zero-valent nanoparticles (Fe, Zn, etc.), noble and transition metal nanoparticles (Fe, Cu, Ag, Au, Pd, etc.), metal oxide nanoparticles (iron oxide, titania, zinc oxide, magnesium oxide aluminum oxide, etc.) and overall nanocomposites of metal or metal oxide were discussed in detail. With the recent progress in wastewater treatment technology, nanomaterials-based water treatment methods are considered as extensive promising technology for wastewater decontamination. However, further investigations are still required to solve the issues regarding practical use of nanomaterials.

The drawback of existing nanomaterials will be required to be resolved for better application of these nanomaterials in water decontamination. First, most of the nanomaterials are not stable and easily aggregate. Moreover, it is generally troublesome to separate the nanoparticle from the system after the treatment process due to their nanosize. The development of nanocomposite materials could be an effective tool to solve this separation problem issue. Furthermore, to devise the facile synthesis procedure, to acquire long-time stability and to solve some other problems regarding nanocomposites, it requires more study in this area. Second, the commercial nanomaterials for heavy metal removal are scarce and more research is needed to obtain nanomaterials for commercial use. Finally, the effect and toxicity to the environment and human health due to extensive use of nanomaterials should be paid attention. There has been some research concentrated on the biological behavior and toxicity of nanoparticles toward human health [70, 114, 128, 225, 290]. The standard assessments of the toxicity of nanomaterials are quite inadequate at present. It is noticed from the previous study that most of the nanomaterials are observed as toxic substances after a certain level [32, 205]. Therefore, extensive study on the toxicity of nanomaterials is essential to ensure the safety for practical application.

The introduction of nanomaterials in the water treatment process is becoming a thriving tool. Moreover, the removal efficiency of contaminants with the above-mentioned nanomaterials is mostly studied in laboratory scale. More data of their application in practical wastewater treatment are inadequate and are badly needed. Present nanotechnology approaches for wastewater treatment seem promising. But, more extensive investigation is necessary to prove their safety in practical use. The metal-based nanomaterials should be low cost and superior to the traditional technologies that are applied for the water treatments. It is not easy to figure out the capabilities of different nanomaterials in practical applications and it requires more investigation to find out improved nanomaterials for the real application toward wastewater decontamination. Hence, the assessment of metal-based nanomaterials on the basis of performance in decontamination of wastewater should be perfected in the future. We visualize that metal-based nanomaterials will become excellent candidates for industrial and public water purification systems as more development is done through cost-effective synthesis and utilizing the environmentally acceptable functional materials.

References

1. Abdullah JA, Al Lafi AG, Amin Y, Alnama T (2018) A Styrofoam-nano manganese oxide based composite: Preparation and application for the treatment of wastewater. Appl Radiat Isot 136:73–81. https://doi.org/10.1016/j.apradiso.2018.02.013
2. Adams CP, Walker KA, Obare SO, Docherty KM (2014) size-dependent antimicrobial effects of novel palladium nanoparticles. PLoS ONE 9:e85981.https://doi.org/10.1371/journal.pone.0085981
3. Adegoke HI, AmooAdekola F, Fatoki OS, Ximba BJ (2014) Adsorption of Cr (VI) on synthetic hematite (α-Fe2O3) nanoparticles of different morphologies. Korean J Chem Eng 31:142–154. https://doi.org/10.1007/s11814-013-0204-7
4. Adeli M, Yamini Y, Faraji M (2017) Removal of copper, nickel and zinc by sodium dodecyl sulphate coated magnetite nanoparticles from water and wastewater samples. Arab J Chem 10:S514–S521. https://doi.org/10.1016/j.arabjc.2012.10.012
5. Afshar A, Sadjadi SAS, Mollahosseini A, Eskandarian MR (2016) Polypyrrole-polyaniline/Fe_3O_4 magnetic nanocomposite for the removal of Pb(II) from aqueous solution. Korean J Chem Eng 33:669–677. https://doi.org/10.1007/s11814-015-0156-1
6. Ahmad T, Guria C, Mandal A (2020) A review of oily wastewater treatment using ultrafiltration membrane: a parametric study to enhance the membrane performance. J Water Process Eng 36:101289.https://doi.org/10.1016/j.jwpe.2020.101289
7. Ahmad T, Wani IA, Lone IH, Ganguly A, Manzoor N, Ahmad A, Ahmed J, Al-Shihri AS (2013) Antifungal activity of gold nanoparticles prepared by solvothermal method. Mater Res Bull 48:12–20. https://doi.org/10.1016/j.materresbull.2012.09.069
8. Ahmed S, Brockgreitens J, Xu K, Abbas A (2017) A Nanoselenium sponge for instantaneous mercury removal to undetectable levels. Adv Funct Mater 27:1606572. https://doi.org/10.1002/adfm.201606572
9. Akhbarizadeh R, Shayestefar MR, Darezereshki E (2014) Competitive removal of metals from wastewater by maghemite nanoparticles: a comparison between simulated wastewater and AMD. Mine Water Environ 33:89–96. https://doi.org/10.1007/s10230-013-0255-3
10. Akinoglu GE, Mir SH, Gatensby R, Rydzek G, Mokarian-Tabari P (2020) Block copolymer derived vertically coupled plasmonic arrays for surface-enhanced raman spectroscopy. ACS Appl Mater Interfaces 12:23410–23416. https://doi.org/10.1021/acsami.0c03300
11. Alice GL, Alexandar APM, Herman SM (2021) Advanced functional nanostructures based on magnetic iron oxide nanomaterials for water remediation: a review. Water Res 190:116693.https://doi.org/10.1016/j.watres.2020.116693
12. Alqudami A, Alhemiary NA, Munassar S (2012) Removal of Pb(II) and Cd(II) ions from water by Fe and Ag nanoparticles prepared using electro-exploding wire technique. Environ Sci Pollut Res 19:2832–2841. https://doi.org/10.1007/s11356-012-0788-1
13. Amin RM, Mohamed MB, Ramadan MA, Verwanger T, Krammer B (2009) Rapid and sensitive microplate assay for screening the effect of silver and gold nanoparticles on bacteria. Nanomed 4:637–643. https://doi.org/10.2217/nnm.09.50
14. Anandan S, Kathiravan K, Murugesan V, Ikuma Y (2009) Anionic (IO 3 -) non-metal doped TiO_2 nanoparticles for the photocatalytic degradation of hazardous pollutant in water. Catal Commun 10:1014–1019. https://doi.org/10.1016/j.catcom.2008.12.054
15. Aredes S, Klein B, Pawlik M (2012) The removal of arsenic from water using natural iron oxide minerals. J Clean Prod 29–30:208–213. https://doi.org/10.1016/j.jclepro.2012.01.029
16. Asharani P, Xinyi N, Hande MP, Valiyaveettil S (2010) DNA damage and p53-mediated growth arrest in human cells treated with platinum nanoparticles. Nanomed 5:51–64. https://doi.org/10.2217/nnm.09.85
17. Badruddoza AZM, Tay ASH, Tan PY, Hidajat K, Uddin MS (2011) Carboxymethyl-β-cyclodextrin conjugated magnetic nanoparticles as nano-adsorbents for removal of copper ions: Synthesis and adsorption studies. J Hazard Mater 185:1177–1186. https://doi.org/10.1016/j.jhazmat.2010.10.029

18. Bagtash M, Yamini Y, Tahmasebi E, Zolgharnein J, Dalirnasab Z (2016) Magnetite nanoparticles coated with tannic acid as a viable sorbent for solid-phase extraction of Cd2+, Co2+ and Cr3+. Microchim Acta 183:449–456. https://doi.org/10.1007/s00604-015-1667-5
19. Banholzer MJ, Millstone JE, Qin L, Mirkin CA (2008) Rationally designed nanostructures for surface-enhanced Raman spectroscopy. Chem Soc Rev 37:885. https://doi.org/10.1039/b710915f
20. Baruah S, Dutta J (2009) Nanotechnology applications in pollution sensing and degradation in agriculture: a review. Environ Chem Lett 7:191–204. https://doi.org/10.1007/s10311-009-0228-8
21. Baruah S, Jaisai M, Dutta J (2012) Development of a visible light active photocatalytic portable water purification unit using ZnO nanorods. Catal Sci Technol 2:918. https://doi.org/10.1039/c2cy20033c
22. Baruah S, Sinha SS, Ghosh B, Pal SK, Raychaudhuri AK, Dutta J (2009) Photoreactivity of ZnO nanoparticles in visible light: Effect of surface states on electron transfer reaction. J Appl Phys 105:074308.https://doi.org/10.1063/1.3100221
23. Bashir S, Moosvi SK, Jan T, Rydzek G, Mir SH, Rizvi MA (2020) Development of polythiophene/prussian red nanocomposite with dielectric, photocatalytic and metal scavenging properties. J Electron Mater 49:4018–4027. https://doi.org/10.1007/s11664-020-08117-7
24. Bayoumi TA, Saleh HM (2018) Characterization of biological waste stabilized by cement during immersion in aqueous media to develop disposal strategies for phytomediated radioactive waste. Prog Nucl Energy 107:83–89. https://doi.org/10.1016/j.pnucene.2018.04.021
25. Baysal A, Kuznek C, Ozcan M (2018) Starch coated titanium dioxide nanoparticles as a challenging sorbent to separate and preconcentrate some heavy metals using graphite furnace atomic absorption spectrometry. Int J Environ Anal Chem 98:45–55. https://doi.org/10.1080/03067319.2018.1427741
26. Behnajady M, Modirshahla N, Hamzavi R (2006) Kinetic study on photocatalytic degradation of C.I. Acid Yellow 23 by ZnO photocatalyst. J Hazard Mater 133:226–232. https://doi.org/10.1016/j.jhazmat.2005.10.022
27. Bhatia S, Verma N (2017) Photocatalytic activity of ZnO nanoparticles with optimization of defects. Mater Res Bull 95:468–476. https://doi.org/10.1016/j.materresbull.2017.08.019
28. Bhaumik M, Noubactep C, Gupta VK, McCrindle RI, Maity A (2015) Polyaniline/Fe0 composite nanofibers: an excellent adsorbent for the removal of arsenic from aqueous solutions. Chem Eng J 271:135–146. https://doi.org/10.1016/j.cej.2015.02.079
29. Bokare V, Jung J, Chang Y-Y, Chang Y-S (2013) Reductive dechlorination of octachlorodibenzo-p-dioxin by nanosized zero-valent zinc: Modeling of rate kinetics and congener profile. J Hazard Mater 250–251:397–402. https://doi.org/10.1016/j.jhazmat.2013.02.020
30. Bootharaju MS, Pradeep T (2012) Understanding the degradation pathway of the pesticide, chlorpyrifos by noble metal nanoparticles. Langmuir 28:2671–2679. https://doi.org/10.1021/la2050515
31. Bowen WR, Welfoot JS (2002) Modelling the performance of membrane nanofiltration—critical assessment and model development. Chem Eng Sci 57:1121–1137. https://doi.org/10.1016/S0009-2509(01)00413-4
32. Boyle D, Clark NJ, Handy RD (2020) Toxicities of copper oxide nanomaterial and copper sulphate in early life stage zebrafish: effects of pH and intermittent pulse exposure. Ecotoxicol Environ Saf 190:109985.https://doi.org/10.1016/j.ecoenv.2019.109985
33. Brame J, Li Q, Alvarez PJJ (2011) Nanotechnology-enabled water treatment and reuse: emerging opportunities and challenges for developing countries. Trends Food Sci Technol 22:618–624. https://doi.org/10.1016/j.tifs.2011.01.004
34. Cai Y, Li C, Wu D, Wang W, Tan F, Wang X, Wong PK, Qiao X (2017) Highly active MgO nanoparticles for simultaneous bacterial inactivation and heavy metal removal from aqueous solution. Chem Eng J 312:158–166. https://doi.org/10.1016/j.cej.2016.11.134

35. Chandran GT, Li X, Ogata A, Penner RM (2017) Electrically transduced sensors based on nanomaterials (2012–2016). Anal Chem 89:249–275. https://doi.org/10.1021/acs.analchem.6b04687
36. Chatzimitakos T, Kallimanis A, Avgeropoulos A, Stalikas CD (2016) Antibacterial, anti-biofouling, and antioxidant prospects of metal-based nanomaterials: water. CLEAN—Soil Air Water 44:794–802. https://doi.org/10.1002/clen.201500366
37. Chen L, Wang G, Mathur GN, Varadan VK (2012) Size and shape dependence of the electrochemical properties of hematite nanoparticles and their applications in lithium ion batteries, in: Varadan, V.K. (Ed.), Presented at the SPIE smart structures and materials + nondestructive evaluation and health monitoring, San Diego, California, p 83441J. https://doi.org/10.1117/12.915597
38. Chen Y, Bagnall DM, Koh H, Park K, Hiraga K, Zhu Z, Yao T (1998) Plasma assisted molecular beam epitaxy of ZnO on *c* -plane sapphire: growth and characterization. J Appl Phys 84:3912–3918. https://doi.org/10.1063/1.368595
39. Chen Y-H, Li F-A (2010) Kinetic study on removal of copper(II) using goethite and hematite nano-photocatalysts. J Colloid Interface Sci 347:277–281. https://doi.org/10.1016/j.jcis.2010.03.050
40. Choi H, Al-Abed SR, Agarwal S (2009) Effects of aging and oxidation of palladized iron embedded in activated carbon on the dechlorination of 2-chlorobiphenyl. Environ Sci Technol 43:4137–4142. https://doi.org/10.1021/es803535b
41. Chouchene B, Chaabane TB, Mozet K, Girot E, Corbel S, Balan L, Medjahdi G, Schneider R (2017) Porous Al-doped ZnO rods with selective adsorption properties. Appl Surf Sci 409:102–110. https://doi.org/10.1016/j.apsusc.2017.03.018
42. Crini G, Lichtfouse E (2019) Advantages and disadvantages of techniques used for wastewater treatment. Environ Chem Lett 17:145–155. https://doi.org/10.1007/s10311-018-0785-9
43. Cundy AB, Hopkinson L, Whitby RLD (2008) Use of iron-based technologies in contaminated land and groundwater remediation: A review. Sci Total Environ 400:42–51. https://doi.org/10.1016/j.scitotenv.2008.07.002
44. Cuong ND, Hoa ND, Hoa TT, Khieu DQ, Quang DT, Quang VV, Hieu NV (2014) Nanoporous hematite nanoparticles: Synthesis and applications for benzylation of benzene and aromatic compounds. J Alloys Compd 582:83–87. https://doi.org/10.1016/j.jallcom.2013.08.057
45. Daneshvar N, Salari D, Khataee AR (2004) Photocatalytic degradation of azo dye acid red 14 in water on ZnO as an alternative catalyst to TiO2. J. Photochem. Photobiol. Chem. 162:317–322. https://doi.org/10.1016/S1010-6030(03)00378-2
46. Dankovich TA, Gray DG (2011) Bactericidal paper impregnated with silver nanoparticles for point-of-use water treatment. Environ Sci Technol 45:1992–1998. https://doi.org/10.1021/es103302t
47. Daou I, Moukrad N, Zegaoui O, Rhazi Filali F (2018) Antimicrobial activity of ZnO-TiO_2 nanomaterials synthesized from three different precursors of ZnO: influence of ZnO/TiO_2 weight ratio. Water Sci Technol 77:1238–1249. https://doi.org/10.2166/wst.2017.647
48. Deng C-H, Gong J-L, Zeng G-M, Zhang P, Song B, Zhang X-G, Liu H-Y, Huan S-Y (2017) Graphene sponge decorated with copper nanoparticles as a novel bactericidal filter for inactivation of Escherichia coli. Chemosphere 184:347–357. https://doi.org/10.1016/j.chemosphere.2017.05.118
49. Diao M, Yao M (2009) Use of zero-valent iron nanoparticles in inactivating microbes. Water Res 43:5243–5251. https://doi.org/10.1016/j.watres.2009.08.051
50. Dickson D, Liu G, Cai Y (2017) Adsorption kinetics and isotherms of arsenite and arsenate on hematite nanoparticles and aggregates. J Environ Manage 186:261–267. https://doi.org/10.1016/j.jenvman.2016.07.068
51. Dimapilis EAS, Hsu C-S, Mendoza RMO, Lu M-C (2018) Zinc oxide nanoparticles for water disinfection. Sustain Environ Res 28:47–56. https://doi.org/10.1016/j.serj.2017.10.001
52. Donat R, Akdogan A, Erdem E, Cetisli H (2005) Thermodynamics of Pb2+ and Ni2+ adsorption onto natural bentonite from aqueous solutions. J Colloid Interface Sci 286:43–52. https://doi.org/10.1016/j.jcis.2005.01.045

53. Dorathi PJ, Kandasamy P (2012) Dechlorination of chlorophenols by zero valent iron impregnated silica. J Environ Sci 24:765–773. https://doi.org/10.1016/S1001-0742(11)60817-6
54. Du P, Carneiro JT, Moulijn JA, Mul G (2008) A novel photocatalytic monolith reactor for multiphase heterogeneous photocatalysis. Appl Catal Gen 334:119–128. https://doi.org/10.1016/j.apcata.2007.09.045
55. Egerton T (2014) UV-Absorption—the primary process in photocatalysis and some practical consequences. Molecules 19:18192–18214. https://doi.org/10.3390/molecules191118192
56. Elmi F, Alinezhad H, Moulana Z, Salehian F, Mohseni Tavakkoli S, Asgharpour F, Fallah H, Elmi MM (2014) The use of antibacterial activity of ZnO nanoparticles in the treatment of municipal wastewater. Water Sci Technol 70:763–770. https://doi.org/10.2166/wst.2014.232
57. Elmi F, Hosseini T, Taleshi MS, Taleshi F (2017) Kinetic and thermodynamic investigation into the lead adsorption process from wastewater through magnetic nanocomposite Fe3O4/CNT. Nanotechnol Environ Eng 2:13. https://doi.org/10.1007/s41204-017-0023-x
58. Els ER, Lorenzen L, Aldrich C (2000) The adsorption of precious metals and base metals on a quaternary ammonium group ion exchange resin. Miner Eng 13:401–414. https://doi.org/10.1016/S0892-6875(00)00022-4
59. Etale A, Tutu H, Drake DC (2016) The effect of silica and maghemite nanoparticles on remediation of Cu(II)-, Mn(II)- and U(VI)-contaminated water by Acutodesmus sp. J Appl Phycol 28:251–260. https://doi.org/10.1007/s10811-015-0555-z
60. Fadel TR, Farrell DF, Friedersdorf LE, Griep MH, Hoover MD, Meador MA, Meyyappan M (2016) Toward the responsible development and commercialization of sensor nanotechnologies. ACS Sens 1:207–216. https://doi.org/10.1021/acssensors.5b00279
61. Falinski MM, Plata DL, Chopra SS, Theis TL, Gilbertson LM, Zimmerman JB (2018) A framework for sustainable nanomaterial selection and design based on performance, hazard, and economic considerations. Nat Nanotechnol 13:708–714. https://doi.org/10.1038/s41565-018-0120-4
62. Fan M, Boonfueng T, Xu Y, Axe L, Tyson TA (2005) Modeling Pb sorption to microporous amorphous oxides as discrete particles and coatings. J Colloid Interface Sci 281:39–48. https://doi.org/10.1016/j.jcis.2004.08.050
63. Fan Y, Liu Z, Wang L, Zhan J (2009) Synthesis of Starch-Stabilized Ag Nanoparticles and Hg2+ Recognition in Aqueous Media. Nanoscale Res Lett 4:1230–1235. https://doi.org/10.1007/s11671-009-9387-6
64. Fang X, Li J, Li X, Pan S, Zhang X, Sun X, Shen J, Han W, Wang L (2017) Internal pore decoration with polydopamine nanoparticle on polymeric ultrafiltration membrane for enhanced heavy metal removal. Chem Eng J 314:38–49. https://doi.org/10.1016/j.cej.2016.12.125
65. FAO, United Nations, 2020. Technical brief on water, sanitation, hygiene and wastewater management to prevent infections and... reduce the spread of antimicrobial resistance. Food & Agriculture Org, S.l
66. Farka Z, Juřík T, Kovář D, Trnková L, Skládal P (2017) Nanoparticle-based immunochemical biosensors and assays: recent advances and challenges. Chem Rev 117:9973–10042. https://doi.org/10.1021/acs.chemrev.7b00037
67. Feng J, Zou L, Wang Y, Li B, He X, Fan Z, Ren Y, Lv Y, Zhang M, Chen D (2015) Synthesis of high surface area, mesoporous MgO nanosheets with excellent adsorption capability for Ni(II) via a distillation treating. J Colloid Interface Sci 438:259–267. https://doi.org/10.1016/j.jcis.2014.10.004
68. Ferdosi E, Bahiraei H, Ghanbari D (2019) Investigation the photocatalytic activity of $CoFe_2O4/ZnO$ and $CoFe_2O_4/ZnO/Ag$ nanocomposites for purification of dye pollutants. Sep Purif Technol 211:35–39. https://doi.org/10.1016/j.seppur.2018.09.054
69. Fernandes T, Soares S, Trindade T, Daniel-da-Silva A (2017) Magnetic hybrid nanosorbents for the uptake of paraquat from water. Nanomaterials 7:68. https://doi.org/10.3390/nano7030068
70. Forbes EA, Posner AM, Quirk JP (1976) The specific adsorption of divalent Cd Co, Cu, Pb, and Zn on goethite. J Soil Sci 27:154–166. https://doi.org/10.1111/j.1365-2389.1976.tb01986.x

71. Franci G, Falanga A, Galdiero S, Palomba L, Rai M, Morelli G, Galdiero M (2015) Silver nanoparticles as potential antibacterial agents. Molecules 20:8856–8874. https://doi.org/10.3390/molecules20058856
72. Fuwad A, Ryu H, Malmstadt N, Kim SM, Jeon T-J (2019) Biomimetic membranes as potential tools for water purification: preceding and future avenues. Desalination 458:97–115. https://doi.org/10.1016/j.desal.2019.02.003
73. Garner KL, Keller AA (2014) Emerging patterns for engineered nanomaterials in the environment: a review of fate and toxicity studies. J Nanoparticle Res 16:2503. https://doi.org/10.1007/s11051-014-2503-2
74. Ge L, Wang W, Peng Z, Tan F, Wang X, Chen J, Qiao X (2018) Facile fabrication of Fe@MgO magnetic nanocomposites for efficient removal of heavy metal ion and dye from water. Powder Technol 326:393–401. https://doi.org/10.1016/j.powtec.2017.12.003
75. Giles DE, Mohapatra M, Issa TB, Anand S, Singh P (2011) Iron and aluminium based adsorption strategies for removing arsenic from water. J Environ Manage 92:3011–3022. https://doi.org/10.1016/j.jenvman.2011.07.018
76. Giwa A, Hasan SW, Yousuf A, Chakraborty S, Johnson DJ, Hilal N (2017) Biomimetic membranes: a critical review of recent progress. Desalination 420:403–424. https://doi.org/10.1016/j.desal.2017.06.025
77. Gómez de la Torre TZ, Ke R, Mezger A, Svedlindh P, Strømme M, Nilsson M (2012) Sensitive detection of spores using volume-amplified magnetic nanobeads. Small 8:2174–2177. https://doi.org/10.1002/smll.201102632
78. Gomez-Solís C, Ballesteros JC, Torres-Martínez LM, Juárez-Ramírez I, Díaz Torres LA, Elvira Zarazua-Morin M, Lee SW (2015) Rapid synthesis of ZnO nano-corncobs from Nital solution and its application in the photodegradation of methyl orange. J Photochem Photobiol Chem 298:49–54. https://doi.org/10.1016/j.jphotochem.2014.10.012
79. Gondal MA, Dastageer MA, Khalil A (2009) Synthesis of nano-WO3 and its catalytic activity for enhanced antimicrobial process for water purification using laser induced photo-catalysis. Catal Commun 11:214–219. https://doi.org/10.1016/j.catcom.2009.10.011
80. Guo H, Xu H, Barnard AS (2013) Can hematite nanoparticles be an environmental indicator? Energy Env. Sci 6:561–569. https://doi.org/10.1039/C2EE23253G
81. Guo S, Sun W, Yang W, Xu Z, Li Q, Shang JK (2015) Synthesis of Mn_3O_4/CeO_2 hybrid nanotubes and their spontaneous formation of a paper-like, free-standing membrane for the removal of arsenite from water. ACS Appl Mater Interfaces 7:26291–26300. https://doi.org/10.1021/acsami.5b08862
82. Gupta VK, Jain R, Nayak A, Agarwal S, Shrivastava M (2011) Removal of the hazardous dye—Tartrazine by photodegradation on titanium dioxide surface. Mater Sci Eng C 31:1062–1067. https://doi.org/10.1016/j.msec.2011.03.006
83. Gurreri L, Tamburini A, Cipollina A, Micale G (2020) Electrodialysis applications in wastewater treatment for environmental protection and resources recovery: a systematic review on progress and perspectives. Membranes 10:146. https://doi.org/10.3390/membranes10070146
84. Gutierrez AM, Dziubla TD, Hilt JZ (2017) Recent advances on iron oxide magnetic nanoparticles as sorbents of organic pollutants in water and wastewater treatment. Rev Environ Health 32:111–117. https://doi.org/10.1515/reveh-2016-0063
85. Haijiao L, Jingkang W, Marco S, Ting W, Ying B, Hongxun H (2016) An overview of nanomaterials for water and wastewater treatment. Adv Mater Sci Eng, Article ID 4964828
86. Hajipour MJ, Fromm KM, Akbar Ashkarran A, Jimenez de Aberasturi D, de Larramendi IR, Rojo T, Serpooshan V, Parak WJ, Mahmoudi M (2012) Antibacterial properties of nanoparticles. Trends Biotechnol 30:499–511. https://doi.org/10.1016/j.tibtech.2012.06.004
87. Hanif H, Waseem A, Kali S, Qureshi NA, Majid M, Iqbal M, Ur-Rehman T, Tahir M, Yousaf S, Iqbal MM, Khan IA, Zafar MI (2020) Environmental risk assessment of diclofenac residues in surface waters and wastewater: a hidden global threat to aquatic ecosystem. Environ Monit Assess 192:204. https://doi.org/10.1007/s10661-020-8151-3
88. Hao Y-M, Man C, Hu Z-B (2010) Effective removal of Cu (II) ions from aqueous solution by amino-functionalized magnetic nanoparticles. J Hazard Mater 184:392–399. https://doi.org/10.1016/j.jhazmat.2010.08.048

89. Hayati B, Maleki A, Najafi F, Daraei H, Gharibi F, McKay G (2016) Synthesis and characterization of PAMAM/CNT nanocomposite as a super-capacity adsorbent for heavy metal (Ni2+, Zn2+, As3+, Co2+) removal from wastewater. J Mol Liq 224:1032–1040. https://doi.org/10.1016/j.molliq.2016.10.053
90. Hayati B, Maleki A, Najafi F, Gharibi F, McKay G, Gupta VK, Harikaranahalli Puttaiah S, Marzban N (2018) Heavy metal adsorption using PAMAM/CNT nanocomposite from aqueous solution in batch and continuous fixed bed systems. Chem Eng J 346:258–270. https://doi.org/10.1016/j.cej.2018.03.172
91. He J, Bardelli F, Gehin A, Silvester E, Charlet L (2016) Novel chitosan goethite bionanocomposite beads for arsenic remediation. Water Res 101:1–9. https://doi.org/10.1016/j.watres.2016.05.032
92. Herrmann J-M, Guillard C (2000) Photocatalytic degradation of pesticides in agricultural used waters. Comptes Rendus Académie Sci. Ser. IIC Chem 3:417–422. https://doi.org/10.1016/S1387-1609(00)01137-3
93. Hlongwane GN, Sekoai PT, Meyyappan M, Moothi K (2019) Simultaneous removal of pollutants from water using nanoparticles: a shift from single pollutant control to multiple pollutant control. Sci Total Environ 656:808–833. https://doi.org/10.1016/j.scitotenv.2018.11.257
94. Homhoul P, Pengpanich S, Hunsom M (2011) Treatment of distillery wastewater by the nano-scale zero-valent iron and the supported nano-scale zero-valent iron. Water Environ Res 83:65–74. https://doi.org/10.2175/106143010X12780288628291
95. Hotze EM, Phenrat T, Lowry GV (2010) Nanoparticle aggregation: challenges to understanding transport and reactivity in the environment. J Environ Qual 39:1909–1924. https://doi.org/10.2134/jeq2009.0462
96. Hua M, Jiang Y, Wu B, Pan B, Zhao X, Zhang Q (2013) Fabrication of a New Hydrous Zr(IV) Oxide-based nanocomposite for Enhanced Pb(II) and Cd(II) removal from waters. ACS Appl Mater Interfaces 5:12135–12142. https://doi.org/10.1021/am404031q
97. Huang J, Xu X, Gu C, Wang W, Geng B, Sun Y, Liu J (2012) Size-controlled synthesis of porous $ZnSnO_3$ cubes and their gas-sensing and photocatalysis properties. Sens Actuators B Chem 171–172:572–579. https://doi.org/10.1016/j.snb.2012.05.036
98. Huang P, Ye Z, Xie W, Chen Q, Li J, Xu Z, Yao M (2013) Rapid magnetic removal of aqueous heavy metals and their relevant mechanisms using nanoscale zero valent iron (nZVI) particles. Water Res 47:4050–4058. https://doi.org/10.1016/j.watres.2013.01.054
99. Huang X, Yang J, Wang J, Bi J, Xie C, Hao H (2018) Design and synthesis of core–shell Fe_3O_4@PTMT composite magnetic microspheres for adsorption of heavy metals from high salinity wastewater. Chemosphere 206:513–521. https://doi.org/10.1016/j.chemosphere.2018.04.184
100. Huang Y, Keller AA (2013) Magnetic nanoparticle adsorbents for emerging organic contaminants. ACS Sustain Chem Eng 1:731–736. https://doi.org/10.1021/sc400047q
101. Huangfu X, Jiang J, Lu X, Wang Y, Liu Y, Pang S-Y, Cheng H, Zhang X, Ma J (2015) Adsorption and Oxidation of Thallium(I) by a Nanosized Manganese Dioxide. Water Air Soil Pollut 226:2272. https://doi.org/10.1007/s11270-014-2272-7
102. Hwang Y-H, Kim D-G, Shin H-S (2011) Mechanism study of nitrate reduction by nano zero valent iron. J Hazard Mater 185:1513–1521. https://doi.org/10.1016/j.jhazmat.2010.10.078
103. Hyder MKMZ, Ochiai B (2018) Selective recovery of Au(III), Pd(II), and Ag(I) from printed circuit boards using cellulose filter paper grafted with polymer chains bearing thiocarbamate moieties. Microsyst Technol 24:683–690. https://doi.org/10.1007/s00542-017-3277-0
104. Hyder MKMZ, Ochiai B (2017) Synthesis of a Selective Scavenger for Ag(I), Pd(II), and Au(III) based on cellulose filter paper grafted with polymer chains bearing thiocarbamate moieties. Chem Lett 46:492–494. https://doi.org/10.1246/cl.160983
105. Ihsanullah A, A., Al-Amer, A.M., Laoui, T., Al-Marri, M.J., Nasser, M.S., Khraisheh, M., Atieh, M.A., (2016) Heavy metal removal from aqueous solution by advanced carbon nanotubes: critical review of adsorption applications. Sep Purif Technol 157:141–161. https://doi.org/10.1016/j.seppur.2015.11.039

106. Islam NU, Jalil K, Shahid M, Rauf A, Muhammad N, Khan A, Shah MR, Khan MA (2019) Green synthesis and biological activities of gold nanoparticles functionalized with Salix alba. Arab J Chem 12:2914–2925. https://doi.org/10.1016/j.arabjc.2015.06.025
107. Jafry HR, Liga MV, Li Q, Barron AR (2011) Simple route to enhanced photocatalytic activity of P25 Titanium Dioxide nanoparticles by silica addition. Environ Sci Technol 45:1563–1568. https://doi.org/10.1021/es102749e
108. Jiang C, Xiao DA (2014) Nanosized Zirconium Dioxide particles as an efficient sorbent for lead removal in waters. Adv Mater Res 926–930:166–169. https://doi.org/10.4028/www.scientific.net/AMR.926-930.166
109. Jiao X, Gutha Y, Zhang W (2017) Application of chitosan/poly(vinyl alcohol)/CuO (CS/PVA/CuO) beads as an adsorbent material for the removal of Pb(II) from aqueous environment. Colloids Surf B Biointerfaces 149:184–195. https://doi.org/10.1016/j.colsurfb.2016.10.024
110. Karthikeyan KT, Nithya A, Jothivenkatachalam K (2017) Photocatalytic and antimicrobial activities of chitosan-TiO 2 nanocomposite. Int J Biol Macromol 104:1762–1773. https://doi.org/10.1016/j.ijbiomac.2017.03.121
111. Kasinathan K, Kennedy J, Elayaperumal M, Henini M, Malik M (2016) Photodegradation of organic pollutants RhB dye using UV simulated sunlight on ceria based TiO_2 nanomaterials for antibacterial applications. Sci Rep 6:38064. https://doi.org/10.1038/srep38064
112. Kazadi Mbamba C, Batstone DJ, Flores-Alsina X, Tait S (2015) A generalised chemical precipitation modelling approach in wastewater treatment applied to calcite. Water Res 68:342–353. https://doi.org/10.1016/j.watres.2014.10.011
113. Kefeni KK, Msagati TAM, Nkambule TTI, Mamba BB (2018) Synthesis and application of hematite nanoparticles for acid mine drainage treatment. J Environ Chem Eng 6:1865–1874. https://doi.org/10.1016/j.jece.2018.02.037
114. Khezami L, Ould M'hamed, M., Lemine, O.M., Bououdina, M., Bessadok-Jemai, A., (2016) Milled goethite nanocrystalline for selective and fast uptake of cadmium ions from aqueous solution. Desalination Water Treat 57:6531–6539. https://doi.org/10.1080/19443994.2015.1010231
115. Khosla A, Shah S, Shiblee MNI, Mir SH, Nagahara LA, Thundat T, Shekar PK, Kawakami M, Furukawa H (2018) Carbon fiber doped thermosetting elastomer for flexible sensors: physical properties and microfabrication. Sci Rep 8:12313. https://doi.org/10.1038/s41598-018-30846-3
116. Kim E-J, Lee C-S, Chang Y-Y, Chang Y-S (2013) Hierarchically structured manganese oxide-coated magnetic nanocomposites for the efficient removal of heavy metal ions from aqueous systems. ACS Appl Mater Interfaces 5:9628–9634. https://doi.org/10.1021/am402615m
117. Kim E-S, Hwang G, Gamal El-Din M, Liu Y (2012) Development of nanosilver and multi-walled carbon nanotubes thin-film nanocomposite membrane for enhanced water treatment. J Membr Sci 394–395:37–48. https://doi.org/10.1016/j.memsci.2011.11.041
118. Kim JH, Joshi MK, Lee J, Park CH, Kim CS (2018) Polydopamine-assisted immobilization of hierarchical zinc oxide nanostructures on electrospun nanofibrous membrane for photocatalysis and antimicrobial activity. J Colloid Interface Sci 513:566–574. https://doi.org/10.1016/j.jcis.2017.11.061
119. Kinemuchi H, Ochiai B (2018) Synthesis of hydrophilic sulfur-containing adsorbents for noble metals having thiocarbonyl group based on a methacrylate bearing dithiocarbonate moieties. Adv Mater Sci Eng 2018:1–8. https://doi.org/10.1155/2018/3729580
120. Kneipp K, Wang Y, Dasari RR, Feld MS (1995) Approach to single molecule detection using surface-enhanced resonance raman scattering (SERRS): a study using rhodamine 6g on colloidal silver. Appl Spectrosc 49:780–784
121. Krishna VD, Wu K, Perez AM, Wang J-P (2016) Giant Magnetoresistance-based biosensor for detection of influenza a virus. Front Microbiol 7.https://doi.org/10.3389/fmicb.2016.00400
122. Kudr J, Haddad Y, Richtera L, Heger Z, Cernak M, Adam V, Zitka O (2017) Magnetic nanoparticles: from design and synthesis to real world applications. Nanomaterials 7:243. https://doi.org/10.3390/nano7090243

123. Kumari P, Alam M, Siddiqi WA (2019) Usage of nanoparticles as adsorbents for waste water treatment: an emerging trend. Sustain Mater Technol 22:e00128.https://doi.org/10.1016/j.susmat.2019.e00128
124. Lawton LA, Robertson PKJ, Cornish BJPA, Jaspars M (1999) Detoxification of Microcystins (Cyanobacterial Hepatotoxins) Using TiO_2 Photocatalytic Oxidation. Environ Sci Technol 33:771–775. https://doi.org/10.1021/es9806682
125. Lee C, Kim JY, Lee WI, Nelson KL, Yoon J, Sedlak DL (2008) Bactericidal Effect of Zero-Valent Iron Nanoparticles on Escherichia coli. Environ Sci Technol 42:4927–4933. https://doi.org/10.1021/es800408u
126. Lee KM, Lai CW, Ngai KS, Juan JC (2016) Recent developments of zinc oxide based photocatalyst in water treatment technology: a review. Water Res 88:428–448. https://doi.org/10.1016/j.watres.2015.09.045
127. Lei Y, Chen F, Luo Y, Zhang L (2014) Three-dimensional magnetic graphene oxide foam/Fe_3O_4 nanocomposite as an efficient absorbent for Cr(VI) removal. J Mater Sci 49:4236–4245. https://doi.org/10.1007/s10853-014-8118-2
128. Leiviskä T, Khalid MK, Sarpola A, Tanskanen J (2017) Removal of vanadium from industrial wastewater using iron sorbents in batch and continuous flow pilot systems. J Environ Manage 190:231–242. https://doi.org/10.1016/j.jenvman.2016.12.063
129. Li Q, Mahendra S, Lyon DY, Brunet L, Liga MV, Li D, Alvarez PJJ (2008) Antimicrobial nanomaterials for water disinfection and microbial control: potential applications and implications. Water Res 42:4591–4602. https://doi.org/10.1016/j.watres.2008.08.015
130. Lisha KP, Maliyekkal SM, Pradeep T (2010) Manganese dioxide nanowhiskers: a potential adsorbent for the removal of Hg(II) from water. Chem Eng J 160:432–439. https://doi.org/10.1016/j.cej.2010.03.031
131. Liu C, Hong T, Li H, Wang L (2018) From club convergence of per capita industrial pollutant emissions to industrial transfer effects: An empirical study across 285 cities in China. Energy Policy 121:300–313. https://doi.org/10.1016/j.enpol.2018.06.039
132. Liu G, Jiang J, Yu R, Yan H, Liang R (2020) Silver nanoparticle-incorporated porous renewable film as low-cost bactericidal and antifouling filter for point-of-use water disinfection. Ind Eng Chem Res 59:10857–10867. https://doi.org/10.1021/acs.iecr.0c00157
133. Liu T, Wang Z-L, Sun Y (2015) Manipulating the morphology of nanoscale zero-valent iron on pumice for removal of heavy metals from wastewater. Chem Eng J 263:55–61. https://doi.org/10.1016/j.cej.2014.11.046
134. Lockwood RA, Chen KY (1973) Adsorption of mercury(II) by hydrous manganese oxides. Environ Sci Technol 7:1028–1034. https://doi.org/10.1021/es60083a006
135. Lofrano G, Carotenuto M, Libralato G, Domingos RF, Markus A, Dini L, Gautam RK, Baldantoni D, Rossi M, Sharma SK, Chattopadhyaya MC, Giugni M, Meric S (2016) Polymer functionalized nanocomposites for metals removal from water and wastewater: an overview. Water Res 92:22–37. https://doi.org/10.1016/j.watres.2016.01.033
136. Lu F, Astruc D (2018) Nanomaterials for removal of toxic elements from water. Coord Chem Rev 356:147–164. https://doi.org/10.1016/j.ccr.2017.11.003
137. Lu H, Wang J, Stoller M, Wang T, Bao Y, Hao H (2016) An overview of nanomaterials for water and wastewater treatment. Adv Mater Sci Eng 2016:1–10. https://doi.org/10.1155/2016/4964828
138. Lv Y, Zhang C, He A, Yang S-J, Wu G-P, Darling SB, Xu Z-K (2017) Photocatalytic nanofiltration membranes with self-cleaning property for wastewater treatment. Adv Funct Mater 27:1700251. https://doi.org/10.1002/adfm.201700251
139. Madhushika HG, Ariyadasa TU, Gunawardena SHP (2019) Biological decolourization of textile industry wastewater by a developed bacterial consortium. Water Sci Technol 80:1910–1918. https://doi.org/10.2166/wst.2020.010
140. Madrakian T, Afkhami A, Zadpour B, Ahmadi M (2015) New synthetic mercaptoethylamino homopolymer-modified maghemite nanoparticles for effective removal of some heavy metal ions from aqueous solution. J Ind Eng Chem 21:1160–1166. https://doi.org/10.1016/j.jiec.2014.05.029

141. Madrakian T, Afkhami A, Zolfigol MA, Ahmadi M, Koukabi N (2012) Application of modified silica coated magnetite nanoparticles for removal of iodine from water samples. Nano-Micro Lett. 4:57–63. https://doi.org/10.1007/BF03353693
142. Madzokere TC, Karthigeyan A (2017) Heavy metal ion effluent discharge containment using Magnesium Oxide (MgO) Nanoparticles. Mater. Today Proc. 4:9–18. https://doi.org/10.1016/j.matpr.2017.01.187
143. Magnet C, Lomenech C, Hurel C, Reilhac P, Giulieri F, Chaze A-M, Persello J, Kuzhir P (2017) Adsorption of nickel ions by oleate-modified magnetic iron oxide nanoparticles. Environ Sci Pollut Res 24:7423–7435. https://doi.org/10.1007/s11356-017-8391-0
144. Mahdavi S, Jalali M, Afkhami A (2015) Heavy metals removal from aqueous solutions by Al_2O_3 nanoparticles modified with natural and chemical modifiers. Clean Technol Environ Policy 17:85–102. https://doi.org/10.1007/s10098-014-0764-1
145. Mahmood MA, Baruah S, Dutta J (2011) Enhanced visible light photocatalysis by manganese doping or rapid crystallization with ZnO nanoparticles. Mater Chem Phys 130:531–535. https://doi.org/10.1016/j.matchemphys.2011.07.018
146. Mahmoud ME, Abdelwahab MS, Abdou AEH (2016) Enhanced removal of lead and cadmium from water by Fe_3O_4-cross linked-*O*-phenylenediamine nano-composite. Sep Sci Technol 51:237–247. https://doi.org/10.1080/01496395.2015.1093505
147. Stoller M, Miranda L, Chianese A (2009) Optimal feed location in a spinning disc reactor for the production of tio2 nanoparticles. Chem Eng Trans 17:993–998. https://doi.org/10.3303/CET0917166
148. Masheane ML, Nthunya LN, Malinga SP, Nxumalo EN, Mamba BB, Mhlanga SD (2017) Synthesis of Fe-Ag/f-MWCNT/PES nanostructured-hybrid membranes for removal of Cr(VI) from water. Sep Purif Technol 184:79–87. https://doi.org/10.1016/j.seppur.2017.04.018
149. Massalimov IA, Il'yasova, R.R., Musavirova, L.R., Samsonov, M.R., Mustafin, A.G., (2014) Use of micrometer hematite particles and nanodispersed goethite as sorbent for heavy metals. Russ J Appl Chem 87:1456–1463. https://doi.org/10.1134/S1070427214100115
150. Mauter MS, Wang Y, Okemgbo KC, Osuji CO, Giannelis EP, Elimelech M (2011) Antifouling ultrafiltration membranes via post-fabrication grafting of biocidal nanomaterials. ACS Appl Mater Interfaces 3:2861–2868. https://doi.org/10.1021/am200522v
151. Mazzocchi RA (2016) Medical sensors—defining a pathway to commercialization. ACS Sens 1:1167–1170. https://doi.org/10.1021/acssensors.6b00553
152. Min L-L, Zhong L-B, Zheng Y-M, Liu Q, Yuan Z-H, Yang L-M (2016) Functionalized chitosan electrospun nanofiber for effective removal of trace arsenate from water. Sci Rep 6:32480. https://doi.org/10.1038/srep32480
153. Mir NA, Haque MM, Khan A, Umar K, Muneer M, Vijayalakshmi S (2012) Semiconductor mediated photocatalysed reaction of two selected organic compounds in aqueous suspensions of titanium dioxide. J Adv Oxid Technol 15.https://doi.org/10.1515/jaots-2012-0218
154. Mir SH, Ebata K, Yanagiya H, Ochiai B (2018) Alignment of Ag nanoparticles with graft copolymer bearing thiocarbonyl moieties. Microsyst Technol 24:605–611. https://doi.org/10.1007/s00542-017-3418-5
155. Mir SH, Hasan PMZ, Danish EY, Aslam M (2020) Pd-induced phase separation in poly(methyl methacrylate) telopolymer: synthesis of nanostructured catalytic Pd nanorods. Colloid Polym Sci. https://doi.org/10.1007/s00396-020-04630-7
156. Mir SH, Nagahara LA, Thundat T, Mokarian-Tabari P, Furukawa H, Khosla A (2018) Review—organic-inorganic hybrid functional materials: an integrated platform for applied technologies. J Electrochem Soc 165:B3137–B3156. https://doi.org/10.1149/2.0191808jes
157. Mir SH, Ochiai B (2017) One-Pot Fabrication of Hollow Polymer@Ag nanospheres for printable translucent conductive coatings. Adv Mater Interfaces 4:1601198. https://doi.org/10.1002/admi.201601198
158. Mir SH, Ochiai B (2016) Development of Hierarchical Polymer@Pd nanowire-network: synthesis and application as highly active recyclable catalyst and printable conductive ink. ChemistryOpen 5:213–218. https://doi.org/10.1002/open.201600009

159. Mir SH, Ochiai B (2016) Fabrication of Polymer-Ag honeycomb hybrid film by metal complexation induced phase separation at the air/water interface. Macromol Mater Eng 301:1026–1031. https://doi.org/10.1002/mame.201600035
160. Mir SH, Rydzek G, Nagahara LA, Khosla A, Mokarian-Tabari P (2019) Review—recent advances in block-copolymer nanostructured subwavelength antireflective surfaces. J Electrochem Soc 167:037502.https://doi.org/10.1149/2.0022003JES
161. Mirrezaie N, Nikazar M, Hasan Zadeh M (2013) Synthesis of Magnetic Nanocomposite Fe_3O_4 Coated Polypyrrole (PPy) for Chromium(VI) Removal. Adv Mater Res 829:649–653. https://doi.org/10.4028/www.scientific.net/AMR.829.649
162. Mishra PK, Saxena A, Rawat AS, Dixit PK, Kumar R, Rai PK (2018) Surfactant-free one-pot synthesis of low-density cerium oxide nanoparticles for adsorptive removal of arsenic species. Environ Prog Sustain Energy 37:221–231. https://doi.org/10.1002/ep.12660
163. Moon G, Kim D, Kim H, Bokare AD, Choi W (2014) Platinum-like behavior of reduced graphene oxide as a cocatalyst on tio $_2$ for the efficient photocatalytic oxidation of arsenite. Environ Sci Technol Lett 1:185–190. https://doi.org/10.1021/ez5000012
164. Morris T, Copeland H, McLinden E, Wilson S, Szulczewski G (2002) The effects of mercury adsorption on the optical response of size-selected gold and silver nanoparticles. Langmuir 18:7261–7264. https://doi.org/10.1021/la020229n
165. Mostafiz F, Islam MM, Saha B, Hossain MdK, Moniruzzaman M, Habibullah-Al-Mamun Md (2020) Bioaccumulation of trace metals in freshwater prawn, Macrobrachium rosenbergii from farmed and wild sources and human health risk assessment in Bangladesh. Environ Sci Pollut Res 27:16426–16438.https://doi.org/10.1007/s11356-020-08028-4
166. Mpenyana-Monyatsi L, Mthombeni NH, Onyango MS, Momba MNB (2012) Cost-effective filter materials coated with silver nanoparticles for the removal of pathogenic bacteria in groundwater. Int J Environ Res Public Health 9:244–271. https://doi.org/10.3390/ijerph9010244
167. Mukherjee J, Ramkumar J, Chandramouleeswaran S, Shukla R, Tyagi AK (2013) Sorption characteristics of nano manganese oxide: efficient sorbent for removal of metal ions from aqueous streams. J Radioanal Nucl Chem 297:49–57. https://doi.org/10.1007/s10967-012-2393-7
168. Munnawar I, Iqbal SS, Anwar MN, Batool M, Tariq S, Faitma N, Khan AL, Khan AU, Nazar U, Jamil T, Ahmad NM (2017) Synergistic effect of Chitosan-Zinc Oxide Hybrid Nanoparticles on antibiofouling and water disinfection of mixed matrix polyethersulfone nanocomposite membranes. Carbohydr Polym 175:661–670. https://doi.org/10.1016/j.carbpol.2017.08.036
169. Munshi GH, Ibrahim AM, Al-Harbi LM (2018) Inspired preparation of zinc oxide nanocatalyst and the photocatalytic activity in the treatment of methyl orange dye and paraquat herbicide. Int J Photoenergy 2018:1–7. https://doi.org/10.1155/2018/5094741
170. Mustapha S, Ndamitso MM, Abdulkareem AS et al (2020) Application of TiO2 and ZnO nanoparticles immobilized on clay in wastewater treatment: a review. Appl Water Sci 10:49. https://doi.org/10.1007/s13201-019-1138-y
171. Nabi D, Aslam I, Qazi IA (2009) Evaluation of the adsorption potential of titanium dioxide nanoparticles for arsenic removal. J Environ Sci 21:402–408. https://doi.org/10.1016/S1001-0742(08)62283-4
172. Nel A, Xia T, Meng H, Wang X, Lin S, Ji Z, Zhang H (2013) Nanomaterial toxicity testing in the 21st century: use of a predictive toxicological approach and high-throughput screening. Acc Chem Res 46:607–621. https://doi.org/10.1021/ar300022h
173. Neumann A, Kaegi R, Voegelin A, Hussam A, Munir AKM, Hug SJ (2013) Arsenic removal with composite iron matrix filters in bangladesh: a field and laboratory study. Environ Sci Technol 47:4544–4554. https://doi.org/10.1021/es305176x
174. Ng SM, Koneswaran M, Narayanaswamy R (2016) A review on fluorescent inorganic nanoparticles for optical sensing applications. RSC Adv 6:21624–21661. https://doi.org/10.1039/C5RA24987B
175. Ngomsik A-F, Bee A, Talbot D, Cote G (2012) Magnetic solid–liquid extraction of Eu(III), La(III), Ni(II) and Co(II) with maghemite nanoparticles. Sep Purif Technol 86:1–8. https://doi.org/10.1016/j.seppur.2011.10.013

176. Norouzian Baghani A, Mahvi AH, Gholami M, Rastkari N, Delikhoon M (2016) One-Pot synthesis, characterization and adsorption studies of amine-functionalized magnetite nanoparticles for removal of Cr (VI) and Ni (II) ions from aqueous solution: kinetic, isotherm and thermodynamic studies. J Environ Health Sci Eng 14:11. https://doi.org/10.1186/s40201-016-0252-0
177. O'Carroll D, Sleep B, Krol M, Boparai H, Kocur C (2013) Nanoscale zero valent iron and bimetallic particles for contaminated site remediation. Adv Water Resour 51:104–122. https://doi.org/10.1016/j.advwatres.2012.02.005
178. Ojea-Jiménez I, López X, Arbiol J, Puntes V (2012) Citrate-coated gold nanoparticles as smart scavengers for Mercury(II) Removal from polluted waters. ACS Nano 6:2253–2260. https://doi.org/10.1021/nn204313a
179. Pan S, Shen H, Xu Q, Luo J, Hu M (2012) Surface mercapto engineered magnetic Fe_3O_4 nanoadsorbent for the removal of mercury from aqueous solutions. J Colloid Interface Sci 365:204–212. https://doi.org/10.1016/j.jcis.2011.09.002
180. Pan Z, Zhu X, Satpathy A, Li W, Fortner JD, Giammar DE (2019) Cr(VI) Adsorption on engineered iron oxide nanoparticles: exploring complexation processes and water chemistry. Environ Sci Technol 53:11913–11921. https://doi.org/10.1021/acs.est.9b03796
181. Patra AK, Dutta A, Bhaumik A (2012) Self-assembled mesoporous γ-Al2O3 spherical nanoparticles and their efficiency for the removal of arsenic from water. J Hazard Mater 201–202:170–177. https://doi.org/10.1016/j.jhazmat.2011.11.056
182. Peng F, Xu T, Wu F, Ma C, Liu Y, Li J, Zhao B, Mao C (2017) Novel biomimetic enzyme for sensitive detection of superoxide anions. Talanta 174:82–91. https://doi.org/10.1016/j.talanta.2017.05.028
183. Perez T, Pasquini D, de Faria Lima A, Rosa EV, Sousa MH, Cerqueira DA, de Morais LC (2019) Efficient removal of lead ions from water by magnetic nanosorbents based on manganese ferrite nanoparticles capped with thin layers of modified biopolymers. J Environ Chem Eng 7:102892.https://doi.org/10.1016/j.jece.2019.102892
184. Peterson J, Green G, Iida K, Caldwell B, Kerrison P, Bernich S, Aoyagi K, Lee SR (2000) Detection of hepatitis c core antigen in the antibody negative "window" phase of hepatitis C infection. Vox Sang 78:80–85. https://doi.org/10.1046/j.1423-0410.2000.7820080.x
185. Peyravi M, Jahanshahi M, Rahimpour A, Javadi A, Hajavi S (2014) Novel thin film nanocomposite membranes incorporated with functionalized TiO_2 nanoparticles for organic solvent nanofiltration. Chem Eng J 241:155–166. https://doi.org/10.1016/j.cej.2013.12.024
186. Pol R, Céspedes F, Gabriel D, Baeza M (2017) Microfluidic lab-on-a-chip platforms for environmental monitoring. TrAC Trends Anal Chem 95:62–68. https://doi.org/10.1016/j.trac.2017.08.001
187. Prabhakar R, Samadder SR (2018) Low cost and easy synthesis of aluminium oxide nanoparticles for arsenite removal from groundwater: a complete batch study. J Mol Liq 250:192–201. https://doi.org/10.1016/j.molliq.2017.11.173
188. Pradhan N, Pal A, Pal T (2002) Silver nanoparticle catalyzed reduction of aromatic nitro compounds. Colloids Surf Physicochem Eng Asp 196:247–257. https://doi.org/10.1016/S0927-7757(01)01040-8
189. Qian J, Gao X, Pan B (2020) Nanoconfinement-mediated water treatment: from fundamental to application. Environ Sci Technol 54:8509–8526. https://doi.org/10.1021/acs.est.0c01065
190. Qiu X, Fang Z, Yan X, Gu F, Jiang F (2012) Emergency remediation of simulated chromium (VI)-polluted river by nanoscale zero-valent iron: Laboratory study and numerical simulation. Chem Eng J 193–194:358–365. https://doi.org/10.1016/j.cej.2012.04.067
191. Qu X, Brame J, Li Q, Alvarez PJJ (2013) Nanotechnology for a safe and sustainable water supply: enabling integrated water treatment and reuse. Acc Chem Res 46:834–843. https://doi.org/10.1021/ar300029v
192. Quek J-A, Lam S-M, Sin J-C, Mohamed AR (2018) Visible light responsive flower-like ZnO in photocatalytic antibacterial mechanism towards Enterococcus faecalis and Micrococcus luteus. J Photochem Photobiol B 187:66–75. https://doi.org/10.1016/j.jphotobiol.2018.07.030

193. Rafiq Z, Nazir R, Durr-e-Shahwar S, M.R., Ali, S., (2014) Utilization of magnesium and zinc oxide nano-adsorbents as potential materials for treatment of copper electroplating industry wastewater. J Environ Chem Eng 2:642–651. https://doi.org/10.1016/j.jece.2013.11.004
194. Rahbar N, Jahangiri A, Boumi S, Khodayar MJ (2014) Mercury removal from aqueous solutions with chitosan-coated magnetite nanoparticles optimized using the box-behnken design. Jundishapur J Nat Pharm Prod 9. https://doi.org/10.17795/jjnpp-15913
195. Rahman MA, Muneer M (2005) Photocatalysed degradation of two selected pesticide derivatives, dichlorvos and phosphamidon, in aqueous suspensions of titanium dioxide. Desalination 181:161–172. https://doi.org/10.1016/j.desal.2005.02.019
196. Rajakumar K, Kirupha SD, Sivanesan S, Sai RL (2014) Effective removal of heavy metal ions using Mn_2O_3 doped polyaniline nanocomposite. J Nanosci Nanotechnol 14:2937–2946. https://doi.org/10.1166/jnn.2014.8628
197. Rajasulochana P, Preethy V (2016) Comparison on efficiency of various techniques in treatment of waste and sewage water—a comprehensive review. Resour-Effic Technol 2:175–184. https://doi.org/10.1016/j.reffit.2016.09.004
198. Rajendran S, Khan MM, Gracia F, Qin J, Gupta VK, Arumainathan S (2016) Ce3+-ion-induced visible-light photocatalytic degradation and electrochemical activity of ZnO/CeO_2 nanocomposite. Sci Rep 6:31641. https://doi.org/10.1038/srep31641
199. Rajput S, Pittman CU, Mohan D (2016) Magnetic magnetite (Fe3O4) nanoparticle synthesis and applications for lead (Pb2+) and chromium (Cr6+) removal from water. J Colloid Interface Sci 468:334–346. https://doi.org/10.1016/j.jcis.2015.12.008
200. Rajput S, Singh LP, Pittman CU, Mohan D (2017) Lead (Pb 2+) and copper (Cu 2+) remediation from water using superparamagnetic maghemite (γ-Fe_2O_3) nanoparticles synthesized by Flame Spray Pyrolysis (FSP). J Colloid Interface Sci 492:176–190. https://doi.org/10.1016/j.jcis.2016.11.095
201. Rambabu K, Bharath G, Fawzi B, Pau LS (2021) Green synthesis of zinc oxide nanoparticles using Phoenix dactylifera waste as bioreductant for effective dye degradation and antibacterial performance in wastewater treatment. J. Hazardous Mater. 402:123560.https://doi.org/10.1016/j.jhazmat.2020.123560
202. Recillas S, Colón J, Casals E, González E, Puntes V, Sánchez A, Font X (2010) Chromium VI adsorption on cerium oxide nanoparticles and morphology changes during the process. J Hazard Mater 184:425–431. https://doi.org/10.1016/j.jhazmat.2010.08.052
203. Recillas S, García A, González E, Casals E, Puntes V, Sánchez A, Font X (2011) Use of CeO_2, TiO_2 and Fe_3O_4 nanoparticles for the removal of lead from water. Desalination 277:213–220. https://doi.org/10.1016/j.desal.2011.04.036
204. Rehman K, Fatima F, Waheed I, Akash MSH (2018) Prevalence of exposure of heavy metals and their impact on health consequences. J Cell Biochem 119:157–184. https://doi.org/10.1002/jcb.26234
205. Renzi M, Blašković A (2019) Ecotoxicity of nano-metal oxides: a case study on daphnia magna. Ecotoxicology 28:878–889. https://doi.org/10.1007/s10646-019-02085-3
206. Richter K, Ayers J (2018) An approach to predicting sediment microbial fuel cell performance in shallow and deep water. Appl Sci 8:2628. https://doi.org/10.3390/app8122628
207. Rodrigues SM, Demokritou P, Dokoozlian N, Hendren CO, Karn B, Mauter MS, Sadik OA, Safarpour M, Unrine JM, Viers J, Welle P, White JC, Wiesner MR, Lowry GV (2017) Nanotechnology for sustainable food production: promising opportunities and scientific challenges. Environ Sci Nano 4:767–781. https://doi.org/10.1039/C6EN00573J
208. Saad AHA, Azzam AM, El-Wakeel ST, Mostafa BB, Abd El-latif MB (2018) Removal of toxic metal ions from wastewater using ZnO@Chitosan core-shell nanocomposite. Environ Nanotechnol Monit Manag 9:67–75. https://doi.org/10.1016/j.enmm.2017.12.004
209. Saadi Z, Saadi R, Fazaeli R (2013) Fixed-bed adsorption dynamics of Pb (II) adsorption from aqueous solution using nanostructured γ-alumina. J Nanostructure Chem 3:48. https://doi.org/10.1186/2193-8865-3-48

210. Sadati Behbahani N, Rostamizadeh K, Yaftian MR, Zamani A, Ahmadi H (2014) Covalently modified magnetite nanoparticles with PEG: preparation and characterization as nano-adsorbent for removal of lead from wastewater. J Environ Health Sci Eng 12:103. https://doi.org/10.1186/2052-336X-12-103
211. Sadovnikov SI, Gusev AI (2017) Recent progress in nanostructured silver sulfide: from synthesis and nonstoichiometry to properties. J Mater Chem A 5:17676–17704. https://doi.org/10.1039/C7TA04949H
212. Saharan P, Chaudhary GR, Mehta SK, Umar A (2014) Removal of water contaminants by iron oxide nanomaterials. J Nanosci Nanotechnol 14:627–643. https://doi.org/10.1166/jnn.2014.9053
213. Santhosh C, Velmurugan V, Jacob G, Jeong SK, Grace AN, Bhatnagar A (2016) Role of nanomaterials in water treatment applications: a review. Chem Eng J 306:1116–1137. https://doi.org/10.1016/j.cej.2016.08.053
214. Sapkota A, Anceno AJ, Baruah S, Shipin OV, Dutta J (2011) Zinc oxide nanorod mediated visible light photoinactivation of model microbes in water. Nanotechnology 22:215703.https://doi.org/10.1088/0957-4484/22/21/215703
215. Saravanan R, Sacari E, Gracia F, Khan MM, Mosquera E, Gupta VK (2016) Conducting PANI stimulated ZnO system for visible light photocatalytic degradation of coloured dyes. J Mol Liq 221:1029–1033. https://doi.org/10.1016/j.molliq.2016.06.074
216. Savage N, Diallo MS (2005) Nanomaterials and water purification: opportunities and challenges. J. Nanoparticle Res 7:331–342. https://doi.org/10.1007/s11051-005-7523-5
217. Schmidtke C, Eggers R, Zierold R, Feld A, Kloust H, Wolter C, Ostermann J, Merkl J-P, Schotten T, Nielsch K, Weller H (2014) Polymer-assisted self-assembly of superparamagnetic iron oxide nanoparticles into well-defined clusters: controlling the collective magnetic properties. Langmuir 30:11190–11196. https://doi.org/10.1021/la5021934
218. Schwermer CU, Krzeminski P, Wennberg AC, Vogelsang C, Uhl W (2018) Removal of antibiotic resistant E. coli in two Norwegian wastewater treatment plants and by nano- and ultra-filtration processes. Water Sci Technol 77:1115–1126. https://doi.org/10.2166/wst.2017.642
219. Sciortino F, Mir SH, Pakdel A, Oruganti A, Abe H, Witecka A, Shri DNA, Rydzek G, Ariga K (2020) Saloplastics as multiresponsive ion exchange reservoirs and catalyst supports. J Mater Chem A 8:17713–17724. https://doi.org/10.1039/D0TA05901C
220. Segura Y, Martínez F, Melero JA, Molina R, Chand R, Bremner DH (2012) Enhancement of the advanced Fenton process (Fe0/H2O2) by ultrasound for the mineralization of phenol. Appl Catal B Environ 113–114:100–106. https://doi.org/10.1016/j.apcatb.2011.11.024
221. Sekoai PT, Ouma CNM, du Preez SP, Modisha P, Engelbrecht N, Bessarabov DG, Ghimire A (2019) Application of nanoparticles in biofuels: an overview. Fuel 237:380–397. https://doi.org/10.1016/j.fuel.2018.10.030
222. Seyedi SM, Rabiee H, Shahabadi SMS, Borghei SM (2017) Synthesis of zero-valent iron nanoparticles via electrical wire explosion for efficient removal of heavy metals: water. Clean: Soil, Air, Water 45:1600139. https://doi.org/10.1002/clen.201600139
223. Shan C, Ma Z, Tong M, Ni J (2015) Removal of Hg(II) by poly(1-vinylimidazole)-grafted Fe_3O_4@SiO_2 magnetic nanoparticles. Water Res 69:252–260. https://doi.org/10.1016/j.watres.2014.11.030
224. Shelby T, Sulthana S, McAfee J, Banerjee T, Santra S (2017) Foodborne pathogen screening using magneto-fluorescent nanosensor: rapid detection of E. Coli O157:H7. J Vis Exp 55821. https://doi.org/10.3791/55821
225. Shen Y, Huang Z, Liu X, Qian J, Xu J, Yang X, Sun A, Ge J (2015) Iron-induced myocardial injury: an alarming side effect of superparamagnetic iron oxide nanoparticles. J Cell Mol Med 19:2032–2035. https://doi.org/10.1111/jcmm.12582
226. Shen Y, Saboe PO, Sines IT, Erbakan M, Kumar M (2014) Biomimetic membranes: a review. J Membr Sci 454:359–381. https://doi.org/10.1016/j.memsci.2013.12.019
227. Shi J, Li H, Lu H, Zhao X (2015) Use of carboxyl functional magnetite nanoparticles as potential sorbents for the removal of heavy metal ions from aqueous solution. J Chem Eng Data 60:2035–2041. https://doi.org/10.1021/je5011196

228. Shipley HJ, Engates KE, Grover VA (2013) Removal of Pb(II), Cd(II), Cu(II), and Zn(II) by hematite nanoparticles: effect of sorbent concentration, pH, temperature, and exhaustion. Environ Sci Pollut Res 20:1727–1736. https://doi.org/10.1007/s11356-012-0984-z
229. Shirin S, Balakrishnan VK (2011) Using chemical reactivity to provide insights into environmental transformations of priority organic substances: the Fe^0-mediated reduction of acid blue 129. Environ Sci Technol 45:10369–10377. https://doi.org/10.1021/es202780r
230. Shokati Poursani A, Nilchi A, Hassani AH, Shariat M, Nouri J (2015) A novel method for synthesis of nano-γ-Al_2O_3: study of adsorption behavior of chromium, nickel, cadmium and lead ions. Int J Environ Sci Technol 12:2003–2014. https://doi.org/10.1007/s13762-014-0740-7
231. Shukla S, Arora V, Jadaun A, Kumar J, Singh N, Jain V (2015) Magnetic removal of Entamoeba cysts from water using chitosan oligosaccharide-coated iron oxide nanoparticles. Int J Nanomedicine 4901.https://doi.org/10.2147/IJN.S77675
232. Singh J, Dutta T, Kim K-H, Rawat M, Samddar P, Kumar P (2018) 'Green' synthesis of metals and their oxide nanoparticles: applications for environmental remediation. J Nanobiotechnology 16:84. https://doi.org/10.1186/s12951-018-0408-4
233. Singh J, Kumar V, Kim K-H, Rawat M (2019) Biogenic synthesis of copper oxide nanoparticles using plant extract and its prodigious potential for photocatalytic degradation of dyes. Environ Res 177:108569.https://doi.org/10.1016/j.envres.2019.108569
234. Sizmur T, Fresno T, Akgül G, Frost H, Moreno-Jiménez E (2017) Biochar modification to enhance sorption of inorganics from water. Bioresour Technol 246:34–47. https://doi.org/10.1016/j.biortech.2017.07.082
235. Skubal LR, Meshkov NK, Rajh T, Thurnauer M (2002) Cadmium removal from water using thiolactic acid-modified titanium dioxide nanoparticles. J Photochem Photobiol Chem 148:393–397. https://doi.org/10.1016/S1010-6030(02)00069-2
236. Sneed MC, Brasted RC, King CV (1956) Comprehensive inorganic chemistry. J Electrochem Soc 103:83C. https://doi.org/10.1149/1.2430269
237. So H-M, Park D-W, Jeon E-K, Kim Y-H, Kim BS, Lee C-K, Choi SY, Kim SC, Chang H, Lee J-O (2008) Detection and titer estimation ofescherichia coli using aptamer-functionalized single-walled carbon-nanotube field-effect transistors. Small 4:197–201. https://doi.org/10.1002/smll.200700664
238. Song J, Kong H, Jang J (2011) Adsorption of heavy metal ions from aqueous solution by polyrhodanine-encapsulated magnetic nanoparticles. J Colloid Interface Sci 359:505–511. https://doi.org/10.1016/j.jcis.2011.04.034
239. Sorbiun M, Shayegan Mehr E, Ramazani A, Taghavi Fardood S (2018) Green synthesis of zinc oxide and copper oxide nanoparticles using aqueous extract of oak fruit hull (Jaft) and comparing their photocatalytic degradation of basic violet 3. Int J Environ Res 12:29–37. https://doi.org/10.1007/s41742-018-0064-4
240. Sorbiun M, Shayegan Mehr E, Ramazani A, Taghavi Fardood S (2018) Biosynthesis of Ag, ZnO and bimetallic Ag/ZnO alloy nanoparticles by aqueous extract of oak fruit hull (Jaft) and investigation of photocatalytic activity of ZnO and bimetallic Ag/ZnO for degradation of basic violet 3 dye. J Mater Sci Mater Electron 29:2806–2814. https://doi.org/10.1007/s10854-017-8209-3
241. Sponza DT, Oztekin R (2016) Treatment of olive mill wastewater by photooxidation with ZrO_2 -doped TiO_2 nanocomposite and its reuse capability. Environ Technol 37:865–879. https://doi.org/10.1080/09593330.2015.1088579
242. Srinivasan NR, Shankar PA, Bandyopadhyaya R (2013) Plasma treated activated carbon impregnated with silver nanoparticles for improved antibacterial effect in water disinfection. Carbon 57:1–10. https://doi.org/10.1016/j.carbon.2013.01.008
243. Sriram G, Bhat MP, Patil P, Uthappa UT, Jung H-Y, Altalhi T, Kumeria T, Aminabhavi TM, Pai RK, Madhuprasad K, M.D., (2017) Paper-based microfluidic analytical devices for colorimetric detection of toxic ions: a review. TrAC Trends Anal Chem 93:212–227. https://doi.org/10.1016/j.trac.2017.06.005

244. Stietiya MH, Wang JJ (2014) Zinc and cadmium adsorption to aluminum oxide nanoparticles affected by naturally occurring ligands. J Environ Qual 43:498–506. https://doi.org/10.2134/jeq2013.07.0263
245. Stoimenov PK, Klinger RL, Marchin GL, Klabunde KJ (2002) Metal oxide nanoparticles as bactericidal agents. Langmuir 18:6679–6686. https://doi.org/10.1021/la0202374
246. Street A, Sustich R, Duncan J, Savage N, Gray G (eds) (2014) Nanotechnology applications for clean water: solutions for improving water quality, 2, ed. Micro & nano technologies series. William Andrew/Elsevier, Amsterdam
247. Su C, Puls RW, Krug TA, Watling MT, O'Hara SK, Quinn JW, Ruiz NE (2012) A two and half-year-performance evaluation of a field test on treatment of source zone tetrachloroethene and its chlorinated daughter products using emulsified zero valent iron nanoparticles. Water Res 46:5071–5084. https://doi.org/10.1016/j.watres.2012.06.051
248. Su H, Zhang D, Antwi P, Xiao L, Liu Z, Deng X, Asumadu-Sakyi AB, Li J (2020) Effects of heavy rare earth element (yttrium) on partial-nitritation process, bacterial activity and structure of responsible microbial communities. Sci Total Environ 705:135797.https://doi.org/10.1016/j.scitotenv.2019.135797
249. Suman K, A., Gera, M., Jain, V.K., (2015) A novel reusable nanocomposite for complete removal of dyes, heavy metals and microbial load from water based on nanocellulose and silver nano-embedded pebbles. Environ Technol 36:706–714. https://doi.org/10.1080/09593330.2014.959066
250. Sun YB, Wang Q, Yang ST, Sheng GD, Guo ZQ (2011) Characterization of nano-iron oxyhydroxides and their application in UO2 2+ removal from aqueous solutions. J Radioanal Nucl Chem 290:643–648. https://doi.org/10.1007/s10967-011-1325-2
251. Sunada K, Minoshima M, Hashimoto K (2012) Highly efficient antiviral and antibacterial activities of solid-state cuprous compounds. J Hazard Mater 235–236:265–270. https://doi.org/10.1016/j.jhazmat.2012.07.052
252. Syngouna VI, Chrysikopoulos CV, Kokkinos P, Tselepi MA, Vantarakis A (2017) Cotransport of human adenoviruses with clay colloids and TiO_2 nanoparticles in saturated porous media: effect of flow velocity. Sci Total Environ 598:160–167. https://doi.org/10.1016/j.scitotenv.2017.04.082
253. Szczepanik B, Rogala P, Słomkiewicz PM, Banaś D, Kubala-Kukuś A, Stabrawa I (2017) Synthesis, characterization and photocatalytic activity of TiO_2 -halloysite and Fe_2O_3-halloysite nanocomposites for photodegradation of chloroanilines in water. Appl Clay Sci 149:118–126. https://doi.org/10.1016/j.clay.2017.08.016
254. Tabesh S, Davar F, Loghman-Estarki MR (2018) Preparation of γ-Al_2O_3 nanoparticles using modified sol-gel method and its use for the adsorption of lead and cadmium ions. J Alloys Compd 730:441–449. https://doi.org/10.1016/j.jallcom.2017.09.246
255. Tadic M, Panjan M, Damnjanovic V, Milosevic I (2014) Magnetic properties of hematite (α-Fe_2O_3) nanoparticles prepared by hydrothermal synthesis method. Appl Surf Sci 320:183–187. https://doi.org/10.1016/j.apsusc.2014.08.193
256. Tahir K, Nazir S, Li B, Ahmad A, Nasir T, Khan AU, Shah SAA, Khan ZUH, Yasin G, Hameed MU (2016) Sapium sebiferum leaf extract mediated synthesis of palladium nanoparticles and in vitro investigation of their bacterial and photocatalytic activities. J Photochem Photobiol B 164:164–173. https://doi.org/10.1016/j.jphotobiol.2016.09.030
257. Tan L, Xu J, Xue X, Lou Z, Zhu J, Baig SA, Xu X (2014) Multifunctional nanocomposite Fe_3O_4 @SiO_2–mPD/SP for selective removal of Pb(II) and Cr(VI) from aqueous solutions. RSC Adv 4:45920–45929. https://doi.org/10.1039/C4RA08040H
258. Tan Y, Chen M, Hao Y (2012) High efficient removal of Pb (II) by amino-functionalized Fe_3O_4 magnetic nano-particles. Chem Eng J 191:104–111. https://doi.org/10.1016/j.cej.2012.02.075
259. Naseem T, Tayyiba D (2021) The role of some important metal oxide nanoparticles for wastewater and antibacterial applications: a review. Environ Chem Ecotoxicol 3:59–75. https://doi.org/10.1016/j.enceco.2020.12.001
260. Tesh SJ, Scott TB (2014) Nano-composites for water remediation: a review. Adv Mater 26:6056–6068. https://doi.org/10.1002/adma.201401376

261. Tratnyek PG, Salter AJ, Nurmi JT, Sarathy V (2010) Environmental applications of zerovalent metals: iron vs. Zinc, in: Erickson LE, Koodali RT, Richards RM (eds), Nanoscale materials in chemistry: environmental applications, ACS symposium series. american chemical society, Washington, DC, pp. 165–178. https://doi.org/10.1021/bk-2010-1045.ch009
262. Tu Y-J, You C-F, Chang C-K (2013) Conversion of waste Mn–Zn dry battery as efficient nano-adsorbents for hazardous metals removal. J Hazard Mater 258–259:102–108. https://doi.org/10.1016/j.jhazmat.2013.04.029
263. Tuutijärvi T, Lu J, Sillanpää M, Chen G (2009) As(V) adsorption on maghemite nanoparticles. J Hazard Mater 166:1415–1420. https://doi.org/10.1016/j.jhazmat.2008.12.069
264. Umar A, Kumar R, Akhtar MS, Kumar G, Kim SH (2015) Growth and properties of well-crystalline cerium oxide (CeO_2) nanoflakes for environmental and sensor applications. J Colloid Interface Sci 454:61–68. https://doi.org/10.1016/j.jcis.2015.04.055
265. Umar K, Parveen T, Khan MA, Ibrahim MNM, Ahmad A, Rafatullah M (2019) Degradation of organic pollutants using metal-doped TiO2 photocatalysts under visible light: a comparative study. Desalination Water Treat 161:275–282. https://doi.org/10.5004/dwt.2019.24298
266. Vanegas DC, Gomes CL, Cavallaro ND, Giraldo-Escobar D, McLamore ES (2017) Emerging biorecognition and transduction schemes for rapid detection of pathogenic bacteria in food: rapid monitoring of pathogens in food…. Compr Rev Food Sci Food Saf 16:1188–1205. https://doi.org/10.1111/1541-4337.12294
267. Venkateswarlu S, Yoon M (2015) Core-shell ferromagnetic nanorod based on amine polymer composite (Fe3O4@DAPF) for fast removal of Pb(II) from aqueous solutions. ACS Appl Mater Interfaces 7:25362–25372. https://doi.org/10.1021/acsami.5b07723
268. Vikesland PJ, Wigginton KR (2010) Nanomaterial enabled biosensors for pathogen monitoring—a review. Environ Sci Technol 44:3656–3669. https://doi.org/10.1021/es903704z
269. Vimala K, Mohan YM, Sivudu KS, Varaprasad K, Ravindra S, Reddy NN, Padma Y, Sreedhar B, MohanaRaju K (2010) Fabrication of porous chitosan films impregnated with silver nanoparticles: a facile approach for superior antibacterial application. Colloids Surf B Biointerf 76:248–258. https://doi.org/10.1016/j.colsurfb.2009.10.044
270. Wan S, Wu J, Zhou S, Wang R, Gao B, He F (2018) Enhanced lead and cadmium removal using biochar-supported hydrated manganese oxide (HMO) nanoparticles: Behavior and mechanism. Sci Total Environ 616–617:1298–1306. https://doi.org/10.1016/j.scitotenv.2017.10.188
271. Wang C, Yu C (2013) Detection of chemical pollutants in water using gold nanoparticles as sensors: a review. Rev Anal Chem 32:1–14. https://doi.org/10.1515/revac-2012-0023
272. Wang J, Wu X, Wang C, Rong Z, Ding H, Li H, Li S, Shao N, Dong P, Xiao R, Wang S (2016) Facile synthesis of Au-coated magnetic nanoparticles and their application in bacteria detection via a SERS method. ACS Appl Mater Interf 8:19958–19967. https://doi.org/10.1021/acsami.6b07528
273. Wang J-C, Lou H-H, Xu Z-H, Cui C-X, Li Z-J, Jiang K, Zhang Y-P, Qu L-B, Shi W (2018) Natural sunlight driven highly efficient photocatalysis for simultaneous degradation of rhodamine B and methyl orange using I/C codoped TiO2 photocatalyst. J Hazard Mater 360:356–363. https://doi.org/10.1016/j.jhazmat.2018.08.008
274. Wang X, Huang K, Chen Y, Liu J, Chen S, Cao J, Mei S, Zhou Y, Jing T (2018) Preparation of dumbbell manganese dioxide/gelatin composites and their application in the removal of lead and cadmium ions. J Hazard Mater 350:46–54. https://doi.org/10.1016/j.jhazmat.2018.02.020
275. Wang X, Zhan C, Kong B, Zhu X, Liu J, Xu W, Cai W, Wang H (2015) Self-curled coral-like γ-Al_2O_3 nanoplates for use as an adsorbent. J Colloid Interf Sci 453:244–251. https://doi.org/10.1016/j.jcis.2015.03.065
276. Wang Y, Liu J, Wang P, Werth CJ, Strathmann TJ (2014) Palladium Nanoparticles encapsulated in core-shell silica: a structured hydrogenation catalyst with enhanced activity for reduction of oxyanion water pollutants. ACS Catal 4:3551–3559. https://doi.org/10.1021/cs500971r
277. Watts MP, Coker VS, Parry SA, Pattrick RAD, Thomas RAP, Kalin R, Lloyd JR (2015) Biogenic nano-magnetite and nano-zero valent iron treatment of alkaline Cr(VI) leachate and

chromite ore processing residue. Appl Geochem 54:27–42. https://doi.org/10.1016/j.apgeochem.2014.12.001
278. Wei H, Hossein Abtahi SM, Vikesland PJ (2015) Plasmonic colorimetric and SERS sensors for environmental analysis. Environ Sci Nano 2:120–135. https://doi.org/10.1039/C4EN00211C
279. Wu Y, Pang H, Liu Y, Wang X, Yu S, Fu D, Chen J, Wang X (2019) Environmental remediation of heavy metal ions by novel-nanomaterials: a review. Environ Pollut 246:608–620. https://doi.org/10.1016/j.envpol.2018.12.076
280. Xu F, Deng S, Xu J, Zhang W, Wu M, Wang B, Huang J, Yu G (2012) Highly active and stable Ni–Fe bimetal prepared by ball milling for catalytic hydrodechlorination of 4-chlorophenol. Environ Sci Technol 46:4576–4582. https://doi.org/10.1021/es203876e
281. Xu J, Tan L, Baig SA, Wu D, Lv X, Xu X (2013) Dechlorination of 2,4-dichlorophenol by nanoscale magnetic Pd/Fe particles: Effects of pH, temperature, common dissolved ions and humic acid. Chem Eng J 231:26–35. https://doi.org/10.1016/j.cej.2013.07.018
282. Yalçınkaya Ö, Kalfa OM, Türker AR (2011) Chelating agent free-solid phase extraction (CAF-SPE) of Co(II), Cu(II) and Cd(II) by new nano hybrid material (ZrO2/B2O3). J Hazard Mater 195:332–339. https://doi.org/10.1016/j.jhazmat.2011.08.048
283. Yan W, Vasic R, Frenkel AI, Koel BE (2012) Intraparticle Reduction of Arsenite (As(III)) by Nanoscale Zerovalent Iron (nZVI) Investigated with In Situ X-ray Absorption Spectroscopy. Environ Sci Technol 46:7018–7026. https://doi.org/10.1021/es2039695
284. Yin J, Deng B (2015) Polymer-matrix nanocomposite membranes for water treatment. J Membr Sci 479:256–275. https://doi.org/10.1016/j.memsci.2014.11.019
285. Yin W, Wu J, Li P, Wang X, Zhu N, Wu P, Yang B (2012) Experimental study of zero-valent iron induced nitrobenzene reduction in groundwater: The effects of pH, iron dosage, oxygen and common dissolved anions. Chem Eng J 184:198–204. https://doi.org/10.1016/j.cej.2012.01.030
286. Youssef AM, Malhat FM (2014) Selective removal of heavy metals from drinking water using titanium dioxide nanowire. Macromol Symp 337:96–101. https://doi.org/10.1002/masy.201450311
287. Zhang B, Wang D (2019) Preparation of biomass activated carbon supported nanoscale zero-valent Iron (Nzvi) and its application in decolorization of methyl orange from aqueous solution. Water 11:1671. https://doi.org/10.3390/w11081671
288. Zhang G, Wang S, Yang F (2012) Efficient adsorption and combined heterogeneous/homogeneous fenton oxidation of amaranth using supported Nano-FeOOH as cathodic catalysts. J Phys Chem C 116:3623–3634. https://doi.org/10.1021/jp210167b
289. Zhang L, Lv P, He Y, Li S, Chen K, Yin S (2020) Purification of chlorine-containing wastewater using solvent extraction. J Clean Prod 273:122863.https://doi.org/10.1016/j.jclepro.2020.122863
290. Zhang L, Wang X, Miao Y, Chen Z, Qiang P, Cui L, Jing H, Guo Y (2016) Magnetic ferroferric oxide nanoparticles induce vascular endothelial cell dysfunction and inflammation by disturbing autophagy. J Hazard Mater 304:186–195. https://doi.org/10.1016/j.jhazmat.2015.10.041
291. Zhang Q, Teng J, Zhang Z, Nie G, Zhao H, Peng Q, Jiao T (2015) Unique and outstanding cadmium sequestration by polystyrene-supported nanosized zirconium hydroxides: a case study. RSC Adv 5:55445–55452. https://doi.org/10.1039/C5RA09628F
292. Zhang X, Lin S, Chen Z, Megharaj M, Naidu R (2011) Kaolinite-supported nanoscale zero-valent iron for removal of Pb2+ from aqueous solution: reactivity, characterization and mechanism. Water Res 45:3481–3488. https://doi.org/10.1016/j.watres.2011.04.010
293. Zhang X, Ma Y, Xi L, Zhu G, Li X, Shi D, Fan J (2019) Highly efficient photocatalytic removal of multiple refractory organic pollutants by BiVO4/CH3COO(BiO) heterostructured nanocomposite. Sci Total Environ 647:245–254. https://doi.org/10.1016/j.scitotenv.2018.07.450
294. Zhang Y, Wu B, Xu H, Liu H, Wang M, He Y, Pan B (2016) Nanomaterials-enabled water and wastewater treatment. NanoImpact 3–4:22–39. https://doi.org/10.1016/j.impact.2016.09.004

295. Zhao F, Peydayesh M, Ying Y, Mezzenga R, Ping J (2020) Transition metal dichalcogenide-silk nanofibril membrane for one-step water purification and precious metal recovery. ACS Appl Mater Interf 12:24521–24530. https://doi.org/10.1021/acsami.0c07846
296. Zhao G, Huang X, Tang Z, Huang Q, Niu F, Wang X (2018) Polymer-based nanocomposites for heavy metal ions removal from aqueous solution: a review. Polym Chem 9:3562–3582. https://doi.org/10.1039/C8PY00484F
297. Zhao M, Xu Y, Zhang C, Rong H, Zeng G (2016) New trends in removing heavy metals from wastewater. Appl Microbiol Biotechnol 100:6509–6518. https://doi.org/10.1007/s00253-016-7646-x

Water Purification by Carbon Quantum Dots

Karthiyayini Sridharan, Vijaya Ilango, and R. Sugaraj Samuel

Abstract Sustainable, sufficient and pure water is vital on earth at present. Of all processes used to purify water, nanotechnology plays an important role. Nanomaterials are available in all stages like membranes in filtration, adsorbents of pollutants, photocatalysts for degradation and for detection of pollutants in water purification. Carbon quantum dots (CQDs), an accidentally invented nanomaterial, like in many applications have supported in water purification in all stages. Hence, it is intended for a mini-review for the water purification techniques with carbon quantum dots (CQDs). The chapter consists of three parts. Many nanotechnological methods are employed in water purification. Among them, nanomembranes play a dominant role in detection, nanofiltration and degradation in water purification. Next, nanomaterials as adsorbents, sensors and photocatalytic activity for water remediation are being investigated largely. Hence, they are discussed in the first part. In the second part, the features of CQDs are considered for its versatile applications. The easy and low-cost preparation of CDQs from easily available materials are investigated from the research published articles. In the third part, an attempt is made to collectively analyze the CQDs contributions as membranes in water purification, to detect the presence of organic, inorganic and dye contamination from water is reviewed. Further, the removal of the harmful and or toxic elements from polluted and impure water is discussed from various research groups.

1 Introduction

Water is essential for every living being. Earth and human body are made up of almost 70% of water. Yet only 2.5% of freshwater is available. Out of which only 0.007% is accessible although there are many sources of water like spring, rivers,

K. Sridharan (✉) · V. Ilango
Birla Institute of Technology and Science Pilani, Dubai Campus, Dubai, UAE

R. S. Samuel
The New College, Chennai, India

E. Lichtfouse et al. (eds.), *Inorganic-Organic Composites for Water and Wastewater Treatment*, Environmental Footprints and Eco-design of Products and Processes,
https://doi.org/10.1007/978-981-16-5928-7_4

etc., being periodically filled by rain. There will be no life if there is no water on earth. Thus, the present challenge is to "Ensure availability and sustainable management of water and sanitation for all". Traditional water purification methods include boiling, filtration, sedimentation and solar radiation. They may be cheap and feasible for rural area, but at present a safe, convenient and more sustainable methods are essential. Hence, nanotechnology is used for sustainable methods for water purification. Nanotechnology can solve the technical challenge faced in almost all the methods of purification of water. The advantages being its size and properties therefrom. For example, titanium oxide is a more effective catalyst in nanoscale than as microscale. It is used to degrade organic pollutants, for example, in water treatment. However, in some cases, manufactured nanoparticle's small size may become more toxic than the normal material. Another notable factor is removal of water contaminants including arsenic, viruses, bacteria, pesticides, mercury, salt pose, etc.

The goal of Nanotechnology in water is to increase the water availability, to be delivered effectively at the required destination and to employ the required implementation to ensure its sustained availability for our future generation.

First a detailed analysis must be continuously collected about the following:

1. Water sources
2. Pollutants
3. Purification techniques

Surface water like river, lake or freshwater marsh; Under river flow, that flows through rocks and sediments under the ground, Groundwater which is a fresh water source available at pore spaces between the soil and rocks, Frozen water like icebergs and glaciers, Desalination which is fresh water converted from these a water are the main sources of water available [https://en.wikipedia.org/wiki/Water_purification].

In broad sense, inorganic, organic and biological water contaminants that are harmful for living beings and environment are the pollutants. The toxic and/or harmful metal ions like arsenic, mercury, lead, cadmium, chromium, zinc, nickel, copper, nitrates, sulfates, phosphates, fluorides, chlorides, selenides, chromates, oxalates, etc., are some pollutants that change the taste and color of water. Organic and pharmaceutical pollutants such as pesticides, fertilizers, radioactive waste, hydrocarbons, plasticizers, biphenyls, phenols, detergents, oils, greases are also beyond consumable or usable limit in water. Biological water contaminants like bacteria, viruses and parasites that cause water-borne diseases must be destroyed before water consumption [10].

Water purification is generally accomplished by a few successive stages consisting of chemical coagulation, flocculation, sedimentation, filtration and disinfection. Selection of the methods to be used for water remediation will depend on the contamination and the application of the remediated water. Further, a cheaper and effective combination of the processes is essential and is the ultimate requirement. There are many water purification techniques by nanotechnology at various processes. Nanotechnology offers nanomaterials synthesized with desirable properties for the necessary treatment like nanofiltrations and membranes, nanoadsorbents, nanocatalysts, disinfection and microbial control for removing the organic and inorganic contaminants and bacteria.

The engineered nanomaterials are customized with certain properties like high specific area, pore sizes to suit hydrophilic, and hydrophobic interactions in nanomembranes and nanoadsorbents useful in filtration and adsorption techniques. Nanomaterial composites with specific functional groups for selective adsorption and interaction with bacteria or other organic or inorganic pollutants. Thus, Nanotechnology is employed for an efficient and economic method for water purification [41].

2 Nanotechnology for Water Purification

The basic methods like boiling, filtration, distillation, chlorination, sedimentation and oxidation were commonly used for water purification for the removal of physical, biological and chemical pollutants. For efficiency and sustainability, nanomaterials and nanotechnology are preferred for water treatment processes.

Water is purified by successive processes so that in each stage some selective pollutants are removed by their respective purification method by filtration, adsorption etc. It is reported [19] that the processes like coagulation/flocculation, precipitation, biodegradation, filtration (sand) and adsorption using activated carbon are conventional methods adopted. The established recovery processes are solvent extraction then evaporation followed by oxidation and electrochemical treatment then membrane separation, membrane bioreactors, ion-exchange and incineration are categorized as. Further, the new emerging processes are advanced oxidation, adsorption onto, nonconventional solids, biosorption, biomass and nanofiltration.

Thus, a combination of physical, chemical and biological processes are being incorporated in complete removal of pollutants for safe water to be available for living beings.

The major nanotechnology methods for water purification are remediation, desalination, filtration, purification, etc. The advantages for using nanomaterials are their surface area, compact volume, strength, stability and durability due to their structure and the volume to surface area ratio. Also, their chemical and biological reactions with the pollutants.

The property of nanomaterials being chemically inert and versatile hydrophilic surface chemistry are suitable for the removal of bacterial pollutants from polluted water. Metal ions such as Cu^{2+}, Fe^{2+}, etc., sulfates, fluorides and organic pollutants are removed by some biobased membranes.

Nanofilter membranes are also found to have good adsorption capacity. The membranes are doped with functionalized groups, say, negatively charged anionic group which are used for the adsorption of the positively charged pollutants. They are also utilized as anti-fouling agent, for removal of salt ions, dechlorination of water

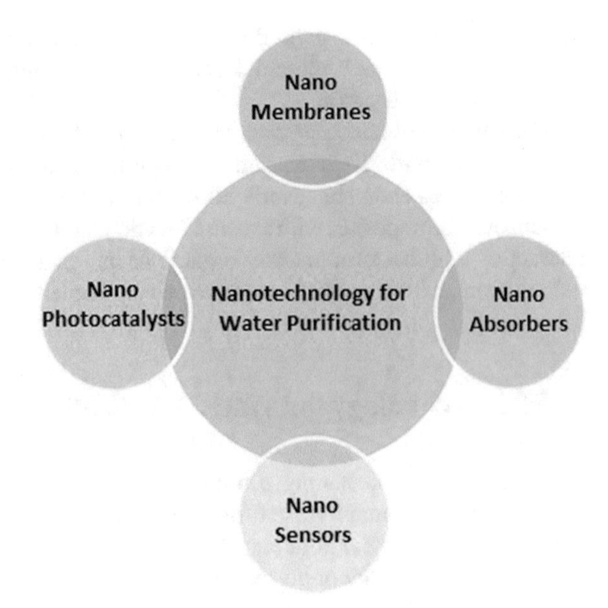

Fig. 1 Nanotechnology for water purifications

and removal of dyes from the polluted water. [*Nanotechnology for water purification-wikipedia, en.wikipedia.org*]. Thus, the major areas of nanotechnology considered for water purifications is given in Fig. 1.

2.1 Nanomembranes for Water Purification

Filtration implies the removal of unwanted constituents by the passage of untreated water through a porous barrier. After the pretreatment stage the suspended solid is removed. Coagulation and flocculation are then done where the dissolved impurities are precipitated through sedimentation. Filtration is done to remove the suspended particles if present from the water. In the conventional water treatment, the dissolved salts like soluble inorganic and organic substances cannot be removed. Hence, nanotechnology is considered for the water treatment. Thus nanomembranes, nanoporous polymers, etc., are used for water remediation. Further, desalination is being considered under filtration method [68].

Nanofiltration membrane process is used on surface water and fresh groundwater, softening purposes and removal of disinfection byproduct precursors such as natural organic matter and synthetic organic matter. There are different kinds of membranes such as microfiltration (MF), ultrafiltration (UF), nanofiltration (NF) and reverse osmosis (RO) membranes that are categorized according to their pore size. Selected molecular organics and salts are detained depending on pore size distribution and complex separation mechanism. Thus, nanofiltration membranes first blocks the pollutants based on their size. Membrane must offer high selectivity along with high permeability. Membrane fouling must be greatly reduced as they will cause severe flux loss due to the blockades in the membranes. This also shortens the lifespan of the membrane [96].

Generally, nanomembrane consists of many layers which also includes a few nanomaterials for the filtration of specific pollutants. Usually, polymers like fluoropolymers, amphiphilic polymers, nanoporous ceramic membrane, carbonized nanomaterials, etc., are utilized. Mainly the organic pollutants, inorganic pollutants and pathogens from viruses and bacteria are either filtered or adsorbed selectively by customized nanomembranes [5]. Antifouling properties are developed on the membranes generally by surface modification. One of the modifications is thin film coating on the membrane surface. The other one is by grafting of polymer materials on the membrane surface. Both these methods are to stop the interactions between the foulant and membrane. The membrane surface must be nonporous, smooth to stop the foulant entrance into the membrane and avoid internal fouling. The membrane surface is charged by some dopants to repel the foulants by electrostatic interactions [57]. Nanocomposite membranes are made from specific structured nanomaterials. They are found to have higher hydrophilicity, thermal stability and increased water permeability. It is mostly used for Reverse osmosis and removal of micropollutants.

Highly porous nanocarbon membranes are found to sustain high salinity, hydrophobicity and can withstand high pressure and vacuum. Homogeneous nanoporosity is found in certain self-assembled membranes for ultra-filtrations. Aquaporins membranes are noted highly selective to reject specific ionic molecules under certain conditions [69]. Nanofiltration consumes less energy and the water productivity is more with high salt rejection. Hence, it is reported that nanofiltration plant is more economical and efficient for desalination of water [88]. Some nanostructured materials are used polymeric and ceramic membranes to enhance their properties for efficient water treatment. Zeolitic nanoparticle on ceramic, isoporous block copolymer and hybrid inorganic–organic nanocomposite membranes are a few examples mentioned [80].

2.2 Adsorption of Pollutants by Nanotechnology

Certain nanomaterials can attract particular substances on to the molecules on their surface that are at close contact with them are known as nanoadsorbents [28]. High adsorption of organic and inorganic pollutants like heavy metals and micropollutants are noticed in these nanoadsorbents. There are nanoadsorbents based on carbon, metal, polymer and zeolites used as adsorbents for selective pollutants. Generally, these nanoadsorbents are more efficient and faster in adsorption process. Therefore, they are used for contaminated or polluted water treatment.

Adsorption is another well-known technique for water remediation. Nanomaterials are also considered as good adsorbent in treating wastewater. For example, catalytic, absorptive, catalytic membrane, bioactive nanoparticles, biomimetic membrane, polymeric and nanocomposite membrane, thin-film composite membrane, etc. are some of the applications where they are used. Carbon nanotubes (CNTs) are used as absorbent for several organic chemicals. Carboxylic, hydroxyl and amide functional groups of Organic compounds which form hydrogen bond with the graphitic CNT surface also donate electrons. Ferrous oxide, TiO_2 and Al_2O_3 are some nanoscale metal oxides which are effective, low-cost adsorbents for heavy metals and radionuclides. Dendrimers (polymeric nanomaterials) are adsorbents used for removal of organics and heavy metals. Powder, beads, or porous granules are various forms of nanoadsorbents available for water treatment [30]. Commercial activated carbon is available as adsorbents. Because of their structural properties with large surface area and porosity, they are generally good adsorbents of organic pollutants and pesticides, aromatic and phenolic derivatives, metals ions like iron, manganese, nitrate and dye molecules that produce different taste or smell in water. They also retain toxic organic materials and cause degradation of bacteria from the adsorbed materials.

Activated carbons being expensive, they are being replaced by CQDs and their composites [10]. Natural adsorbents such as sawdust, wood, etc., are available for water treatment. Several materials are customized to develop a particular structure and properties for selective adsorbents. Materials like zeolites are produced specifically as adsorbents. Adsorbents that are good in interaction with bioorganisms are manufactured for the adsorption of chitosan, fungi or bacterial biomass.

The removal of pollutants from wastewaters by a variety of conventional and nonconventional adsorbents are done by various mechanisms. Researchers have classified the mechanisms of adsorption as physisorption, chemisorption, ion exchange and precipitation. Physisorption is the process where an adsorbate bound to the surface by weak van der Waals forces before being removed from the wastewater. Chemisorption is the process where an adsorbate is tethered through either covalent bonding or due to electrostatic attraction.

Organic resins, activated alumina, zeolites and sand are a few of the adsorbents equally competent as commercial activated carbon. There are innumerable nonconventional efficient adsorbents available. To list, agricultural solid waste, industrial by-products such as red mud, natural materials such as clays, biosorbents such as chitosan are not only efficient solid adsorbents but also very economical. Recently cellulose and chitosan are used in many applications as adsorbents. However, the effective adsorbents are not utilized in industrial level as yet [19]. As the adsorbate is accumulated on the adsorbent surface, adsorbent is a surface process. Carbon-based nanomaterials, metal oxide-based nanomaterials, Carbon, metal oxide hybrid nanomaterials and polymer-based nanomaterials are widely used as adsorbents for wastewater treatment specifically for the removal of heavy metals and dye removal.

Carbon-based nanomaterials like activated carbon, CNTs, fullerenes, and graphene are widely used because of their non-toxicity, structural stability and high adsorption [25]. They are also available in abundance and are prepared easily. Metal oxide-based nanomaterials like Fe_3O_4, MnO_2, TiO_2, MgO, CdO and ZnO are used as adsorbents for the removal of heavy metal ions and dyes. High surface area, specific affinity, low solubility with relatively no environmental impact are the characteristics responsible for them to be utilized as adsorbents.

Retention and permeability are the properties to be considered in nanofiltration membranes. They depend on the electric charge and the valency of the solute and the solution. Monovalent ions are more permeable than the multi-valent ions. Similarly, the retention rate depends on the specific cations and anions. The membrane pore size is another important factor in membrane processes. Beside nanofiltration process, reverse osmosis techniques are performed using nanomembranes [85].

The common difficulties on nanomembrane technology are scaling and fouling issues which are being overcome by specific composites and CQD doped membranes.

2.3 Role of Nanomaterials as Sensors for Water Remediation

Nanomaterials are used as sensors for water quality monitoring due to its unique optical and electronic properties [75]. The particles containing a core and a shell are the core–shell nanocomposite materials. They were prepared from metal nanoparticles as precursors following bi-functional molecules or co-precipitation method. Their customized physical and chemical properties over their single-component enable them to be utilized in many applications. These synthesized nanomaterials are used as sensors because of their tunable optical properties by their ratio of core-to-shell thickness.

Nanosensors play an important role in detecting pollutants from essential materials like air and water in environment. SnO^2 and reduced graphene oxide together are used for electrochemical detection of ultra-trace heavy metal ions in drinking water. A pH-based sensor synthesized from ZnO nanorods was used to find the pH in water. A bismuth porous carbon nanocomposite showed high sensitivity in detecting Pb^{2+} and Cd^{2+} ions in tap drinking water and wastewater even at a very low concentration levels [67]. The sensitivity and selectivity of nanomaterials and nanostructures are the key factors to employ them as sensors. Inorganic quantum dots, graphene oxide, QCD and metallic clusters are excellent fluorescent materials. Further their high photostability, high quantum efficiency, size-dependent fluorescence emission peaks enable them to be used as fluorescent sensors for the detection of heavy metals. QCDs linked with Au NPs are good sensors of Hg^{2+} in river water with excellent selectivity over other metal ions. ARhodamine B-Au NP-based probe with nanometal surface energy transfer is reported [54] as a sensor for detecting mercury in water.

2.4 Nanomaterials as Photocatalysts for Water Purification

Photocatalysis is the enhancement of a photoreaction by the presence of a catalyst. The photocatalyst is the light that is absorbed. The photocatalytic activity (PCA) is first to create electron-hole pairs by the ability of the catalyst, which in turn generate free radicals (e.g., hydroxyl radicals: OH^{-}) that are able to undergo secondary reactions. They are then used to break down organic molecules, organisms and inorganic molecules in various applications like removal of air pollution, in building materials for self-cleaning surfaces and in water purification [http://protecsolutions.com.tr/uploads/41bf0b835349bea49bd2d50fc0327678.pdf].

The techniques from nanotechnology water treatment using nanostructured catalytic membranes, nanosorbents and nanophotocatalyst are both eco-friendly and efficient, but the investment and energy in implementing them are enormous.

Organic pollutants such as Congo red, azo dyes, phenol aromatic base pollutants, toluene, dichlorophenol trichlorobenzene, chlorinated ethene, etc. are removed using TiO_2 based nanotubes effectively from wastewater. Titanium dioxide (TiO_2) is considered as good nanophotocatalyst material due to toxic-free property, chemical stability, easy availability and low cost. Photocatalysts like ZnO, which are used to eliminate pollutants in wastewater are also effectively reused. Pd incorporated ZnO nanomaterial is used for the removal of Escherichia coli (E.Coli) from wastewater due to their high photocatalytic activity [4]. TiO_2 activity is by the oxidation of pollutants by the formation of hydroxyl radicals. The photogenerated e^--h^+ pairs are separated which are responsible for the transformation of pollutants. The ability to decompose the pollutants is exceptional. Its photocatalytic activity can be performed under sunlight. It is environmentally safe, non toxic and of low cost.

Generally, light is absorbed, excites the electrons from conduction band to valence band [9]. TiO_2 photocatalytic activity causes adsorption of organics to its surface. Then the degradation of the adsorbents occurs. For an efficient water purification through TiO_2-based photocatalytic process, proper light absorption to enable catalyst action is necessary. Also, recombination of photogeneration of e^--h^+ pairs reduces the photocatalytic efficiency. Hence, e^--h^+ separation must be increased. For the effective adsorption and degradation of organics by TiO_2, its surface area is made large and active. The TiO_2may also be doped appropriately for an effective photocatalytic activity. Nanostructures of metal oxide semiconductors as photocatalysts are efficient in the removal of biological contaminants from polluted water [87]. Nanostructured photocatalytic adsorption is higher for nanocatalysts of large surface-to-volume ratio. Being excited by sunlight they can be employed everywhere to degrade organic, inorganic and microbial contaminants by their effective redox process. Nanoparticles of TiO_2 and ZnO catalysis are examples for groundwater remediation of chemical contaminants. Ternary oxide zinc stannate photocatalyst is another good photocatalyst for water purification. The addition of hydrogen peroxide (H_2O_2) being an electron acceptor on potassium persulfate ($K_2S_2O_8$) and TiO_2enhances the degradation rate of the pollutants. Complete mineralization of carbofuran was observed by addition of total organic carbon (TOC) analyzer. Cymoxanil, dimethoate, methomyl, oxamyl, pyrimethanil and telone are some pesticides that are completely destroyed by complete mineralization. Bacteria including Escherichia coli, Salmonella typhimurium, Pseudomonas aeruginosa, and Enterobacter cloacae were successfully degraded by TiO_2 nanoparticles (Degussa P25) under solar light irradiation.

3 Understanding Carbon Quantum Dots

3.1 Properties of Carbon Quantum Dots

Carbon dots are nanoparticles with all the dimensions measured within the nanoscale (no dimensions are larger than 100 nm). Carbon quantum dots, also recognized as C-dots, have the molecular geometry and physical properties comparable with that of graphene oxide. They vary from graphene oxide in terms of dimensions, being quasispherical nanoparticles with width below 10 nm [49]. The assembly of carbon quantum dot structure is represented in Fig. 2.

The structure collectively can be viewed as a muster of functional groups comprising of oxygen atoms essentially C=O, −OH and C–O reinforced on the outside of a solitary layer graphene sheet. The structure and properties of the graphene sheets are altered due to the accumulation of oxygenated functional groups and particle sizes to exhibit the quantum internment properties. Also, the electron-transfer and inventory properties of CQDs can be functional to distinct generation of electrons by light.

The ample functional groups (–NH_2, −COOH, OH, etc.) on the exterior of CQDs can be functioned as dynamic binding site with transition metal ions. Principally, CQDs can be employed as resourceful electrocatalysts when blended with other inorganic compounds, such as metal phosphates, metal sulfides and layered double hydroxides (LDHs), etc. CQDs contrast to the naturally occurring rare precious metals are economical, easily accessible, provide more active catalytic sites, improved structural stability and enhanced electronic conductivity [31, 90]. The modest chemical structure of carbon dots recommends that they can be produced from organic or eco-friendly waste, providing a stimulating prospect to renovate

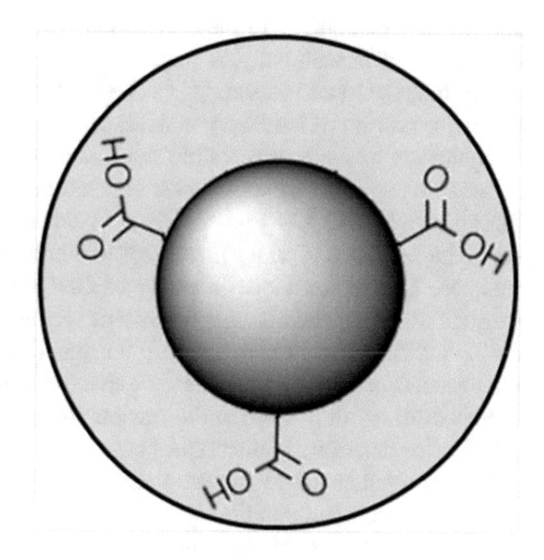

Fig. 2 Carbon quantum dot structure

waste into progressive efficient nanomaterials for attaining a globular economy and justifiable development. Several waste matter have been turned into carbon dots such as orange peels, fish scales, waste paper, onion waste and hair [22]. By varying the concentration of precursors, reaction time, solution pH and reaction temperature, the physicochemical properties of CQDs can be modified. Consequently, CQDs are appropriate in the versatile applications of energy conversion, optoelectronic devices, bio-imaging and water purification methods.

Various studies have stated the usage of CQDs in wastewater treatment or for wastewater monitoring. For instance, CQDs have been used to discard uranium and also in tracking the dangerous materials in industrial waste. CQDs can also be functional as adsorbents for the removal of benzopyrene and Cd(II) in environmental water samples. Additional study has stated the use of CQDs as an antimicrobial agent for degradation of bacteria.

CQDs have fascinated the attention in biomedical applications, as they possess exceptional biocompatibility, good solubility in aqueous solution and necessary optical properties apart from wastewater treatment. Manufacture of CQDs from eco-friendly materials (carbon wastes and plant-based sources) should be fortified due to its low cost, extensive accessibility, low toxicity and sustainability. Also, this can decrease waste production and the consumption of chemical substances [89]. Due to the emerging complex environmental problems, carbon-based materials, especially CQDs, have sparked a lot of interest in the field of energy conversion and storage (Lim 2015).

Contamination of heavy metal is identified to be the origin of toxic effects on the human health and environment. The discharge of overwhelming metal in water sources could be a challenging issue confronted by the mankind in the twenty-first century. Even when present in trace amount in natural systems these metal ions will have the poisonous effects and they are nonbiodegradable. Extremely low concentration of these metal ions in the environment can affect the signaling mechanisms and cellular machinery in the body. This will cause damage to bones, teeth, liver and organs of human body. A major portion of inorganic pollutants in the form of anions ($Cr_2O_7^{2-}$) and cations (Hg^{2+}, Fe^{3+} Co^{2+}, Ni^{2+}, Pb^{2+} and Cd^{2+}) in water form the major portion of the heavy metal ions.

Synthesis of water-dispersible fluorescent CQDs (2–5 nm size) from naturally obtainable cabbage as the carbon source has been established. The results indicated the potential of CQDs from cabbage in advancement of chemosensor probes to sense the existence of heavy metal ions in innumerable environmental and biological samples. The fluorescence intensity of CDQs obtained by this method could be substantially quenched in the presence of various metal ions such as Hg^{2+}, Fe^{3+} and Pb^{2+}. Additional work is mandatory for quantitative estimation of metal ions since the nanosensor prepared lacks selectivity. Sustainable carbon-based photosensitive probes offer an alternative to the noxious semiconductor QDs that are presently in practice for detecting applications [1].

The components and structure of CQDs govern their diverse properties. CQD surface expose biocompatibility and excellent solubility in water due to numerous carboxyl moieties. CQDs are also appropriate for surface passivation with various biological, polymeric, inorganic or organic materials and chemical modification. Because of this surface passivation, the CDQs can improve the fluorescence properties as well as physical properties (Lim 2015). Carbon quantum dots are group of carbon nanostructures which have established widespread attention due to their exceptional properties including chemical inertness, high resistance to photobleaching, non-toxicity, good biocompatibility, remarkable photostability, high solubility in water [78].

Quantum-sized semiconductor nanoparticles have been established as a significant class of photoactive nanomaterials for a diversity of scope and implementation. There has been wide research on the photoresponse, photoinduced charge separation and electron transfer processes for the use of semiconductor quantum dots in light energy conversion and related areas. The electron-donating abilities of the photoexcited carbon dots were also verified in the photoreduction of Ag(I) to Ag [98]. Compared to the traditional semiconductor quantum dots, the fluorescent CDs possess excessive advantages comprising their exceptional stable chemical properties and optical properties. CDs replace semiconductor quantum dots in biological imaging since CDs are eco-friendly with low toxicity characteristics.

Fluorescent nanomaterials have gained a lot of attention as potential competitors to conventional fluorescent dye probe in recent years due to the high demand for fluorescent probe in chemical sensing, biological monitoring, and other related fields. Traditional fluorescent dye flaws such as low stability, low fluorescence intensity and rapid photobleaching can be overcome by fluorescent nanomaterials with its quantum size effect.

Because of their low toxicity, changeable surface functional groups, exceptional biocompatibility, CDs have become a hotspot of drug delivery science [35]. From both the view of fundamental research and real-world application Photoluminescence (PL) is one of the most captivating features of CQDs [16, 42, 61, 104]. For CDQs, the diverse reliance of the intensity and emission wavelength is one of the consistent features of the photoluminescence.

The distinct dependency of the emission wavelength and intensity is a typical feature of the PL for CQDs. The optical array of nanoparticles of different sizes or CQDs with different emissive traps on the surface may be the cause of this unusual phenomenon (Li 2018).

The large and excitation-dependent PL emission spectrum reflects the difference in particle size and PL emission [84, 107] CQD emission activity under 470 nm wavelength irradiation with different concentrations was investigated. As the concentration of CQDs was increased, the PL intensity of the solution first increased and then decreased [89].

The size-dependent optical absorption of CQDs is one of the most intriguing properties. The PL is a well-known sign of quantum confinement. Zhao et al. stressed the importance of size in determining the excitation wavelength of CQDs. When compared with particles of the same size, PL was with more varying emissive trap sites [51, 66]. The fluorescence properties of CQDs depends on excitation of certain

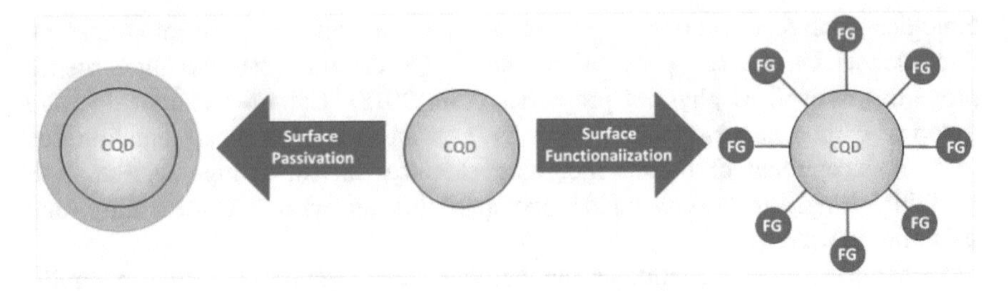

Fig. 3 Two modifications of CQD

range of wavelength. This makes it more viable for many applications. CQDs fluorescence properties depend on their functional groups, sizes and crystalline degrees. Barely CQDs are less in fluorescence quantum yield. But it can be intensely enhanced by functionalization, passivation and chemical alteration by other molecules. The precursor of carbon dots also plays a crucial role because of its core and surface structure originate from the initial raw material [24].

3.2 Passivation and Functionalization

Passivation is a thin protective or insulating layer formed around CQDs. Often this layer prevents the impurities bonding with CQDs as its surface is highly reactive due to the presence of oxygen and carbon thus improving the photoluminescent intensity. This covering is generally done by polymer or organic molecules which are non-radiative by the visible radiations. Hence, the CQDs have their usual luminescence. Further, because of this protection, CQDs become stable and hence will have long life.

Functionalization is to create functional groups such as hydroxyl, carboxyl and carbonyl, for enhancing various features of CQDs. Generally, the functional groups are attached during the synthesis itself. It is by acid treatment or hydrothermal process for the attachment of sulfur-, amines, selenium-, phosphorus-, boron-, and oxygen-containing or by attaching boron, amines containing by Solvothermal method [40]. The two modifications, passivation and functionalization on the surface of CQD are depicted in Fig. 3.

3.3 Some Methods of CQD Synthesis

Synthesis of Carbon Quantum dots can be broadly classified as Top-Down Method and Bottom-Up Method.

The macromolecule is crushed or disseminated into small-sized CQDs using chemical or physical methods in the top-down process, while the bottom-up approach refers to the chemical polymerization and carbonization of a sequence of small molecules into CQDs.

Top-Down Methods

Arc-discharge:

In this approach, C-dots was developed from crude carbon nanotube soot (residue) [29]. Next step the residue was oxidized with 3.3 M HNO_3 to bring in carboxyl groups. Later the resulted substance was extracted with NaOH/basic solution of pH 8.4. A stable dark colored suspension was observed. The extracted matter was purified by Gel electrophoresis. The arc-discharge soot method comprises a number of complex segments and the CNPs produced by the arc-discharge process have minimal yield. But it was hard to purify the segments. In addition, since they were mixed with oxygen externally, the fluorescent NPs obtained by the oxidized carbon nanotube had the ability to gather when dispersed in water. It was coated in a thin layer of carbon, indicating that it was more widespread division.

Laser ablation method:

By laser treatment of graphite fragments in polymer solution Carbon quantum dots (C-dots) with typical sizes of about 3, 8 and 13 nm were produced [26]. The size control of C-dots can be understood by regulating the laser pulse width. Carbon dots obtained show excitation wavelength and size-dependent photoluminescence behavior.

The cause may be the effect of laser pulse width on C-dot nucleation and growth conditions. As opposed to short-pulse-width lasers, the long-pulse-width laser will be better suited in controlling the size and morphology of nanostructures in different material systems.

Electrochemical method:

Carbon Quantum Dots (CQDs) are captivating incredible attention due to their, biocompatibility, optical properties, low toxicity, extensive applicability and water dispersibility. By the electrochemical oxidation of a graphite electrode in alkaline alcohols, CQDs with an average width of (4.0 ± 0.2) nm and high crystallinity were produced. CQDs dispersion formed was colorless, then changed to bright yellow when stored in ambient conditions. This color distinction appeared to be due to oxygenation of surface species over time based on fluorescence spectroscopy, Fourier transform infrared spectroscopy (FTIR), UV–Vis absorption, high-resolution transmission electron microscopy (HRTEM) and X-ray photoelectron spectroscopy (XPS).

Carbon quantum dots are produced by electrochemically by oxidizing graphite electrodes in alkaline alcohols [53] and are used for ferric ion detection and cell imaging.

Bottom-up Methods

Thermolysis-Chemical route:

Most common synthesis method for fluorescent carbon dot is by thermal decomposition of citric acid. Several intermediates are formed and they also produce fluorescent species during the process. The reaction pathway is very complicated, and the specifics are still unknown. In the current work [71], By combining, Xray photoelectron spectroscopy, infrared analysis and liquid chromatography/mass spectroscopy (LC/MS) with the modification of the absorption, emission, and optical properties, the formation of fluorescent C-dots from citric acid was studied as a function of reaction time. As shown by the decay time study, the reaction intermediates have been identified and found to generate two key emissive species, in the green and blue, at various stages. C-dots were also synthesized from the intermediates through thermal decomposition, with an emission limit of around 450 nm. To improve the control and repeatability of C-dots synthesis, a better understanding of the process is needed.

In a Teflon-lined autoclave, fresh pepper was suspended in water and heated at 180 °C for 5 h [101]. In the next step to obtain purified CDQs, large particles were separated by centrifugation and the filtrate was dialyzed with water.

Microwave synthesis:

Fluorescent CQDs were initially produced from roasted chickpeas by a microwave-supported pyrolysis [6]. It has been characterized using fourier transform infrared spectroscopy (FTIR) spectroscopy, transmission electron microscopy (TEM), UV–vis absorption spectroscopy, X-ray diffraction (XRD) technique and Fluorescence spectroscopy. Roasted chickpea was used as the carbon source. This entire synthesis is eco-friendly and the CDQs produced have shown beneficial properties such as good water solubility, excellent photostability and high fluorescence intensity.

Under UV light, the CQDs emit a blue fluorescence (at 365 nm). CQDs were prepared in 120 s using a microwave oven (350 watts). The effect of different metal ions on the fluorescence intensity of CQDs was studied to see if it could be used to detect metal ions.

One-pot microwave treatment of lotus root (LR) yielded fluorescent nitrogen-doped CQDs with a nitrogen content of 5.23%, Gu et al. [11] without the use of any other surface passivation agents.

Hydrothermal method:

In hydrothermal method the carbon dots are synthesized with external heating from precursors such as birch bark soot, blueberry, blackbird cherry, redcurrant, glucose, citric acid and cowberry [36]. The photoluminescence of these carbon dots at various excitation wavelengths was studied. It is shown that as the excitation wavelengths are increased, the luminescence spectra intensity peaks change. The discovery of this effect opens up new possibilities for the creation of composite materials made of carbon dots, starch, and epoxy that can turn blue LED light to green. In a Teflon-lined autoclave, fresh chopped rose-heart radish was mixed with water and heated to 180 C for 3 h [70]. A dialysis membrane was used to isolate the solution.

Pyrolysis carbonization 800 °C. Sweet potato was carbonized by hydrothermal treatment [77]. The microstructure and chemical composition of the CQDs were studied. A simple hydrothermal method was used to make highly luminescent carbon dots (C-dots) directly from lemon juice [7]. To control the luminescence of CQDs, different temperatures, time, precursor aging and diluted solvents were used.

A one-step hydrothermal method was used to make CQDs from waste tea leaves and peanut shells [112]. CQDs with high quantum yield and good stability were studied in terms of their synthetic conditions, structure and optical properties.

The use of cowberry, blackbird cherry, blueberry, redcurrant and birch bark soot as precursors in a hydrothermal method with external heating was investigated [36]. It is shown that there is an effect of the shift of luminescence spectra intensity peaks with the increasing of excitation wavelengths.

Solvothermal method:

Carbon quantum dots (CQDs) were prepared via a hydrothermal method starting from phosphoric acid, urea and citric acid in dimethylformamide solution by doping with phosphorus and nitrogen [37]. CQDs were characterized for the surface composition, optical properties, energy levels, size and morphology. Cyclic voltammetry was used to differentiate the LUMO and HOMO levels of the doped CQDs. Ferric (III) ions are found to quench the fluorescence. The selective coordination of Fe^{3+} by the surface functional groups on the CQDs can be attributed due to the quenching mechanism, as calculated by energy level measurements and absorption spectra. CQDs have been demonstrated to be viable fluorescent probes for determining Fe^{3+} with high selectivity and sensitivity.

Template method:

Developing C-dots by calcination in appropriate mesoporous silicon spheres or patterns, and etching to remove supports and produce nano-sized C-dots are the two stages of this process. Zong [114] determined a method for using silica mesoporous spheres as rough models. After this, the silica spheres were saturated with mixed solution of complex salts and citric acid. Following that, mesoporous supports were calcined and expelled and the photostability of subsequent C-dots, as well as mono-dispersion, verified that they had remarkable luminescence properties. Yang [99] reported a soft-hard template approach for developing consistently structural PL C-dots. The obtained C-dots' composition, tunable size, and crystalline degrees were found to have extra high stability properties after template elimination, passivation and carbonization. The challenge of forming aggregates has been effectively removed [29].

Other Methods

Carbonization procedure:

Carbon quantum dots (CQDs) are formed by the gathering of the central aromatic ring system of asphaltene molecules due to π–π interaction. The direct carbonization of dispersed carbonaceous microcrystals in mesophase pitch is viewed as a simple method for the synthesis of CQDs [23].

Regulating the nucleation temperature for mesophase formation changes the size of the as-prepared CQDs. Excitation-independent fluorescent behavior was observed in the CQDs, with a quantum yield of up to 87%. CDQs were successfully used for the fluorescent detection of Fe^{3+} ions with high sensitivity and specificity.

The outcome of this method furnished the ascendable production of CQDs at low cost. In addition to this, it also gives clear evidence to realize the hardening of asphaltene at nanoscale.

Synthesis of S–N–C-dots on a large scale was done using etching and carbonization of hair fiber by sulfuric acid [82]. It was shown that the S content of the as-formulated S–N–C-dots increased with reaction temperature, while the N content remained almost same.

CQDs nanohybrids:

Recent studies on the preparation of novel hybrids consisting of CQDs and inorganic nanoparticle cores (e.g., titania, silica, iron oxide and zinc oxide) have been attempted [91]. The hybrids formed, mix the fluorescence properties of the CQDs with the mechanical, magnetic and optical or properties of the oxide cores. These hybrids could be used as photocatalysts or magneto-optical biolabeling agents. Bidentate TiO_2/vitamin-C (VC) complexes were used to make TiO_2/CQDs composites using a hydrothermal process.

The hydrogen production from photocatalytic water splitting catalyzed by the TiO_2/CQDs nanohybrids was explored. The impact of hydrothermal temperature, reaction time, and vitamin C concentration was also investigated. The photocatalytic efficiency of these nanohybrids can be attributed to the hydrothermal treatment's collaborative effects, as well as the appropriate electron transfer capability during the transformation of CQDs.

4 Carbon Quantum Dots (CQDs) for Water Purification

CQDs are known to exhibit strong photoluminescence, nontoxic unlike metal-based quantum dots with exceptional optical and fluorescence characteristics with high quantum yield. Also known for high thermal and optical photostability, tunable excitation and emission, with easy surface functionalization [73].

Many researchers have reported [89] that two absorption peaks due to $\pi \rightarrow \pi^*$ transition of C=C and $n \rightarrow \pi^*$ transition of C=O from CQD. They are further tuned by either doping or by CQD-based composites. Photoluminescence is another useful property ofCQDs, dependent on the emission wavelength and intensity. It is attributed due to the surface defects resulting from sp^2 and sp^3 hybridized carbon. The various functional groups (–COOH, –OH, –NH2, etc.) on the surface of CQDs are another fascinating feature with which the interaction takes place with the organic and inorganic ions. Due to good water solubility, chemical stability, easy preparation from cheap and easily available materials, innumerable applications of CQDs are being reported. It is intended to review some of them for the water purification in this part of the chapter.

CQDs are synthesized by each research group by their requirement for their application. The synthesized CQDs were then characterized by a few of the following methods to confirm its properties.

UV molecules on absorption of ultraviolet or visible light excite the bonding and non-bonding electrons to higher anti-bonding molecular orbitals. The electron that absorbs longer wavelength of light is more easily excited (i.e., lower energy gap between the HOMO and the LUMO). π–π*, n–π*, σ–σ*, and n–σ* are four possible types of transitions. Their probable transitions are as σ–σ* > n–σ* > π–π* > n–π*.

PL—Photoluminescence is light emission by the absorption of photons from any form of matter. It is one of many forms of luminescence caused by photoexcitation.

TEM—In transmission electron microscopy, an image of the specimen is formed by a beam of electrons transmitted through it. The thickness of specimen is most often less than 100 nm thick or a suspension on a grid.

XRD—The atomic and molecular structure of a crystal is found using X-ray crystallography. The incident X-rays diffract through the specimen in the required direction to get its crystalline structure.

Raman spectroscopy is used to determine vibrational modes of molecules. Besides, the rotational and other low-frequency modes of vibrations of the systems are observed. It usually provides a structural fingerprint for identifying the molecules.

IR—chemical substances or functional groups in material in the form of solid, liquid, or gas are identified using infrared spectroscopy. It is found by the interaction of matter with infrared radiation through absorption, emission, or reflection.

XPS—X-ray photoelectron spectroscopy is performed using photoelectric effect which is a surface-sensitive technique. In a material, it is performed to identify the surface and inside elements and their chemical state. Further, the overall electronic structure and density of the electronic states are determined. This technique also shows the elements that are bonded to the material elements. Generally, this technique gives both the in line profiling, in depth profiling of the elemental composition across the surface, or when paired with ion-beam etching.

4.1 Carbon Quantum Dots as Membranes

Nanofiltration membranes play a vital role in water remediation. One of the drawbacks of the conventional water filtration method is that dissolved salts and some soluble inorganic and organic substances cannot be removed. So, nanotechnology

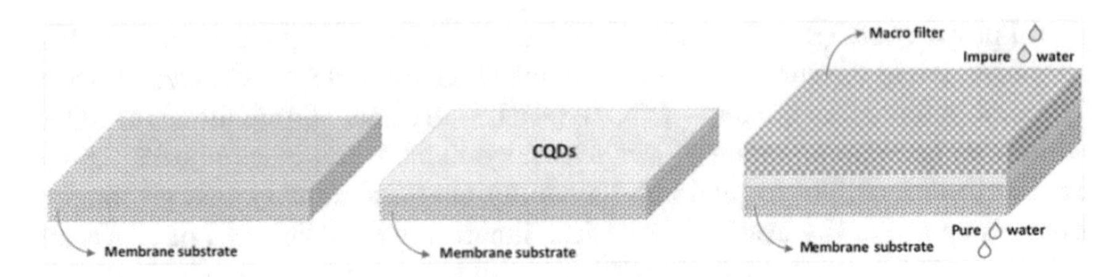

Fig. 4 Membrane structure with CQDs embedded as a layer

is preferred for the water treatment. This is done by using nanoporous polymers, nanomembranes, etc., which typically have pore sizes that range from 1 to 50 nm, filtering out almost all the bacteria and many harmful substances. Desalination is also categorized under this method [68]. The general structure of the membrane consists of several layers. In a simple membrane structure, CDQs are evenly spread as a layer between dense top layer and porous sublayer as shown in Fig. 4.

The application of membranes has been significantly enhanced by the addition of CQDs in thin-film nanocomposite (TFN) membranes. The water flux and power density are enhanced. The purity of water was very high, due to permeability through these membranes. The toxic ions were almost nil as the water purity reported was 99%. TFN membrane surface becomes more negatively charged when CQDs are incorporated in selective layers. Hence, the anions are very effectively reduced. There is a remarkable improvement in antifouling properties on the selective layer. This is also due to the electrostatic repulsions between foulants and the membrane surface. The efficient and large surface area, large interstitial space with many functional groups are the main reason for the highest fluxes. The presence of specific hydrophilic groups in CQDs also causes reduction of non-specific adsorption, thereby increasing the selectivity in the required adsorption of pollutants. These membranes are used in reverse osmosis application because of enhanced water permeability, high permeate flux and antifouling capability.

All the processes like desalination performance, porosity, permeability, hydrophilicity, selectivity, etc., are enhanced by the incorporation of CQDs onto membrane surfaces of the selective layers.

One research group [89] have published that three types of element doped CQDs were synthesized (carboxyl, amino and sulfur) to study about anti-fouling. Another study for the removal of selenium and arsenic by CQDs modified TFN membranes using sodium ion (Na-CQD).

Membrane selectivity and permeability are major challenges of the membrane technology. The energy consumption in the power-driven membrane, fouling adds to further disadvantages. Therefore, prior to ultrafiltration or reverse osmosis (RO) some pretreatment have been employed. Improvements are also focused on mechanical strength, physicochemical properties of membranes and structures further to selectivity and permeability. Beside polymers many composites were developed, using many nanomaterials and it offers solutions to an extent due to the selective properties of all the materials in the membrane. Membrane surface hydrophilicity, heat production, compaction are other difficulties encountered. Some nanomaterials have selective properties to overcome these issues. Further, the polymeric membranes are modified by incorporating nano zero-valent iron (nZVI) and noble metals supported on nZVI for reductive degradation of the microbial contaminants.

Although many approaches have been employed to improve the membrane technology, Carbon quantum dots (CQDs) with its optical properties, excellent hydrophilicity, low toxicity, environmental friendliness, and low cost was first embedded into the polyamide layer during interfacial polymerization and novel thin-film composite membranes produced are used for pressure retarded osmosis application.

Similarly, forward osmosis, membrane distillation, pressure retarted osmosis are other nanomembrane technology, improved drastically by CQD modified membranes.

The membrane plays an important role in pressure retarded osmosis (PRO) and a high-performance PRO membranes were desired. For the first time, thin-film composite membranes with carbon quantum dots incorporation, was developed for osmotic power generation [15]. Both the original CQDs (O-CQD) and Na^+–functionalized CQDs (NaCQD) were synthesized and reported that enhanced water flux and power density were found in the incorporation of Na^+–functionalized CQDs.

Forward osmosis (FO) is preferred due to its less energy consumption, low fouling and low cost. In FO, a semipermeable membrane is used in the water separation process to separate the dissolved solutes from water. Natural energy in the form of osmotic pressure is used in FO to transport water through the membrane where the dissolved solutes are retained on the other side.

The water flux (J_w) and reverse salt (J_s) flux were calculated using the following equations:

$$J_w = \frac{\Delta m}{A_m \Delta t}$$

$$J_s = \frac{\Delta(C_t V_t)}{J_w A_m \Delta t}$$

where Δm is the weight in grams change of the draw solution, A_m is the effective membrane area in square *cm*, C_t and V_t are the salt concentration in mol/L and feed solution volume at the end of experiment, respectively, and Δt is the measured time period in minutes.

One of the main drawbacks is membrane fouling due to osmotic driving force that may even lead to some toxic effects in FO membrane [44]. Carbon quantum dots due to hydrophilicity and surface passivation properties reveal high osmotic pressure and hence allow high water flux.

CQDs are synthesized by hydrothermal method from tulsi leaves as a draw solute [13]. The osmotic pressure difference between the feed and the draw solution was used to calculate the Water flux through the FO membrane. Higher water flux and lower reverse solute flux were got with deionized water (DI) as feed and gave maximum water flux with synthetic wastewater.

A thin-film composite membrane was synthesized from oil palm biomass-based CQD which was derived from activated carbon inserted into polysulfone-selective layers, for forward osmosis. The membrane surface hydrophilicity and porosity in CQDs-PSF got significantly enhanced, which in turn improved membrane permeability. It had been found that the porosity reduces water resistance during transport. The following equation was used to determine the porosity of the membrane.

$$\text{Porosity} = \frac{W_w - W_d}{\rho A t}$$

where W_w is the mass of wet membrane in g, W_d is the mass of dry membrane in g, A is the active membrane surface area in cm^2, ρ is the density of pure water in g/cm^3, and t is the thickness of membrane measured in μ_m [47].

It is also understood that increasing hydrophilicity allows water, but not foulants. But disruption of foulants occurs. The polymer membranes are contaminated by microorganisms or bacteria since they are hydrophobic and hence cause fouling. Thus, incorporating CQDs on the polysulfone (CQDs-PSF) membrane is substituted in FO. Mostly this composite membrane prevented bacterial activity largely and the antifouling property enhanced water flux.

Carbon quantum dots (CQDs) are fabricated from citric acid and the polydopamine (PDA) layer transplanted on the surface of poly(ether sulfone) substrate (PES) membranes are immobilized onto through covalent bonding [110]. This CQD-PDA layer was found to inactivate E. coli and S. aureus, and results in anti-biofouling. The CQD surface is negatively charged. Hence, unreacted carboxyl groups which are also the negatively charged bacteria are electrostatically repelled. Further CQD due to the smaller size and with oxygen-containing functional groups could have damaged or killed the bacteria by oxidative stress. Thus, the CQD modified PRO membranes not only showed antifouling properties for osmotic power generation, but also indicated higher recovery after backwash. This is an essential requirement of the membrane particularly when natural water streams are used as the feed solution.

Biogas slurry is treated with reverse osmosis (RO). Biogas slurry is mainly made up small molecular-weight amino acids. Nanofiltration membranes were one option for RO, but there are two disadvantages. One is that since both the amino acids and nanofiltration containing negatively charged particles, RO becomes ineffective. The second disadvantage is that due to low molecular-weight amino acid, only a strong NF membrane must be used for higher retention. But this will greatly reduce water flux. Hence, a conventional TFC membrane with CQD layer between substrate is fabricated [100]. This is not only the best for separation of small molecules from biogas slurry but also enhances the water flux. Polyethersulfone (PES) ultrafiltration membranes were used as substrates. Then these substrates were immersed in an aqueous solution containing CQDs and 48.8 g deionized water for 10 min to activate the –COOH groups. The modified substrate was named as "CQDs/PES".

The intermediate layer of CQD increases the water transmission due to increased hydrophilicity.

Seawater desalination is mainly done by forward osmosis which is a large scale process. A suitable draw solute is essential for a high-performance. So, high osmotic pressure must be generated and low solute leakage is required. Some organic and inorganic draw solutes, like sodium chloride, ethanol, etc., have been considered but with drawbacks like reverse draw permeation, energy consumption and fouling. Thus, causes damage to membranes. Hence, Na+-functionalized carbon quantum dots was employed as draw solutes in FO for seawater desalination [20]. Being rich in ionic species Na_CQDs, their aqueous dispersion shows high osmotic pressure than that of seawater. Further, Na_CQDs demonstrate high water fluxwith very low reverse draw solute permeation. However, in the FO desalination process using Na_CQDs as draw solutes, the energy consumption is yet to be explored.

Antifouling and anti-microbial properties of CQDs' are the essential properties for the production of a multi-functional composite material. Due to its solubility in polar solvents like water, it is even more viable for membranes applications. Many membrane modifications with CQDs were done for all purposes like reverse osmosis, forward osmosis, seawater desalination, with enhanced water flux and low energy consumption.

CQDs for membrane applications are mostly synthesized by heating of citric acid by various methods like solvothermal, hydrothermal and microwave-assisted pyrolysis. This technique is simple, and the yield is high. The size and functional groups on CQD surfaces are generally controlled by pyrolysis temperature and time.

Membranes by CQDs are developed as thin-film nanocomposite (TFN) membranes or as CQD/polymer composite membranes or as Membranes with CQDs on top of substrates. TFN membranes modified with CQDs were found to have high hydrophilicity, larger permeability, selective solute retention and good anti-fouling property.

CQDs were found to disperse well in polymer matrix to form CQD/polymer composite membranes with high stability and strength. These membranes are formed through chemical bonds between CQDs and host polymer matrix. So, there is uniformity in properties throughout the membrane. They also exhibited high hydrophilicity and larger permeability.

Membranes with CQDs on top of substrates are obtained when CQDs are coated on various membrane surfaces. Even in these membranes, covalent bond exists between CQDs oxygen groups and amine groups at the membrane surfaces [111]. Enhanced antifouling resistance was noticed. Due to the fluorescence property of CQDs, the surface effects are prominent. Especially the interactions between the membranes and foulants were studied to enhance the antibacterial properties of the membranes. One such revelation was the electrostatic repulsion between the bacteria and membrane surface as both of them are negatively charged.

CQDS modified membranes were also found to have greater resistance for chlorine. This requirement is essential since in polymer composite membranes, the polymer layers like polyamide layer, are generally degraded by chlorine. Thereby, it increases the lifetime of such membranes.

The interesting properties of CQDs interaction with foulants, not only make it an antibacterial but also as a sensor. Although there were many advantages in membranes modified by CQDs like water recovery, reuse, desalination, clean energy production etc., there are also some drawbacks. First drawback is to understand the properties and features of CQDs according to their synthesis. A systematic study on CQDs on synthesis, size, shape, and surface dispersion and property may make an efficient agent for membrane applications. Further, CQDs being small and water-soluble, the thorough study becomes essential before it is made available for commercial use although they are found to be biocompatible and nontoxic.

A research group [12], reported the construction of a composite membrane which also possess self-cleaning ability. CQD membrane is reported to have high permeability and selectivity. A Poly(ether sulfone) (PES) UF substrate is precoated with polydopamine (PDA). This membrane is introduced into CQDs solution to form selective layer to construct "PDA–CQDs–TFC" membrane. The purpose of the PDA–CQDs layer is to prevent the solute penetration and reduces concentration polarization (CP) issues. Thickness of 30 nm and smooth surface of PDA–CQDs–TFC membrane are the reasons for it to be great on mitigation of concentration polarization. It is found that the salt rejection of divalent ions and the water permeability are higher for PDA–CQDs–TFC membrane. The structure of the membrane with large voids is the cause of large permeability and salt rejection. The organic molecules that are adsorbed onto the membrane surface are then degraded by the efficient photocatalytic effect of CQD layer. MB and orange II are the dyes that were degraded.

Thin-film nanocomposite (TFN) membranes with three functionalized CQDs namely, carboxylic CQD (CCQD), amino CQD (NCQD) and sulfonated CQD (SCQD) were synthesized by some researchers [83]. CCQD was prepared directly by pyrolyzing citric acid (CA), NCQD by low-temperature pyrolysis of CA in the presence of branched polyethyleneimine (BPEI) and SCQD by pyrolysis method using CA and Poly(sodium 4-styrene sulfonate) as precursors and TFN-CCQD, TFN-NCQD and TFN-SCQD, are the respective TFN membranes formed by interfacial polymerization. Due to their loose structures, water fluxes increased linearly with the applied pressure. In separation performance, when tested with Na_2SO_4, $MgSO_4$, $MgCl_2$, NaCl salts, the sequence of rejection was $Na_2SO_4 > MgSO_4 > MgCl_2 > NaCl$ showing the membranes are negatively charged polyamide layer. The antifouling properties were tested with bovine serum albumin (BSA) which was found to have improved due to hydrophilicity from CQD as the adsorption of hydrophobic BSA by the membranes were reduced. The researchers also reported that better retention of divalent cations by the membranes is due to the adhesion strength between amino groups and polyamide matrix.

CQDs were produced from citric acid by hydrothermal method. Sodium-ion modified carbon quantum dot (Na-CQD) is prepared by alkalization process. TFN membranes with varying Na-CQD loading were analyzed for the nanofiltration process for water purification. The hydrophilicity of polyamide (PA) layers was found to be enhanced due to Na-CQDs in the membrane. As the Na-CQD loading was increased, salt rejections was observed to increase, due to pore size being smaller and pore size distribution. The modified TFN membranes, Na-CQD, was then experimented for simultaneous removal of Se and As in single ion and mixed ion solutions. Se rejection was found higher in mixed solution which was suggested due to pH value. But As rejections are found almost same for both single and mixed ion solutions and were also suggested due to the close pH values. Na-CQD loading of 0.05 wt.% membrane was reported [102] to be superior with high pure water permeability and salt rejections. It had been concluded that by the addition of Na-CQDs the polyethersulfone polymer membrane showed better fouling resistance against proteins. Because of these improved anti-fouling properties, this Na-CQD membrane is effective in water remediation.

CQD synthesized from lemon juice by hydrothermal treatment. It was embedded with Ag nanoparticles to produce silver/carbon quantum dot (Ag/CQD) membrane. The structure of the membrane consisted of dense top layer and porous sublayer. Of the varying wt.% of Ag/CQD, 0.5 wt.% of Ag/CQD was reported to have high permeability which is attributed due to the hydrogen bonding of composite CQD and water which increased the water flow through the membrane. In the analysis of tartrazine dye, removal 0.7 wt.% of Ag/CQD had the highest rejection rate which is due to the electrostatic repulsive force between the membrane and foulants. The irreversible fouling and reversible fouling were seen to be high for no or low loading of Ag/CQD. It was reported [33] that due to low electronegativity, the tartrazine foulants were not repelled and hence easily entered into the membrane pores. Thus, the size of the pores and the electronegativity are important in both permeability and antifouling of membranes. The nanomembrane applications associated with CQDs are listed in Table 1.

Table 1 CQD/composite in nanomembrane applications

	CQD/composite	Enhanced/detected/degraded	References
Membrane	Na-CQD	Selenium and arsenic	Rani et al. [89]
	NaCQD	Flux and power density	Gai et al. [15]
	CQD	High osmotic pressure	Doshi and Mungray [13]
	CQDs-PSF	Bacterial activity	Mahat et al. [47]
	CQD-PDA	E. coli and S. aureus	Zhao et al. [110]
	CQDs/PES	Biogas slurry	Yang et al. [100]
	NaCQD	FO	Guo et al. [20]
	PDA–CQDs–TFC	MB and orange II	Shao et al. [12]
	TFN-NCQD and TFN-SCQD	Na_2SO_4, $MgSO_4$, $MgCl_2$, NaCl salts	Sun and Wu [83]
	Na-CQDs	Se and As	Hea et al. [102]
	Ag/CQD	High permeability	Gan et al. [33]

4.2 Carbon Quantum Dots as Sensors

Tunable fluorescence property of CQDs and the simple method to modify the surface state of CQDs for selectivity have made them very attractive as sensors. Being biocompatible and nontoxic are added advantages. In a mixture of different ions or chemical species, the specific ions or chemical compounds could be detected by them. The size of the CQDs plays a role in the photoluminescent properties. They give UV, visible and near infra-red light emission for small (1.2 nm), medium (~3 nm) and large (~3.8 nm) size, respectively [21].

Many mechanisms make CQDs possible sensors. 1. The excitation or deexcitation of electron between the CQD and another species. 2. In resonance energy transfer (RET) mechanism energy is transferred between two light-sensitive molecules. The energy is transferred from a donor initially in its excited state to an acceptor, where CQD may be either a donor or an acceptor. 3. In photoinduced electron transfer (PET) is an electron from an excited state is transferred from a donor to an acceptor. Thus, a charge separation is generated or may be considered as a redox reaction taking place in excited state as denoted in Fig. 5.

In inner filter effect (IFE) an absorber in the detection system absobs the excitation and/or emission light. Also, the observed fluorescence intensity is proportional to the intensity of the exciting light. This inner filter effect is used for sensing applications by the fluorescent carbon dots.

It is also been reported that fluorescence emission of the CQDs is being caused by surface by the incorporated material with specific functional groups.

CQDs are directly used as sensors through their fluorescent signals. They can also be coupled with some receptors to reflect the signals. CQDs may also be integrated with other similar sensor materials [95].

Polystyrene sulfonate-coated CQDs (PSS-CQDs) were used as hydrogeological sensor where surface functionalization and fluorescence emission of the CQDs are the parameters used to study the flow of water.

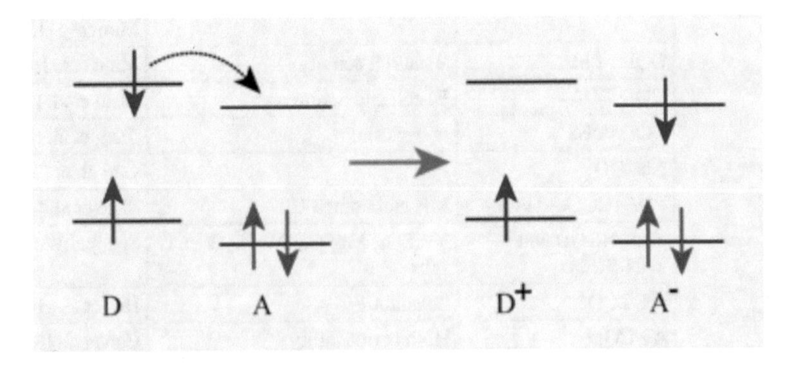

Fig. 5 Photoinduced electron transfer

PSS-CQDs showed excellent fluorescence property under various physical conditions like temperature, pressure, etc. It was tested in real humic situations as tracers and found to be detected easily without any background effects. PSS-CQDs also revealed less stickiness in different types of soil and sand samples when tested in real ground experiments.

The velocity of laminar groundwater flow is proportional to the ground slope. The hydraulic gradient can be calculated between two points using Darcy's law as follows.

$$Vn = kj$$

where n is effective porosity of an aquifer, Vis pore velocity of groundwater, k is hydraulic conductivity of an aquifer and j is hydraulic gradient.

The borehole tracer method is employed, where the groundwater velocity is then found from the filtration velocity of the groundwater and concentration. This procedure is used in groundwater velocity analysis.

Finally, it was found that PSS-CQDs were promising hydrogeological tracers for groundwater studies [86]. CQDs are found to have excellent selectivity toward the target. For example, N-doped CQDs are appropriate sensors for Fe^{3+} ions detection. CQDs changed the colors of the solution to indicate their presence. It is attributed due to the quenching of the fluorescence intensity. Also, it was able to selectively sense Fe^{3+} ions although there were about 15 metal ions present in the solution. Similarly, nitrogen and sulfur-doped CQDs was able to detect Hg^{2+} in lake water and tap water [46].

CQDs may be tuned with different functional groups to be used as sensors selectively. Even the synthesis of the CQDs plays a role in the selectivity or sensing performance. Synthesis of CQDs from human hair was used for detection of chloroform ($CHCl_3$) in water. CQDs were produced by two different methods, using microwave and by conventional thermal treatment method [22]. The presence of chloroform was indicated by enhanced fluorescence. CQDs by the conventional method showed high selectivity to chloroform. In chlorine-treated drinking water, the nitrogen–oxygen functional group on CQDS was found to be an effective sensor of chlorine.

The nitrogen-doped CQDs have been synthesized from lotus root by microwave treatment method [11]. This LR-CQDs were found to be an active and efficient selective fluorescent sensor for Hg^{2+}. This was tested for the detection of Hg^{2+} in tap water and found to be equally effective as it was tested for biomedical applications. Thus, it can be considered as a possible sensor of Hg^{2+} in environmental samples also.

CQDs were synthesized from ginkgo leaves as precursors by hydrothermal method. Various water samples with different Hg^{2+} concentrations were analyzed [110]. The hydrophilic-OH, $-COOH$ and $-NH_2$ functional groups in CQDs were responsible for its uniform dissolution in aqueous solution which was essential for Hg^{2+} detection. CQDs photoluminescence intensity was tested in the presence of several metal ions. And it was found that there was a clear decrease in the photoluminescence intensity showing that CQD is an effective and selective sensor for Hg^{2+}.

Amine-capped CQDs were produced and used for the detection of picric acid (PA) in aqueous solution through fluorescent intensity [58]. It was found to be very effective and highly selective. This was evident when common reagents like ethanol, phenol, acetone, etc., showed little effect of fluorescence emission. The selectivity is found because of the electrostatic interaction between PA in water and amino groups on the CQDs surface. The fluorescent intensity decreased when the PA concentration was increased. When the fluorescence quenching experiment was performed in water and ethanol as a function of PA concentration, fluorescent quenching was more pronounced for water. It was suggested that polarity of the water being higher, there was a strong electrostatic interaction between PA and CQDs thereby increasing the fluorescence quenching in water.

Researchers [38] have used doped CQDs for the detection of picric acid (PA). Doped CQDs were an effective fluorescent sensor. It was found to be a highly selective probe by Fluorescence resonance energy transfer mechanism. Nitrogen and sulfur co-doped CQDs (NS-CQDs) were prepared by hydrothermal method. From the water sample collected from tap water, the concentration of PA, was measured. Thus, NS-CQDs as a sensor, was sensitive, selective and also fast in detection of PA in water. Fluorescence intensity was directly related to the concentrations of PA.

Fluorescence quenching is one of the efficient methods used for the detection of Hypochlorite ions (ClO^-). Recently researchers [32] synthesized nitrogen-fluorine-co-doped carbon quantum dots (NFCDs) by one-pot hydrothermal method. Strong green fluorescence light was emitted by NFCDs under UV light. It was found to have good stability in water. As soon as ClO^- was added, the fluorescence intensity was quenched immediately and had no effect on pH factor of solution. When examined in the presence of many other ions, fluorescence quenching was absent for other elements. This is evident for the perfect selectivity, without any practical interference due to other elements. It was revealed to be an ideal probe for ClO^- detection. The wide range of detection and low limit of detection proved it an excellent probe for the ClO^- detection in all water samples for essential utilities.

Polyamine functionalized CQD by capping with branched poly- (ethylenimine) (BPEI) with excellent fluorescence activity was used to sense free copper ions [103]. Amino groups at the surface of the BPEI-CQDs react with Cu^{2+} to form cupric amine and this may attribute to the fluorescence quenching of CQD. It is found to be very sensitive and selective among many other ions to detect Cu^{2+} especially in river water. Further its wide range of response makes it a favorable probe for Cu^{2+} detection in environmental water samples.

Nitrogen-doped carbon quantum dots (NCQDs) from tartaric acid and L-arginine as the precursors was prepared by solvothermal method and was used to detect Hg^{2+} or Fe^{3+} from tap water and river water [113]. The fluorescence intensities of Hg^{2+} or Fe^{3+} at different concentrations in a buffer solution were recorded at room temperature. NCQDs showed different colors in solution and solid state. pH range was optimized from 3 to 7 for the detection Hg^{2+} and Fe^{3+}. When Hg^{2+} and Fe^{3+}

coexisted, addition of thiourea was necessary for their detection. It was suggested that it may be due to stronger binding affinity of Hg^{2+} to thiourea than to NCQDs. The recoveries of Hg^{2+} and Fe^{3+} were reported to be in the range of 86.50–115.05% with relative standard deviation less than 5.99%. The results obtained on the detection of Hg^{2+} and Fe^{3+} were done for the tap water and river water. It proved to be an apt probe for its detection.

N and S co-doped carbon quantum dots (N, S-CQDs) was synthesized by hydrothermal method using L-cysteine as the single precursor. The N, S-CQDs proved to be a good sensor for detection of Hg^{2+} ions in DI water and a real water sample [92]. In the selectivity analysis of this detection system for Hg^{2+} ions with different concentrations, no remarkable decrease is observed by the addition of other metal ions into the N, S-CQDs solution. Hence, good selectivity of Hg^{2+} ions by this system was confirmed. Photoluminescent intensity was found to decrease when the concentration of Hg^{2+} ions was increased in the real lake water sample. This was done with various minerals and organics coexisting in the lake water. Another group [43] prepared similar Nitrogen and sulfur co-doped CQDs (N,S/C-dots)from of L-cysteine and citric acid, from hair fiber, gentamycin sulfate and rice as precursors for the detection of Hg^{2+}. At Hg^{2+} concentration of chelating agent 300 μM, the fluorescence intensity was completely quenched. Detection of Hg^{2+} was analyzed in the lake water and tap water. It was reported to have showed good detection and recovery. Electron-transfer rate and coordination interaction between N,S/C-dots and Hg^{2+} were responsible for quenching which detected the presence of Hg^{2+}. Thus, N,S/C-dots were reported to have high sensitivity and ion selectivity as a sensor.

CQDs were synthesized from glycerin, ethylene glycol and cellulose microcrystalline as precursors by hydrothermal process [3]. The fluorescent CQDs are then used to determine contaminants such as carboxylated multi-walled carbon nanotubes (c-MWCNTs), single-walled carbon nanotubes (SWCNTs), humic acids (HA) and graphene-based products. The preparation of CQD affects the fluorescent features. pH dependence on the fluorescence of CQD was also investigated. The researchers reported that CDs synthesized from cellulose microcrystalline was suitable in detecting contaminants in river water. Further hydrogen bonding interactions were responsible in detecting the contaminants. The poor dependance on pH made it possible to sense over a wide range of pH value. It was found that no appreciable change in quenching due to high concentration of NaCl. This may be attributed due to the weak interaction between ions of NaCl and carboxylic groups of MWCNTs. Thus, CQDs are found to be an effective sensor of c-MWCNTs in water samples.

One-pot solvothermal method was employed using BBr_3 as the boron source and hydroquinone as the precursor and Fluorescent B-doped carbon quantum dots (BCQDs) was synthesized. It was reported [97] that it was used as a sensor to detect hydrogen peroxide (H_2O_2) and glucose. The detection of H_2O_2 was indicated by the complete quenching of fluorescence emission of BCQD. It is due to the electron transfer between the boron from BCQD and H_2O_2. The electrons were given by H_2O_2 to boron to form stable B–O coordination bonds. This was responsible for fluorescence quenching. The fluorescence intensity was found to decrease by the increase of H_2O_2 concentration. It was reported that glucose oxidase (GOx) is the catalyst for the oxidation of glucose to form gluconic acid and H_2O_2. Thus, the glucose concentration in the presence of GOx was found to quench the BCQD fluorescence thereby indicating its presence.

One research group [22] synthesized CQDs from cabbage by hydrothermal method. Such CQDs were found to have –OH and –COOH groups on its surface. This photo fluorescent CQDs with these functional groups were used as sensors to detect Pb^{2+}, Hg^{2+}, Cd^{2+}, $Cr^{3+}$and Fe^{3+} metal ions. A mixture of the CQDs and heavy metal ion solutions prepared were explored for fluorescent sensing from aqueous solution. The fluorescence intensity measurements showed that the intensity was completely quenched in case of Fe^{3+} ions. For Pb^{2+} and Hg^{2+} there was significant decrease in the intensity. However, Cd^{2+} and Cr^{3+} showed negligible quenching. The group reported that further studies must be done on CQD as sensor for food, water and health.

CQDs were synthesized from 1,3-phenylenediamineand citric acid as precursors by one-step ultrasonic vibrating method with m-phenylenediamine and citric acid as raw materials. $Cr_2O_7^{2-}$ of various concentrations with CQDs solution were prepared and their fluorescence spectra was studied for sensing dichromate ions. River water and drinking water were directly used for the study. pH value was adjusted to 5 as optimum value. The result displayed a linear correlation between the quenching of intensity to the concentration of $Cr_2O_7^{2-}$. The quenching effect of CQDs on $Cr_2O7_2^-$ ion was distinctly displayed. The test was done in the presence of $Cr_2O_7^{2-}$, MnO_4^-, NO^{2-}, F^-, Cl^-, Br^-, I^-, S^{2-}, SO_4^{2-}, $S_2O_3^{2-}$, $S_2O_8^{2-}$, $(SCN)^-$, CO_3^{2-}, PO_4^{3-}, Fe^{2+}, Fe^{3+}, Cu^{2+}, K^+, Ca^{2+}, Na^+, Mg^{2+}, Mn^{2+}, Zn^{2+}, Al^{3+} and H_2O_2. The results displayed that the quenching effect of CDs on $Cr_2O_7^{2-}$ ion was highly selective. The recoveries ranged from 96 to 98% in drinking water. The research group [18] stated ion-selective recognition in real samples is convenient, response fast and detected accurately. Hence, a nanoplatform for sensing dichromate ion with high accuracy and reliability was claimed by them.

CQDs were synthesized from whey as precursor for detecting selenite, (Se(IV)) in water by functionalized CQD (fGCQD) [63]. It is found that –OH, –COOH are the functional groups on the surface. The detection was done by analyzing the photoluminescence intensity. It is found that the presence of Se (IV) had considerably quenched the fluorescent intensity. It is due to the interaction between amine groups and Se (IV) which reduced almost all the vacancies and produce no radiation by the recombination of the charge carriers. The detection meets the requirement of routine selenium analysis in potable water. In the selectivity analysis, the detection was done in the presence of Ni^{2+}, Se^{6+}, Cl^-, Cu^{2+}, As^{3+}, As^{5+}, Pb^{2+}, Br^-, NO^{3-}, NO^{2-} and F^- ions. 70% quenching was found due to selenium, showing reasonable selectivity although slight quenching is found due to Pb^{2+}, may be possible by the complex formation of H–Pb^{2+}. Thus, fGCQDs probes have high sensitivity and selectivity of selenium against real water samples collected from lake, ground and well.

Carbon quantum dots were synthesized citric acid one-pot hydrothermal method. Then using urea as nitrogen source nitrogen-doped carbon dots (N-CDs) were prepared for detection of Hg^{2+} from water. The N-CDs concentration was fixed as 8 μg mL-1in this detection experiment. A strong PL peak at 440 nm was exhibited by N-CDs alone. Addition of 40 mM Hg^{2+}, drastically quenched the fluorescence indicating the presence of Hg^{2+}. This is attributed due to the interaction between Hg^{2+} with the carboxyl or hydroxyl group of N-CDs. Non-radiative recombination by the photo charge carriers was responsible for the fluorescence quenching. The detection was reported [109] to be highly sensitive and selective. It was investigated in both tap and mineral water samples. It was found linear over a range of 0–50 nM. Further, the limit of detection was 2.88 nM for tap water and 2.87 nM for mineral water.

A review on CQDs as an optical sensor for detecting heavy metals using the optical characteristics of CQDs was reported. The various mechanisms attributed for the investigations were described below.

- Fluorescence quenching, inner filter effects (IFEs)—Here the non-radiative recombination of the photo-excited charge carriers cause quenching of fluorescence emission. Cu (II) and Fe (III) ions are detected using this mechanism.
- Inner filter effects (IFEs)—here an emission is caused when CQDs are combined with IFE, by the absorption spectra from an absorbent. Determination of Cr(VI) is done by this mechanism. Fluorescence enhancement is also caused by binding an analyte to CQDs. The example of this method is the detection of Ag+ions. When Ag+ions are added to CQDs, an enhancement in fluorescence occurs and this phenomenon is used for detection.
- Photo induced electron transfer (PET)—here the photoelectron excites the analyte to form a complex with the receptor. This excited electron interferes with the excited CQD. Hence, there will be a change in its phosphorescent signal. The metal detection is done on this basis like in the detection of Cu^{2+}.
- Phosphorescence, ratiometric dual emission, and fluorescence resonance energy transfer (FRET)—is a tool used for determining the distance between two fluorophores. It is a process of non-radiative transfer of energy. The energy is transferred from an excited, donor, molecular fluorophore to another acceptor, fluorophore. Hence, there will be quenching of the donor and excitation of the acceptor unit which may either produce a radiative or non-radiative photon. CQD-labeled oligodeoxyribonucleotide and graphene oxide are used for detection of Hg (II) in water.
- Activation/enhancement, surface-enhanced Raman scattering (SERS)—This mechanism allows detection of the molecules by the adsorption of metals on rough metallicsurfaces. By passivation on CQDs, some inorganic pollutants may be detected.
- Mercury, arsenic, selenium, iron, copper, lead, cadmium, zinc and chromium are some of the metals reported [64]. These are detected using CQDs, modified CQDs and composite CQDs from water by following any one of the mechanisms explained above.

CQDs were synthesized by hydrothermal process from orange peel, paulownia leaves, ginkgo biloba leaves, and magnolia flower and denoted as OP-CQDs, PL-CQDs, GB-CQDs, and MF-CQDs, respectively. It is used for Fe^{3+} ions detection in pond water. When applied to detect Fe^{3+} ions, fluorescence quenching is recorded in PL spectrum. It is reported [89] that the interaction of amino groups, carboxyl groups and hydroxyl groups on CQDs surface was responsible for the detection. The aggregation of CQDs was caused due to strong attraction/ interaction between Fe^{3+} ions, NH_2, $COOH^-$ and OH^- functional groups of CQDs. This prompted the non-radiative recombination of photo charge carriers caused by the fluorescence quenching. The selective determination of Fe^{3+} was effective in the presence of Ag^+, K^+, Na^+, Pb^{2+}, Fe^{3+}, Fe^{2+}, Al^{3+}, Cd^{2+}, Zn^{2+}, Cu^{2+}, Mg^{2+}, Ca^{2+}, Cr^{3+}, Ba^{2+} and Hg^{2+} ions. Thus, a highly sensitive and selective sensor for Fe^{3+} was successfully applied with satisfactory recoveries of 94%–108%.

Boron nitrogen co-doped carbon quantum dots (B, N-CDs) were synthesized by hydrothermal method for the fluorescence detection of Cr(VI) in river water samples by UV–Visible absorption spectrum [105]. The optimized concentration of B, N-CDs for maximum response is (0.01–0.10 mg/L). The pH and contact time are (5–11) and (0–10 min), respectively. The selectivity was excellent when the analysis was performed with Al^{3+}, Ca^{2+}, Cr^{3+}, Cu^{2+}, Fe^{2+}, Fe^{3+}, Hg^{2+}, K^+, Mg^{2+}, Mn^{2+}, Pb^{2+} and Zn^{2+}ions. The static quenching mechanism and IFE were the mechanisms reported for fluorescence quenching in the detection.

Nitrogen-doped carbon quantum dots (N-CQDs) were synthesized by hydrothermal method using citric acid and nitrogen from ethylenediamine (EDA) as source for the detection of nitrite from tap water. In the optimized synthesis of N-CQDs, 3 g of citric acid and 3 ml of EDA were used. The optimum value of pH is chosen as 2. In the analysis of the fluorescence spectra of N-CQDs it was reported [14] to have quenched by nitrite in 15 min. Further, it was selectively quenched when analyzed in the presence of Na^+, Co^{2+}, Ba^{2+}, Ni^{2+}, Hg^{2+}, Cr^{3+}, Pb^{2+}, Cu^{2+}, Zn^{2+}, Fe^{3+}, F^-, Cl^-, Br^-, I^-, $PO_4{}^{3-}$, $HPO_4{}^{2-}$, $H_2PO_4{}^-$, $SO_3{}^{2-}$, $CO_3{}^{2-}$ and $NO_3{}^-$ ions. The reaction of amide group of CQDs with nitrous acid formed N-nitroso compounds, which was stated to be responsible of fluorescence quenching.

Carbon quantum dots with Nitrogen-doped (N-CQDs) was synthesized by hydrothermal method, carbon quantum dots from orange peel (ON-CQDs) and carbon quantum dots from watermelon peel (WN-CQDs) and used for the detection of oxytetracycline (OTC) with different concentrations in tap water and lake water [17]. In the fluorescence analysis, fluorescence quenching responses observed were 0.973 μmol L − 1 for orange peel carbon quantum dots (ON-CQDs) and 0.077 μmol L − 1 for watermelon peel carbon quantum dots (WN-CQDs)) under the optimal reaction conditions. Both the fluorescent probes were found to be both sensitive and selective detectors of OTC.

CQDs were synthesized using various concentrations of citric acid and ammonium dihydrogen phosphate. An optimum CQD was synthesized by two-step microwave method and used for polarization fluorescence analysis (FPA) for copper (2+) cations detection in water samples. For comparison, the fluorescence quenching was analyzed with FPA quenching. The fluorescence polarization versus angular displacement is measured as it is a process of absorption to emission of photon. Fluorescence polarization shows particle size changes due to absorption of copper at CQDs surface, at constant temperature and viscosity. The FPA measurement of CQDs was stated to be effective when compared with fluorescence quenching to determine

Table 2 CQD application as nanosensor

	CQD/composite	Enhanced/detected/degraded	Reference
Sensor	PSS-CQDs	Hydrogeological tracers	Warsi et al. [86]
	Sulfur-doped CQDs, N-CQDs	Fe^{3+}, Hg^{2+}	Molaei [45]
	CQDs	Chloroform	Singh et al. [22]
	Nitrogen-doped CQDs	Hg^{2+}	Gu et al. [11]
	Amine-capped CQDs	Picric acid	Niu et al. [58]
	NS-CQDs	Picric acid	Khan et al. [38]
	NFCDs	ClO^-	Guo et al. [20]
	BPEI-CQDs	Cu^{2+}	Dong et al. [103]
	NCQDs	Fe^{3+}, Hg^{2+}	Zhu et al. [113]
	N, S-CQDs	Hg^{2+}	Wei et al. [92]
	CQDs	c-MWCNTs	Cayuela et al. [3]
	BCQDs	H_2O_2, Glucose	Shan et al. [99]
	CQDs	Fe^{3+}, Hg^{2+}, Pb^{2+}	Singh et al. [22]
	N, S/C-dots	Hg^{2+}	Li et al. [43]
	CQDs	$Cr_2O_7{}^{2-}$	Qiaoa et al. [18]
	fCQDs	Selenium	Devi et al. [64]
	OP-CQDs, GB-CQDs, PL-CQDs and MF-CQDs	Fe^{3+}	Wang et al. [89]
	B, N-CDs	Cr(VI)	Wang et al. [105]
	N-CQDs	Nitrite	Feng et al. [14]
	ON-CQDs, WN-CQDs	OTC	Gao et al. [17]
	CQDs	Cu^{2+}	Yakusheva et al. [2]

copper cations in water. Thus, polarization fluorescence analysis (FPA) is considered yet another method for various applications [2]. The nanosensor applications associated with CQDs are listed in Table 2.

4.3 Carbon Quantum Dots for Removal of Pollutants

Photocatalysis is a favorable method for water remediation. Both solar and UV light are used for pollutant degradation. It is an oxidation process, a sustainable and an environmentally friendly technique for water purification [93].

CQDs exhibits photoluminescence. It is an excellent fluorescent material and effective photocatalyst upon irradiation by UV light. CQDs are also sensitive toward visible light that results in the enhancement of charge carriers and photocatalytic performance. CQDs have also been shown to undergo electron transfer in different

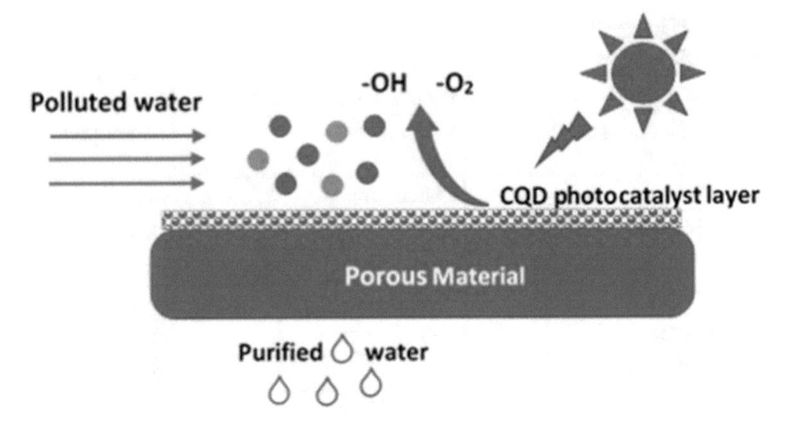

Fig. 6 General mechanism of CQD as photocatalyst

situations. It is capable of reducing the recombination of photogenerated charges. All these factors make it suitable for wastewater treatment in organic and inorganic pollutants removal from water. The general mechanism of photocatalyst is indicated in Fig. 6.

A composite made from CQDs and NH_2-MIL-125, a kind of metal-organic framework, (CQDs/NH_2-MIL-125), photocatalyst is used for almost 100% degradation of Rhodamine B (RhB) by using UV-Vis irradiation. RhB degradation under visible light irradiation was done by CQDs/$La_2Ti_2O_7$ composite as photo-catalyst. Degradation of methylene blue (MB) by the photocatalyst formed by CQDs on the zirconia surface (ZrO_2) under a UV irradiation has been reported. Similarly, degradation of 93% acid violet 43 and p-nitrophenol by the photocatalyst CQDs onto the surface of pyrogenic nanoparticles (P25) was reported. CQDs/TiO_2 composite nanofibers is used for photocatalytic degradation of methylene blue (MB) dyes.

CQDs are also used as a fluorescent probe for sensing Hg (II) and Cr^{6+} ions in water samples. Nitrogen and phosphorus-doped CQDs are used for the detection Fe^{3+}ions. Fluorescence quenching of Fe^{3+} by nitrogen-doped-CQDs (NCQDs) is also reported.

CQDs can be used for the removal of removing organic and inorganic pollutants Cd^{2+} and Pb^{2+} ions from wastewater by absorption treatment. N-CQDs were successfully incorporated for this treatment. Another adsorbent, polyethyleneimine-functionalized CQDs onto the magnetic materials ($MnFe_2O_4$) to produce a nanocomposite (PECQDs/$MnFe_2O_4$) is applied for the removal of uranium. Another research group [89] has used three types of iron oxide adsorbents (Fe_3O_4, C11-Fe_3O_4 and CQDs/C11-Fe_3O_4) and reported the removal of benzo[a]pyrene (BaP) in wastewater samples. The NCQDs with iron oxide (Fe_3O_4) are used as an adsorbent for removal of lead. The composite of CQDs with zinc-aluminium layered double hydroxide (CQDs/ZnAl-LDH) is used for the adsorption of cadmium.

CQDs are also recognized as a good disinfection of wastewater besides waste water monitoring. CQDs are also used for decontamination of Cd (II) and removal of benzopyrene from water because of its photocatalytic property, dispersed well in water and good antimicrobial property. The antibacterial activity of CQDs has been reported for the degradation of E. coli and S. aureus. The analysis of antibacterial property of CQDs-TiO_2 is also reported. NCQDs are found to develop antimicrobial activity against Staphylococcus. NCQDs are first found to interact with the bacteria, Staphylococcus. Then they are destroyed by UV light irradiation.

Sunlight being sustainable and a renewable source of energy, semiconductor photocatalysis was an attractive process for water remediation [8]. But there were some modifications essential for this simple process. One such requirement was that the photo energy must be greater than the bandgap energy. This was generally not feasible in visible light as the principle of photocatalysis is the excitation of electron-hole through the bandgap by light energy. So, as a requirement, their electronic and optical properties were initially modified by nanomaterials. Their properties due to their size and structure, were more suitable for water purification.

The invent of CQDs made it more suitable material as a photocatalyst for water purification. Increase in the size of CQDs decreases the bandgap and hence results in a change in its luminescence. It may further be changed when the CQDs are modified with other functional groups like hydroxyl, etc. CQDs are noted for their wide tunable photoluminescence effect, surface effect, quantum confinement effect (QCE), and edge effect. A good absorbance in UV region is noticed in CQDs due to π-π^* transition of C=C bonds and n-π^* transition of C=O bonds which may be enhanced to UV–visible region by additional functional groups.

Up-conversion photoluminescence is a process in which emission of light by sequential absorption of two or more photons. The emitted light is of shorter wavelength than that of the excitation wavelength. It is an anti-Stokes type emission. This optical property is also noticed in CQDs.

All these properties are the requirement of photocatalytic reactions, like transport of charge carriers and surface redox reactions. CQDs is capable photocatalyst for degradation of pollutants in water remediation [74]. In photocatalysis, the pollutant molecules are broken down to harmless substances due to photo-oxidative reactions. Hence, efficient, visible light photocatalyst was necessary for degrading the pollutants in water. Although electrons are excited easily in narrow bandgaps by visible light, recombination among the charge carriers also take place at a faster rate. Therefore, UV light was preferred, but it also demanded more protection in the case of water remediation.

All the above requirements were fulfilled by the fluorescent CQDs. They are amorphous, sp^3 hybridized and fluoresce due to quantum confinement. Thus, researchers focused CQDs modified semiconductor as catalysts. Further advantages like photostability and recyclability were also found in CQDs modified photocatalysts. CQDs are very effective in photocatalysis process capable of optical absorbance in both UV and visible region. Also, it is excellent in the case of both narrow and wide bandgaps. CQDs are also active in up-conversion photoluminescence. It is good in surface effects with abundant surface functional groups.

The interaction between the π-π^* transition of CQD and the aromatic ring of organic pollutants are responsible for the abundant absorption of organic pollutants. The enhanced adsorption of metal ions is due to their interaction with the functional groups on the CQDs surface.

The photocatalytic activity of many wide bandgap (Eg $\geq$ 3 eV) semiconductors modified with CQDs has been improved by reducing the recombination rate or increasing the charge separations. For example, CQDs modified TiO_2 photocatalyst to emit charge carriers from longer wavelength light, which are reemitted as shorter wavelength light by CQD was studied. CQDs/TiO_2 photocatalyst was reported for the degradation of pollutant phenol under UV light irradiation due to π–π interaction of CQDs and Phenol. CQDs modified ZnO photocatalysts were known for enhanced transfer of e^--h^+ pairs due to increased separation of photogenerated charge. N-doped ZnO-CQD photocatalyst composites were found to degrade malachite green (MG), methylene blue (MB) and fluorescein dyes under daylight irradiation.

The wide range capability of CQDs as catalyst improved the charge efficiency, reduced photocorrosion in waste water remediation. CQDs modified ZnS photocatalyst degrade dyes. Degradation of Tetracycline (TC), Bisphenol A (BPA), and Rhodamine B (Rh B) under visible light was observed by CQDs modified BiOBr. CQDs modified $KNbO_3$ photocatalyst removed crystal violet dye under the illumination of visible light.

The CQDs modified photocatalyst reduces the recombination of photogenerated electrons and holes narrow band gap (Eg < 3) photocatalysts and thereby improve photocatalytic activity. For example, CQDs modified CdS are reported for degradation of dye Rhodamine B (Rh B) under visible light.

CQDs were found to have modified heterojunction photocatalysts. For example, CQDs modified Bi_2MoO_6 photocatalyst are reported for degradation of Rhodamine B (RhB) and Methylene Blue (MB) below visible light irradiation [76]. CQDs/BiOCl/BiOBr reported for degradation of rhodamine B (RhB), tetracycline hydrochloride (TC), ciprofloxacin (CIP) and bisphenol A (BPA) under visible light irradiation.

A nitrogen-doped CQDs (N-CQD) from grass was synthesized with good absorbance of UV and visible radiation with 2.5 eV band gap. Its photocatalytic activity was then tested on Acid Blue, Acid Red, Methylene blue, Eriochrome Black T, Methyl orange and Eosin Ydyes with the degradation time observed were 30 min, 30 min, 90 min, immediately, immediately and 90 min, respectively. The synthesized N-CQD was found to decompose the dye molecules. It required more time to decompose dyes of higher concentration. It is attributed that more time is required to produce enough amount of reactive oxide species (ROS) for degradation [46]. Synthesized N-CQD can decompose under visible radiation because of a low band gap (1.94 eV) beside ROS produced from UV radiation to the water. From the study of surface adsorption ability of the N-CQD, it was observed that the heavy metal ions like Cd^{2+} and Pb^{2+} were removed from the water. Thus, surface adsorption activity of synthesized N-CQD may be adopted in removing organic and inorganic pollutants from the drinking water.

CQDs were synthesized from m-phenylelnediamine (MPDA) by hydrothermal heating method. A chitosan/carboxymethyl cellulose (CMC) hydrogel/CQD composite was prepared and used for water desalination and water remediation [72]. The hydrogel/CQD composite was made to absorb water through its porous structure. The water-swollen hydrogel/CQD composite were then irradiated by solar illumination for evaporation. Water is then recollected by condensation. This procedure is repeated for 50 cycles. Even after so many purification cycles, the C-dot/gel composite is found recyclable and stable. CQD/hydrogel systemis also used for removal of metal ions, detergents and organic pollutants from water. Cu^{2+}, Ni^{2+}, Ag^{+}, and Cd^{2+}, sodium dodecyl sulfate and rhodamine6G got absorbed by CQD/hydrogel composite due to the solar energy illumination. The amount of light absorbance depends on the concentration of CQD in CQD/hydrogel composite. This enhancement was more in the infrared region where heat energy is predominant. Thus, this inexpensive, biodegradable CQD/hydrogel composite contributes more in desalination and purification of water.

CQDs were synthesized from petroleum coke by hydrothermal method and doped with Chitosan to form Chitosan-CQDs (CH–CQD) membrane [39]. It was found to have good the adsorption capacity. The negatively charged CQDs interact electrostatically with positively charged chitosan. A uniform composite membrane is obtained. CQDs are doped with oxygen, nitrogen and sulfur. The removal efficiency of Cd^{2+} ions by CH–CQDs in the presence of Cd^{2+}, Zn^{2+} and Pb^{2+} ions showed it to be very effective in selectivity. Its extraction of Cd^{2+} was very fast under UV irradiation at $pH = 8$. Although several mechanisms are proposed, no confirmed reason was concluded.

CQDs modified $ZnSn(OH)_6$ (ZSH) composites were synthesized by hydrothermal method [65]. It was found to be formed as microspheres with uniform solid, hollow and yolk-shell structures. When photocatalytic activities of ZSH@CQDs composites were studied, degradation of organic pollutant, Rhodamine B(RhB) dye under visible light illumination was observed. When analyzed as photocatalytic disinfection, complete inactivation of Staphylococcus aureus was noticed. Up-conversion effect of CQDs is suggested for the enhanced photocatalytic activity of the composite. Also, the increase in electrons for capturing photogenerated electrons and reduce the recombination rate of charge carriers were the reasons indicated by the authors.

Fe_3O_4/N-Carbon quantum dots were synthesized by chemical method. It was used for adsorption and extraction of Pb^{2+} ions from tap water and seawater samples [50]. The adsorption capacity was high with low toxicity and short separation time. The reasons attributed were high surface area, good solubility and the polar functional groups present on the surface of the CQDs. The adsorption was studied by varying the sample pH, extraction time and amount of the sorbent. The optimized values chosen were pH of 6.5, sorption time of 10 min, and sorbent amount of 25.1 mg. Fe_3O_4/N-CQDs showed high selectivity toward Pb^{2+} among many other metal ions. The multiple usage of the adsorbent showed the stability of it in the desorption solution.

N-doped carbon quantum dots (NCQDs) was synthesized by hydrothermal method [34]. N-doped carbon quantum dots/TiO_2 (NCQDs/TiO_2) hybrid composites were then obtained by ultrasonic dispersion of TiO_2 in NCQDs of various concentrations like 0.2, 1, 2, 4 or 10) mL. Then NCQDs/TiO_2 is used for the photocatalytic degradation of Methylene blue (MB) under visible light irradiation. The photodegradation of MB by 1NCQDs/TiO_2 was more than 86.9% under visible light irradiation found much higher than other concentrations of NCQDs. The photocatalytic activity is enhanced by electron transfer and subsequent electron-MB+ separation efficiency. The NCQDs are then responsible for the electrons to combine with oxygen and complete the photodegradation process.

First, CQDs is synthesized by hydrothermal method. CQDs/$KNbO_3$ composites were then prepared by dispersion of $KNbO_3$ in CQD. The degradation of crystal violet dye as target organic pollutant in water, under visible light irradiation with simultaneous evolution of hydrogen was observed [106]. CQDs is responsible for photoabsorption in visible light range and increased the electron-hole (e^-–h^+) pairs for enhanced photocatalytic effect. The mass proportion of CQDs and $KNbO_3$ was 1.5/0.5 and degradation of crystal violet dye is found to be maximum. The hydrogen evolution amount is the highest for this mass proportion. This composite can be recycled for further photocatalytic effect. Thus, degradation of organic pollutants associated by evolution of hydrogen was done using this CQDs/$KNbO_3$, according to the characteristics of CQDs on $KNbO_3$.

The CQDs were synthesized by microwave irradiation method from starch. To produce CQDs@PAFPnanobiosorbent, the synthesized CQD is added to the polymer solution of anthranilic acid-formaldehyde-phthalic acid (PAFP). The CQDs@PAFPnanobiosorbent was used for the removal of uranium (VI) in three different samples of tap water, seawater and wastewater [48]. These samples were optimized to pH 5.0. CQDs@PAFP was mixed with these samples and subjected to microwave irradiation. The surface of PAFP being coated with CQDs, increased the adsorption capability due to the increase in the available functional groups. The researchers explained the mechanism using the pseudo-second-order model. CQDs@PAFPnanobiosorbent are reported to be reused for repeated application cycles with the recycling reagent being HCl. The sorption mechanism was designated due to the pseudo-second-order model and closely tailored with Freundlich model. The reusability of different sorbents such as CQDs@PAFPnanobiosorbent is a significant economical factor. As said above, HCl is the recycling reagent for the regeneration of the CQDs@PAFPnanobiosorbent.

CQDs synthesized by hydrothermal method from cellulose, which is obtained from wood [79]. The preparation materials are the same for both CQDs and nanofibrillated cellulose (NFC) aerogels. Due to the presence of functional groups on its surface, NFC is a good carrier for composite materials. It is used as a framework to hold CQDs. CQDs/NFC composite aerogels is formed with ease by intermolecular forces or hydrogen bonds because of the large amount of hydroxyl groups present in both. Both the CQDs/NFC composite was then used for the adsorption and detection of heavy metal ions (Cr^{3+}) in water. The detection analysis sustained the CQDs fluorescence and the porous structure of the aerogels. A pH of 6 was considered for a maximum adsorption performance. In the analysis of adsorption, it was found to increase rapidly and stabilized with time. It also increased with the rise in ion concentration of the solution due to the porous structure. Thus, through the measurement of the fluorescence effect of CQDs, using the applications of aerogels and CQDs for the adsorption of heavy metal ions in water, is observed.

Cationic CQDs (L-CQDs) were synthesized by the pyrolysis of ionic liquid and citric acid. L-CQDs/ZnO composites were prepared by hydrothermal method [27]. The absorption of L-CQDs/ZnO composites lies in the visible region. The separation of electron-hole pairs by photogeneration and the fast transfer of electrons by the L-CQDs/ZnO composites enhances the photocatalytic activity. It was reported that L-CQD/ZnO composites degraded phenol more effectively than ZnO. 3%- L-CQDs/ZnOwas found as the best photocatalyst when varied and analyzed between 1 and 5% L-CQDs/ZnO. In the recyclability test of the composite, it proved as a stable photocatalyst with no prominent decrease in the degradation rate of phenol. Thus, it confirms that it is a possible measure for wastewater purification by degradation.

Disposal of dyes is a major pollutant of the main water sources like lakes, rivers, and oceans. These organic pollutants are removed by photocatalytic degradation of CQDs. Direct and indirect degradations are involved in water purification from dyes. Indirect dye degradation employs redox reactions in the conduction bands (CBs) and valence bands (VBs). In direct degradation, the dye themselves are made to absorb light and get excited where they are degraded by redox reactions. In both the mechanisms, CQDs are responsible for both reducing the recombination rate of photo-generated electron-hole pairs. CQDs/2D TiO_2nanaocomposites prepared as sheets by hydrothermal process are used for effective degradation of RhB under visible irradiation [81].

CQDs have been synthesized through hydrothermal process from rice husks [59]. Using Ethylenediamine (EDA) and ascorbic acid as functionalization agents to produce amino and carboxyl functionalized CQDs. They are denoted as NCQD and CCQD. Heavy metal ion (cadmium nitrate) with different concentrations with functionalized CQDs solutions were analyzed for Photoluminescence (PL) intensities. Carboxyl (CCQD-2.5), and amino-functionalized CQDs (NCQD-10.0) amount of EDA and ascorbic acid in ml were being found as the optimum samples for the study of detection and degradation. A linear relation was observed in optical response for both functionalized CQDs for different Cd^{2+} ion concentrations. The electrostatic attractive force between Cd^{2+} cations and anionic functional group of CQDs were responsible for the PL intensity quenching. Further, Cd^{2+} ions improved the electron transfer and reduced the recombination of charges and thereby the quenching was enhanced. The carboxyl functionalized CQDs quenching effect was found more than that of amino-functionalized CQDs.

In Cd^{2+} ions removal, both amino and carboxyl functionalized CQDs were found to be effective. More number of CQDs created more active sites for heavy metal ion adhesion thereby making their removal possible. They have reported that amino-functionalized CQDs removal was higher than that of carboxyl functionalized CQDs.

It is due to adsorptive sites available were increased for the interaction between negatively charged metal cations and CQDs. So, the diffusion of metal ions into these sites of CQDs' surface led to removal of more metal ions. The reduction potential in cadmium solution further increased the removal of Cd^{2+}ions from contaminated water by CQDs.

Another research group [60] synthesized carbon quantum dots/zinc aluminum layered double hydroxide (CQD/ZnAl-LDH) composite by one-pot hydrothermal method for adsorption of Cd(II) ions from water. It is attributed due to the electrostatic attractive force between positively charged ZnAl-LDH nanoplates and negatively charged CQDs. It is found that as the pH is increased from 2 to 5 adsorption of Cd(II) initially increases and reaches a plateau at above pH 5. It is also reported that the adsorption was completed within 20 min, as all the functional groups' site of CQDs/ZnAl-LDH adsorbent have been occupied by Cd(II). The pseudo-second-order kinetic model was indicated by adsorption and confirmed by chemical adsorption.

CQDs were synthesized by hydrothermal method from tapioca for the removal of lead from water [56]. Solution of different concentrations of lead (1l) nitrate with CQDs are prepared and the absorbance spectra was analyzed. Batch adsorption was used to find the amount of lead adsorbed by CQD using adsorption isotherm. Adsorption isotherm gives the equilibrium relationship between CDQs as adsorbent and lead as adsorbate. From the adsorption isotherm, the optimum adsorption was at 260 min with the lead removal efficiency of 80.6%. Adsorption process fits well with both Langmuir and Freundlich models. Further, chemisorption reaction the functional groups on the surface of CQDs and between the lead ions is largely considered for adsorption.

The CQD composite or CQD modified photocatalysts are noted for their increased photocatalytic activity. It is reported that the OH• and O_2 radicals are mainly responsible for this enhancement in the degradation of dye from water through photocatalytic activity. Photogeneratedelectron-hole pairs recombination is decreased by CQD. The functional groups like aldehyde and carbonyl groups then interact with the pollutants to form hydroxyl and oxide ions which cause the degradation of dyes.

The photocatalytic activity of TiO_2-CQDs was stated as an example by a research group [52]. The oxygen on the TiO_2-CQD surface interact with the photoelectrons to form oxide ion. Similarly, hydroxyl ion is formed from holes and water. These are responsible for the degradation of methylene blue (MB).

CQDs/$NaBiO_3$hybrid materials were synthesized through hydrothermal method for the reduction of Cr(VI) from contaminated water [94]. Among different concentrations, 6 wt.% CQDs/$NaBiO_3$ showed 97.7% after 90 min under irradiation of visible light. The optimal pH reported was equal to 3. High specific surface area with the increased number of active sites in CQDs/$NaBiO_3$ enhanced the Cr(VI) adsorption. The enhanced light absorption, increased photo carriers generation, the band gap reduction and recombination reduction supported the effort of absorption. It was reported that the reduction efficiency of Cr(VI) was stable at around 95% even after three consecutive recycling.

N-doped ZnO/fulvic acid (FA)/carbon quantum dot (CQD) nanocomposite (N-ZnO/FA/CQD) was synthesized by hydrothermal method from Saffron for catalytic degradation of MB by ultrasonic irradiation [55]. The polluted water with MB was irradiated with ultrasonic irradiation of 20 kHz, 100 W. It is reported that the degradation process of MB was 94%. When it was also examined on rhodamine B(RhB) in the presence of catalyst, the absorbance was found to be reduced. The production of −OH and −H by ultrasonics were responsible for degradation of MB. The optimum pH of the solution was 8 and the effective contact time was reported to be 50 min.

A flower-like αFeOOH hybridized with carbon quantum dots (CQD@FeOOH) is synthesized for degradation of refractory phenol. αFeOOH and CQD@FeOOH under Fenton, photocatalysis, and photo-Fenton conditions were analyzed for the phenol degradation. When αFeOOH is modified by CQDs, the degradation is significant under all conditions mentioned above. 100% degradation in a contact time of 50 min under photo-Fenton conditions is reported. The degradation mechanism is explained on the basis of electron spin resonance (ESR) spectra, photoluminescence (PL) spectral analysis and free radical trapping. Thus, the photo-Fenton reactions are considered for the degradation by CQDs modified compounds [62]. The applications associated with CQDs for pollutant degradations are listed in Table 3.

CQDs and its composite were synthesized to suit their application. It is then characterized by one or few methods. Next, the examinations are done on a comparative basis to emphasize the importance of CQD in water purification. It is intended to mention the CQD/composite in detection and degradation rather than the complete procedure by each research group. The optical properties exhibited by carbon dots were separately discussed. The fluorescent emission/absorption spectra of CQD extends in UV and visible region with the maximum wavelength ranging from about 300 to 500 nm. In general, attributes of CQD in relevant applications are found to depend on the synthesis and the material used for the synthesis. Thus, CQDs with other nanomaterials play a vital role as a sensor, photocatalyst and membrane in water purification.

5 Conclusion

One of the important reasons for the study of CQDs is the facile method of synthesis from easily available inexpensive materials as depicted in Fig. 7. In this chapter, a mini-review of water purification of CQDs is summarized. CQDs, being a nanomaterial, the role of nanotechnology in various process of water remediation is introduced. CQDs are considered in nanomembranes, nanosensors and degradation through photocatalysts. Hence, the general process of these in water remediation is discussed initially. The CQDs with its versatile properties and easy production from naturally occurring, cheap raw materials, its properties and synthesis have been discussed briefly in introduction before its applications in water purification. Initially, these fascinating CQDs being biocompatible were investigated intensively in bio

Table 3 CQD application for pollutant removal

	CQD/composite	Pollutant	References
Degrader	CQDs/NH_2-MIL-125 CQDs/$La_2Ti_2O_7$ NCQDs PECQDs/$MnFe_2O_4$ CQDs/C11-Fe_3O_4) CQDs/ZnAl-LDH	RhB MB Cd^{2+} and Pb^{2+} Uranium BaP Cadmium	Wang et al. [89]
	CQDs/BiOCl/BiOBr	RhB, TC, CIP, BPA	Sharma et al. [76]
	N-CQD	Acid Blue, Acid Red, Eosin Y, Eriochrome Black T, Methyl orange and MB dyes, Cd^{2+} and Pb^{2+}	Sabet and Mahdavi [46]
	CQD/hydrogel	Cu^{2+}, Ni^{2+}, Ag^{+}, and Cd^{2+}, sodium dodecyl sulfate and rhodamine 6G	Singh et al. [72]
	CH–CQD	Cd^{2+}	Jlassi et al. [39]
	ZSH@CQDs	RhB	Zhang et al. [108]
	Fe_3O_4/ N-CQDs	Pb^{2+}	Mashkani et al. [50]
	NCQDs/TiO_2	MB	Zhang et al. [34]
	CQDs/$KNbO_3$	Organic (Violet dye)	Qu et al. [106]
	CQDs@PAFP	Uranium(VI)	Mahmoud et al. [48]
	CQDs/NFC	Cr^{3+}	Song et al. [79]
	L-CQD/ZnO	Phenol	Liang et al. [27]
	CQDs	RhB	Phanga and Tan [81]
	NCQD and CCQD	Cd^{2+}	Abidin et al. [59]
	CQD/ZnAl-LDH	Cd(II)	Rahmaniana et al. [60]
	CQDs	Pb	Pudza et al. [56]
	TiO_2-CQDs	MB	Pirsaheb et al. [52]
	CQDs/$NaBiO_3$	Cr(VI)	Wu et al. [94]
	N-ZnO/FA/CQD	MB	Moalem-Banhangi et al. [55]
	CQD@FeOOH	Phenol	Wu et al. [62]

and energy applications. Recently, the adventure has started in water purifications as nanomembranes, nanosensors and adsorbent photocatalysts. It had been intended to review the advent in these fields and grouped accordingly for easy reference, application and for future development.

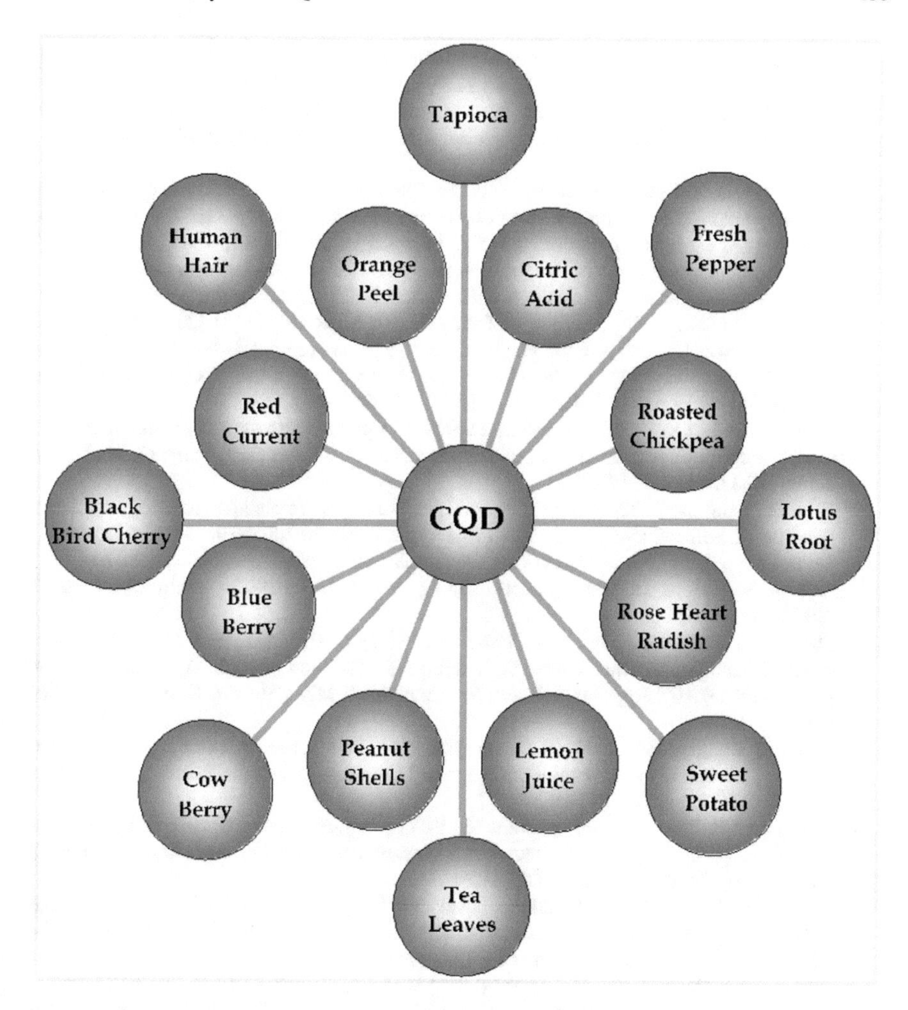

Fig. 7 A few easily available and low-cost raw materials for CQD synthesis

References

1. Singh A, Eftekhar E, Scott J, Kaur J, Yambem S, Leusch F, Wellings R, Gould T, Ostrikov K (Ken), Sonar P, Li Q (2020) Carbon dots derived from human hair for ppb level chloroform sensing in water. Sustain Mater Technol 25:e00159. https://doi.org/10.1016/j.susmat.2020.e00159
2. Yakusheva A, Muratov DS, Arkhipov D, Karunakaran G, Eremin SA, Kuznetsov D (2020) Water-soluble carbon quantum dots modified by amino groups for polarization fluorescence detection of copper (II) ion in aqueous media. Processes 8:1573. https://doi.org/10.3390/pr8121573
3. Cayuela A, Laura Soriano M, Valcárcel M (2015) Photoluminescent carbon dot sensor for carboxylated multiwalled carbon nanotube detection in river water. Sens Actuators B Chem 207, Part A, 596–601. ISSN 0925-4005. https://doi.org/10.1016/j.snb.2014.10.102

4. Yaqoob AA, Parveen T, Umar K, Mohamad Ibrahim MN (2020) Role of nanomaterials in the treatment of wastewater: a review. Water 12:495. https://doi.org/10.3390/w12020495
5. AZoNano (2007) Nanomembranes, an overview of nanomembranes including definition. Applications and Fabrication Techniques. https://www.azonano.com/article.aspx?ArticleID=1834
6. Başoğlu A, Ocak Ü, Gümrükçüoğlu A (2020) Synthesis of microwave-assisted fluorescence carbon quantum dots using roasted–chickpeas and its applications for sensitive and selective detection of Fe^{3+} ions. J Fluoresc 30:515–526.https://doi.org/10.1007/s10895-019-02428-7
7. Hoan BT, Tam PD, Pham VH (2019)Green synthesis of highly luminescent carbon quantum dots from lemon juice. J Nanotechnol 2019. Article ID 2852816, 9 pages
8. Belver C, Bedia J, Gómez-Avilés A, Peñas-Garzón M, Rodriguez JJ (2019) Chapter 22-Semiconductor photocatalysis for water purification, nanoscale materials in water purification, micro and nano technologies, edited by Thomas S, Pasquini D, Leu S-Y, Gopakumar DA, pp 581–65. https://doi.org/10.1016/B978-0-12-813926-4.00028-8
9. Wang C, Liu H, Qu Y (2013) TiO2-based photocatalytic process for purification of polluted water: bridging fundamentals to applications. J Nanomater. Article ID 319637, 14. https://doi.org/10.1155/2013/319637
10. Crini G, Lichtfouse E (2019) Advantages and disadvantages of techniques used for wastewater treatment. Environ Chem Lett 17:145–155. https://doi.org/10.1007/s10311-018-0785-9
11. Gu D, Shang S, Yu Q, Jie S (2016) Green synthesis of nitrogen-doped carbon dots from lotus root for Hg(II) ions detection and cell imaging. Appl Surf Sci 390:38–42https://doi.org/10.1016/j.apsusc.2016.08.012
12. Shao D-D, Yang W-J, Xiao H-F, Wang Z-Y, Zhou C, Cao X-L, Sun S-P (2020) Self-cleaning nanofiltration membranes by coordinated regulation of carbon quantum dots and polydopamine. ACS Appl Mater Interfaces 12:580–590. https://doi.org/10.1021/acsami.9b16704
13. Doshi K, Mungray AA (2020) Bio-route synthesis of carbon quantum dots from tulsi leaves and its application as a draw solution in forward Osmosis. J Environ Chem Eng. https://doi.org/10.1016/j.jece.2020.104174
14. Feng Z, Li Z, Zhang X, Shi Y, Zhou N (2017) Nitrogen-doped carbon quantum dots as fluorescent probes for sensitive and selective detection of nitrite. Molecules 22(12):2061. https://doi.org/10.3390/molecules22122061
15. Gai W, Zhao DL, Chung TS (2018) Novel thin film composite hollow fiber membranes incorporated with carbon quantum dots for osmotic power generation. J Membr Sci 551:94–102. https://doi.org/10.1016/j.memsci.2018.01.034
16. Gan Z, Wu X, Zhou G, Shen J, Chu PK (2013) Is there real upconversion photoluminescence from graphene quantum dots? Adv Opt Mater 1:554–558. https://doi.org/10.1002/adom.201300152
17. Gao R, Wu Z, Wang L, Liu J, Deng Y, Xiao Z, Fang J, Liang YG (2020) Preparation of fluorescent nitrogen-doped carbon quantum dots for sensitive detection of oxytetracycline in environmental samples. Nanomaterials 10:1561. https://doi.org/10.3390/nano10081561
18. Qiaoa G, Lua D, Tangb Y, Gaoc J, Wang Q (2019) Smart choice of carbon dots as a dual-mode onsite nanoplatform for the trace level detection of $Cr_2O_7^{2-}$. Dyes Pigm 163:102–110. https://doi.org/10.1016/j.dyepig.2018.11.049
19. Crini G, Lichtfouse E, Wilson L, Morin-Crini N (2018) Green adsorbents for pollutant removal, vol 18. Springer Nature, pp 23–71. Environmental chemistry for a sustainable world, 978-3-319-92111-2. https://doi.org/10.1007/978-3-319-92111-2_2.hal-02065600
20. Guo CX, Zhao D, Zhao Q, Wang P, Lu X (2014) Na+-functionalized carbon quantum dots: a new draw solute in forward osmosis for seawater desalination. Chem Commun 50(55):7318. https://doi.org/10.1039/c4cc01603c
21. Li H, He X, Kang Z, Huang H, Liu Y, Liu J, Lian S, Tsang CHA, Yang X, Lee ST (2010) Water-soluble fluorescent carbon quantum dots and photocatalyst design. Angew Chem Int Ed 49:4430–4434. https://doi.org/10.1002/anie.200906154

22. Singh H, Bamrah A, Khatri M et al (2020) One-pot hydrothermal synthesis and characterization of carbon quantum dots (CQDs). Mater Today Proc 28, Part 3, pp 1891–1894. ISSN 2214-7853.https://doi.org/10.1016/j.matpr.2020.05.297
23. Wang H, Ning G, He X, Ma X, Yang F, Xu Z, Zhao S, Xu C, Li Y (2018) Carbon quantum dots derived by direct carbonization of carbonaceous microcrystals in mesophase pitch. Nanoscale 10. https://doi.org/10.1039/C8NR07385F
24. Lin H, Huang J, Ding L (2019) Preparation of carbon dots with high-fluorescence quantum yield and their application in dopamine fluorescence probe and cellular imaging, Hindawi. J Nanomater 2019. Article ID 5037243, 9 pages. https://doi.org/10.1155/2019/5037243
25. Sadegh H, Ali GAM, Gupta VK, Makhlouf ASH, Shahryari-ghoshekandi R, Nadagouda MN, Sillanpää M, Megiel E (2017) The role of nanomaterials as effective adsorbents and their applications in wastewater treatment. J Nanostruct Chem 7(1):1–14. https://doi.org/10.1007/s40097-017-0219-4. https://link.springer.com/content/pdf/10.1007%2Fs40097-017-0219-4.pdf
26. Hu S, Liu J, Yang J et al (2011) Laser synthesis and size tailor of carbon quantum dots. J Nanopart Res 13:7247–7252. https://doi.org/10.1007/s11051-011-0638-y
27. Liang H, Tai X, Du Z, Yin Y (2020) Enhanced photocatalytic activity of ZnO sensitized by carbon quantum dots and application in phenol wastewater. Opt Mater 100:109674https://doi.org/10.1016/j.optmat.2020.109674
28. Gehrke I, Geiser A, Somborn-Schulz A (2015) Innovations in nanotechnology for water treatment. Nanotechnol Sci Appl. 8:1–17. Published online 2015 Jan 6. https://doi.org/10.2147/NSA.S43773. PMCID: PMC4294021
29. Singh I, Arora R, Dhiman H, Pahwa R (2018) Carbon quantum dots: synthesis, characterization and biomedical applications. Turk J Pharm Sci 15(2):219–230. https://doi.org/10.4274/tjps.63497
30. Gopalakrishnan I, Sugaraj Samuel R, Sridharan K (2018) Chapter 11, Nanomaterials-based adsorbents for water and wastewater treatments, emerging trends of nanotechnology in environment and sustainability. In Sridharan K (ed) Springer briefs in environmental science. https://doi.org/10.1007/978-3-319-71327-4_11
31. Song J, Wang X, Chang C-T (2014) Preparation and characterization of graphene oxide. J Nanomater 2014. Article ID 276143, 6 pages. https://doi.org/10.1155/2014/276143
32. Guo J, Ye S, Li H, Song J, Qu J (2020) One-pot synthesized nitrogen-fluorine-codoped carbon quantum dots for ClO^- ions detection in water samples. Dyes Pigm 175:108178. https://doi.org/10.1016/j.dyepig.2019.108178
33. Gan JY, Chong WC, Sim LC, Koo CH, Pang YL, Mahmoudi E, Mohammad AW (2020) Novel carbon quantum dots/silver blended polysulfone membrane with improved properties and enhanced performance in tartrazine dye removal. Membranes 10:175. https://doi.org/10.3390/membranes10080175
34. Zhang J, Zhang X, Dong S, Zhou X, Dong S (2016) N-doped carbon quantum dots/TiO_2 hybrid composites with enhanced visible light driven photocatalytic activity toward dye wastewater degradation and mechanism insight. J Photochem Photobiol, A 325:104–110. https://doi.org/10.1016/j.jphotochem.2016.04.012
35. Zuo J, Jiang T, Zhao X, Xiong X, Xiao S, Zhu Z (2015) Preparation and application of fluorescent carbon dots, vol 2015. Article ID 787862. https://doi.org/10.1155/2015/787862
36. Kapitonov AN, Egorova MN, Tomskaya AE, Smagulova SA, Alekseev AA (2018) Hydrothermal synthesis of carbon dots and their luminescence. In AIP conference proceedings 2041, p 030003. https://doi.org/10.1063/1.5079363
37. Omer KM, Tofiq DI, Hassan AQ (2018) Solvothermal synthesis of phosphorus and nitrogen doped carbon quantum dots as a fluorescent probe for iron(III). Microchim Acta 185(10):466. https://doi.org/10.1007/s00604-018-3002-4
38. Khan ZMSH, Saifi S, Shumaila AZ, Khan SA, Zulfequar M (2019) A facile one step hydrothermal synthesis of carbon quantum dots for label -free fluorescence sensing approach to detect picric acid in aqueous solution. J Photochem Photobiol A Chem. https://doi.org/10.1016/j.jphotochem.2019.112201

39. Jlassi K, Eid K, Sliem MH, Abdullah AM, Chehimi MM, Krupa I (2020) Rational synthesis, characterization, and application of environmentally friendly (polymer–carbon dot) hybrid composite film for fast and efficient UV-assisted Cd^{2+} removal from water. Environ Sci Eur 32:12. https://doi.org/10.1186/s12302-020-0292-z
40. Dimos K (2016) Carbon quantum dots: surface passivation and functionalization. Curr Org Chem 20(6). https://doi.org/10.2174/1385272819666150730220948
41. Kunduru KR, Nazarkovsky M, Farah S, Pawar RP, Basu A, Domb AJ (2017) 2-Nanotechnology for water purification: applications of nanotechnology methods in wastewater treatment. In Grumezescu AM (ed) Water purification. Academic Press, pp 33–74. ISBN 9780128043004. https://doi.org/10.1016/B978-0-12-804300-4.00002-2
42. Lan M, Zhao S, Zhang Z, Yan L, Guo L, Niu G et al (2017) Two photon-excited near-infrared emissive carbon dots as multifunctional agents for fluorescence imaging and photothermal therapy. Nano Res 10:3113–3123. https://doi.org/10.1007/s12274-017-1528-0
43. Li L, Yu B, You T (2015) Nitrogen and sulfur co-doped carbondots for highly selective and sensitive detection of Hg(II) ions. Biosens Bioelectron 74:263–269https://doi.org/10.1016/j.bios.2015.06.050
44. Linares RV, Li Z, Abu-Ghdaib M, Wei CH, Amy G, Vrouwenvelder JS (2013) Water harvesting from municipal wastewater via osmotic gradient: an evaluation of process performance. J Membr Sci 447:50–56. https://doi.org/10.1016/j.memsci.2013.07.018
45. Molaei MJ (2020) Principles, mechanisms, and application of carbon quantum dots in sensors: a review. Anal Methods. https://doi.org/10.1039/C9AY02696G.Pages1-78
46. Sabet M, Mahdavi K (2019) Green synthesis of high photoluminescence nitrogen-doped carbon quantum dots from grass via a simple hydrothermal method for removing organic and inorganic water pollutions. Appl Surf Sci 463:283–329. https://doi.org/10.1016/j.apsusc.2018.08.223
47. Mahat NA, Shamsudin SA, Jullok N, Ma'Radzi AH (2020) Carbon quantum dots embedded polysulfone membranes for antibacterial performance in the process of forward osmosis. Desalination 493:114618. https://doi.org/10.1016/j.desal.2020.114618
48. Mahmoud ME, Fekry NA, Abdelfattah AM (2020) Removal of uranium (VI) from water by the action of microwave-rapid green synthesized carbon quantum dots from starch-water system and supported onto polymeric matrix. J Hazard Mater, 122770https://doi.org/10.1016/j.jhazmat.2020.122770
49. d'Amora M, Giordani S (2018) 7-carbon nanomaterials for nanomedicine. In Ciofani G (ed) In micro and nano technologies, smart nanoparticles for biomedicine. Elsevier, pp 103–113. ISBN 9780128141564. https://doi.org/10.1016/B978-0-12-814156-4.00007-0
50. Mashkani M, Mehdinia A, Jabbari A, Bide Y, Reza Nabid M (2018) Preconcentration and extraction of lead ions in vegetable and water samples by N-doped carbon quantum dot conjugated with Fe3O4 as a green and facial adsorbent. Food Chem 239:1019–1026. ISSN 0308-8146. https://doi.org/10.1016/j.foodchem.2017.07.042
51. Farshbaf M, Davaran S, Rahimi F, Annabi N, Salehi R, Akbarzadeh A (2018) Carbon quantum dots: recent progresses on synthesis, surface modification and applications. Artif Cells Nanomedi Biotechnol 46(7):1331–1348https://doi.org/10.1080/21691401.2017.1377725
52. Pirsaheb M, Asadi A, Sillanpää M, Farhadian N (2018) Application of carbon quantum dots to increase the activity of conventional photocatalysts: a systematic review. J Mol Liq 271:857–871. ISSN 0167-7322. https://doi.org/10.1016/j.molliq.2018.09.064
53. Liu M, Xu Y, Niu F, Gooding JJ, Liu J (2016) Carbon quantum dots directly generated from electrochemical oxidation of graphite electrodes in alkaline alcohols and the applications for specific ferric ion detection and cell imaging. Analyst 141:2657–2664.https://doi.org/10.1039/C5AN02231B
54. Li M, Gou H, Al-Ogaidi I, Wu N (2013) Nanostructured sensors for detection of heavy metals: a review. ACS Sustain Chem Eng 1(7):713–723. ISSN 2168-0485.https://doi.org/10.1021/sc400019a
55. Moalem-Banhangi M, Ghaeni N, Ghasemi S (2020) Saffron derived carbon quantum dot/N-doped ZnO/fulvic acid nanocomposite for sonocatalytic degradation of methylene blue. Synth Met 116626. ISSN 0379-6779.https://doi.org/10.1016/j.synthmet.2020.116626

56. Pudza MY, Abidin ZZ, Rashid SA, Yasin FM, Noor ASM, Issa MA (2020) Eco-friendly sustainable fluorescent carbon dots for the adsorption of heavy metal ions in aqueous environment. Nanomaterials 10:315.https://doi.org/10.3390/nano10020315
57. Shahkaramipour N, Tran TN, Ramanan S, Lin H (2017) Review-membranes with surface-enhanced antifouling properties for water purification. Membranes 7:13. https://doi.org/10.3390/membranes7010013
58. Niu Q, Gao K, Lin Z, Wu W (2013) Amine-capped carbon dots as a nanosensor for sensitive and selective detection of picric acid in aqueous solution via electrostatic interaction. Anal Methods 5(21):6228–6233. https://doi.org/10.1039/C3AY41275J
59. Abidin NHZ, Wongso V, Hui KC, Cho K, Sambudi NS, Ang WL, Saad B (2020) The effect of functionalization on rice-husks derived carbon quantum dots properties and cadmium removal. J Water Process Eng 38:101634https://doi.org/10.1016/j.jwpe.2020.101634
60. Rahmaniana O, Dinarib M, Abdolmaleki MK (2018) Carbon quantum dots/layered double hydroxide hybrid for fast and efficient decontamination of Cd(II): The adsorption kinetics and isotherms. Appl Surf Sci 428:272–279. https://doi.org/10.1016/j.apsusc.2017.09.152
61. Peng H, Travas-Sejdic J (2009) Simple aqueous solution route to luminescent carbogenic dots from carbohydrates. Chem Mater 21:5563–5565. https://doi.org/10.1021/cm901593y
62. Wu P, Zhou C, Li Y, Zhang M, Tao P, Liu Q, Cui W (2021) Flower-like FeOOH hybridized with carbon quantum dots for efficient photo-Fenton degradation of organic pollutants. Appl Surf Sci 540:148362.https://doi.org/10.1016/j.apsusc.2020.148362
63. Devi P, Kaur G, Thakur A, Kaur N, Grewal A, Kumar P (2017) Waste derivitized blue luminescent carbon quantum dots for selenite sensing in water. Talanta 170:49–55. ISSN 0039-9140.https://doi.org/10.1016/j.talanta.2017.03.069
64. Devi P, Rajput P, Thakur A, Kim K-H, Kumar P (2019) Recent advances in carbon quantum dot-based sensing of heavy metals in water. Trends Anal Chem 114:171–195. https://doi.org/10.1016/j.trac.2019.03.003
65. Zhang Q, Zhang X, Bao L, Wu Y, Jiang L, Zheng Y, Wang Y, Chen Y (2019) The application of green-synthesis-derived carbon quantum dots to bioimaging and the analysis of mercury(II). J Anal Methods Chem 2019. Article ID 8183134, 9 pages. https://doi.org/10.1155/2019/8183134
66. Zhao Q-L, Zhang Z-L, Huang B-H, Peng J, Zhang M, Pang D-W (2008) Facile preparation of low cytotoxicity fluorescent carbon nanocrystals by electrooxidation of graphite. Chem Commun (41):5116–5118. https://doi.org/10.1039/B812420E
67. Abdel-Karim R, Reda Y, Abdel-Fattah A (2020) Review—nanostructured materials-based nanosensors. J Electrochem Soc 16(7):037554. https://doi.org/10.1149/1945-7111/ab67aa
68. Sugaraj Samuel R, Sridharan K (2018) Water remediation by nanofiltration and catalytic degradation, emerging trends of nanotechnology in environment and sustainability. In Sridharan K (ed) Springer briefs in environmental science. https://doi.org/10.1007/978-3-319-71327-4_12
69. Dongre RS (2018) Rationally fabricated nanomaterials for desalination and water purification, novel nanomaterials-synthesis and applications, George Z. Kyzas and Athanasios C. Mitropoulos, IntechOpen. https://doi.org/10.5772/intechopen.74738
70. Das R, Bandyopadhyay R, Pramanik P (2018) Carbon quantum dots from natural resource: a review. Mater Today Chem 8:96–109. ISSN 2468-5194.https://doi.org/10.1016/j.mtchem.2018.03.003
71. Ludmerczki R, Mura S, Carbonaro CM, Mandity IM, Carraro M, Senes N, Garroni S, Granozzi G, Calvillo L, Marras S, Malfatti L (2019) Carbon dots from citric acid and its intermediates formed by thermal decomposition. Chem A Eur J 25(51):11963–11974https://doi.org/10.1002/chem.201902497
72. Singh S, Shauloff N, Jelinek R (2019) Solar-enabled water remediation via recyclable carbon dot/hydrogel composites. ACS Sustain Chem Eng 7:13186–13194. https://doi.org/10.1021/acssuschemeng.9b02342
73. Sagbas S, Sahiner N, (2019) 22-Carbon dots: preparation, properties, and application, In Khan A, Jawaid M, Inamuddin, Asiri AM, Woodhead publishing series in composites science

and engineering, nanocarbon and its composites. Woodhead Publishing, pp 651–676. ISBN 9780081025093. https://doi.org/10.1016/B978-0-08-102509-3.00022-5
74. Cong S, Zhao Z (2017) Carbon quantum dots: a component of efficient visible light photocatalysts, visible-light photocatalysis of carbon-based materials, Yunjin Yao, Intech Open. https://doi.org/10.5772/intechopen.70801. https://www.intechopen.com/books/visible-light-photocatalysis-of-carbon-based-materials/carbon-quantum-dots-a-component-of-efficient-visible-light-photocatalysts
75. Thatai S, Khurana P, Kumar D (2014) Role of advanced materials as nanosensors in water treatment, biosensor nanotechnology, Editors, Tiwari A, Turner APF. Wiley, pp 315–344. ISBN 978-1-118-77351-2
76. Sharma S, Dutta V, Singh P, Raizada P, Rahmani-Sani A, Hosseini-Bandegharaei A, Thakur VK (2019) Carbon quantum dot supported semiconductor photocatalysts for efficient degradation of organic pollutants in water: a review. J Clean Prod 228:755–769. https://doi.org/10.1016/j.jclepro.2019.04.292
77. Shen J, Shang S, Chen X, Cai Y (2017) Facile synthesis of fluorescence carbon dots from sweet potato for Fe^{3+} sensing and cell imaging. Mater Sci Eng C 76:856–864. https://doi.org/10.1016/j.msec.2017.03.178
78. Iravani S, Varma RS (2020) Green synthesis, biomedical and biotechnological applications of carbon and graphene quantum dots. A Rev Environ Chem Lett 18:703–727. https://doi.org/10.1007/s10311-020-00984-0
79. Song Z, Chen X, Gong X, Gao X, Dai Q, Nguyen TT, Guo M (2020) Luminescent carbon quantum dots/nanofibrillated cellulose composite aerogel for monitoring adsorption of heavy metal ions in water. Opt Mater 100:109642. https://doi.org/10.1016/j.optmat.2019.109642
80. Soutter W (2020) Nanomembranes in water treatment. AZoNano. 27 September. https://www.azonano.com/article.aspx?ArticleID=3170
81. Phanga SJ, Tan L-L (2019) Recent advances in carbon quantum dot (CQD)-based two dimensional materials for photocatalytic applications. Catal Sci Technol 9:5882https://doi.org/10.1039/c9cy01452g
82. Sun D, Ban R, Zhang P-H, Wu G-H, Zhang J-R, Zhu J-J (2013) Hair fiber as a precursor for synthesizing of sulfur-and nitrogen-co-doped carbon dots with tunable luminescence properties. Carbon 64:424–434. https://doi.org/10.1016/j.Carbon2013.07.095
83. Sun H, Wu P (2018) Tuning the functional groups of carbon quantum dots in thin film nanocomposite membranes for nanofiltration. J Membr Sci 564:394–403. https://doi.org/10.1016/j.memsci.2018.07.044
84. Sun Y-P, Zhou B, Lin Y, Wang W, Fernando KAS, Pathak P et al (2006) Quantum-sized carbon dots for bright and colorful photoluminescence. J Am Chem Soc 128:7756–7757. https://doi.org/10.1021/ja062677d
85. Baruah S, Khan MN, Dutta J (2016) Perspectives and applications of nanotechnology in water treatment. Environ Chem Lett 14:1–14. https://doi.org/10.1007/s10311-015-0542-2
86. Warsi T, Bhattacharjee L, Thangamani S, Jat SK, Mohanta K, Bhattacharjee RR, Ramaswamy R, Manikyamba C, Rao TV (2020) Emergence of robust carbon quantum dots as nano-tracer for groundwater studies. Diam Relat Mater 103:107701. https://doi.org/10.1016/j.diamond.2020.107701
87. Rani UA, Ng LY, Ng CY, Mahmoudi E (2020) A review of carbon quantum dots and their applications in wastewater treatment. Adv Colloid Interface Sci 278:102124. https://doi.org/10.1016/j.cis.2020.102124
88. Wafi MK, Hussain N, El-Sharief Abdalla O et al (2019) Nanofiltration as a cost-saving desalination process. SN Appl Sci 1:751. https://doi.org/10.1007/s42452-019-0775-y
89. Wang C, Shi H, Yang M, Yan Y, Liu E, Ji Z, Fan J (2019) Facile synthesis of novel carbon quantum dots from biomass waste for highly sensitive detection of iron ions. Mater Res Bull. https://doi.org/10.1016/j.materresbull.2019.110730
90. Wang X, Feng Y, Dong P, Huang J (2019) A mini review on carbon quantum dots: preparation, properties, and electrocatalytic application. Front Chem 7:671. https://doi.org/10.3389/fchem.2019.00671

91. Youfu W, Aiguo H (2014) Carbon quantum dots: synthesis, properties and applications. J Mater Chem C 2:6921.https://doi.org/10.1039/c4tc00988f
92. Wei J, Liu B, Zhang X, Song C (2018) One-pot synthesis of N, S co-doped photoluminescent carbon quantum dots for Hg^{2+} ion detection. New Carbon Mater 33(4):333–340. https://doi.org/10.1016/s1872-5805(18)60343-9
93. Koe WS, Lee JW, Chong WC, Pang YL, Sim LC (2020) An overview of photocatalytic degradation: photocatalysts, mechanisms, and development of photocatalytic membrane. Environ Sci Pollut Res 27, 2522–2565. https://doi.org/10.1007/s11356-019-07193-5
94. Wu Y, Chen C, He S, Zhao X, Huang S, Zeng G, You Y, Cao Y, Niu L (2021) In situ preparation of visible-light-driven carbon quantum dots/$NaBiO_3$ hybrid materials for the photoreduction of Cr(VI). J Environ Sci (China) 99:100–109. https://doi.org/10.1016/j.jes.2020.06.016. Epub 2020 Jul 2. PMID: 33183687
95. Sun X, Lei Y (2017) Fluorescent carbon dots and their sensing applications. TrAC Trends Anal Chem 89:163–180. ISSN 0165-9936. https://doi.org/10.1016/j.trac.2017.02.001
96. Qu X, Alvarez PJJ, Li Q (2013) Applications of nanotechnology in water and wastewater treatment. Water Res 47(12):3931–3946. ISSN 0043-1354. https://doi.org/10.1016/j.watres.2012.09.058
97. Shan X, Chai L, Ma J, Qian Z, Chen J, Feng H (2014) B-doped carbon quantum dots as a sensitive fluorescence probe for hydrogen peroxide and glucose detection. Analyst 139:2322. https://doi.org/10.1039/c3an02222f
98. Wang X, Cao L, Lu F, Meziani MJ, Li H, Qi G, Zhou B, Harruff BA, Kermarrec F, Sun Y-P (2009) Photoinduced electron transfers with carbon dots. Chem Commun (25):3774.https://doi.org/10.1039/b906252a
99. Yang Y, Wu D, Han S, Hu P, Liu R (2013) Bottom-up fabrication of photoluminescent carbon dots with uniform morphology via a soft-hard template approach. Chem Comm (Camb) 49:4920–4922. https://doi.org/10.1039/C3CC38815H
100. Yang W-J, Shao D-D, Zhou Z, Xia Q-C, Chen J, Cao X-L, Sun S-P (2019) Carbon quantum dots (CQDs) nanofiltration membranes towards efficient biogas slurry valorization. Chem Eng J, 123993.https://doi.org/10.1016/j.cej.2019.123993
101. Yin B, Deng J, Peng X, Long Q, Zhao J, Lu Q, Chen Q, Li H, Tang H, Zhang Y, Yao S. (2013) Green synthesis of carbon dots with down- and up-conversion fluorescent properties for sensitive detection of hypochlorite with a dual-readout assay. Analyst 138(21):6551–7. https://doi.org/10.1039/c3an01003a. PMID: 23982153
102. Hea Y, Zhaob DL, Chung T-S (2018) Na+ functionalized carbon quantum dot incorporated thin-film nanocomposite membranes for selenium and arsenic removal. J Membr Sci 564:483–491. https://doi.org/10.1016/j.memsci.2018.07.031
103. Dong Y, Wang R, Li G, Chen C, Chi Y, Chen G (2012) Polyamine-functionalized carbon quantum dots as fluorescent probes for selective and sensitive detection of copper ions. Chen Anal Chem 84:6220–6224. https://doi.org/10.1021/ac3012126
104. Yuan J-M, Zhao R, Wu Z-J, Li W, Yang X-G (2018) Graphene oxide quantum dots exfoliated from carbon fibers by microwave irradiation: two photoluminescence centers and self-assembly behavior. Small 14:1703714. https://doi.org/10.1002/smll.201703714
105. Wang Y, Hu X, Li W, Huang X, Li Z, Zhang W, Zhang X, Zou X, Shi J (2020) Preparation of boron nitrogen co-doped carbon quantum dots for rapid detection of Cr(VI). Spectrochim Acta Part A Mol Biomol Spectrosc 243:118807. https://doi.org/10.1016/j.saa.2020.118807
106. Qu Z, Wang J, Tang J, Shu X, Liu X, Zhang Z, Wang J (2018) Carbon quantum dots/KNbO3 hybrid composites with enhanced visible-light driven photocatalytic activity toward dye waste-water degradation and hydrogen production. Mol Catal 445:1–11. https://doi.org/10.1016/j.mcat.2017.11.002
107. Zhang Q, Sun X, Ruan H, Yin K, Li H (2017) Production of yellow emitting carbon quantum dots from fullerene carbon soot. Sci China Mater 60:141–150. https://doi.org/10.1007/s40843-016-5160-9
108. Zhang Y, Wang L, Yang M, Wang J, Shi J (2019) Carbon quantum dots sensitized ZnSn(OH)6 for visible light -driven photocatalytic water purification. Appl Surf Sci 466:515–524. https://doi.org/10.1016/j.apsusc.2018.10.087

109. Zhang Y, He YH, Cui PP, Feng XT, Chen L, Yang YZ, Liu XG (2015) Water-soluble, nitrogen-doped fluorescent carbon dots for highly sensitive and selective detection of Hg2+ in aqueous solution. RSC Adv 5:40393–40401
110. Zhao DL, Das S, Chung T-S (2017) Carbon Quantum Dots Grafted Antifouling Membranes for Osmotic Power Generation via Pressure-Retarded Osmosis Process. Environ Sci Technol 51(23):14016–14023. https://doi.org/10.1021/acs.est.7b04190
111. Zhao DL, Chung T-S (2018) Applications of carbon quantum dots (CQDs) in membrane technologies: a review. Water Res. https://doi.org/10.1016/j.watres.2018.09.040
112. Jing Z, Fengyuan Z, Xiaona Y, Peirong C, Yue S, Liang Z, Dongdong M, Fei K (2019) Waste utilization of synthetic carbon quantum dots based on tea and peanut shell. J Nanomater 2019:1–7. https://doi.org/10.1155/2019/7965756s
113. Zhu J, Chu H, Wang T, Wang C, Wei Y (2020) Fluorescent probe based nitrogen doped carbon quantum dots with solid-state fluorescence for the detection of Hg^{2+} and Fe^{3+} in aqueous solution. Microchem J 158:105142. https://doi.org/10.1016/j.microc.2020.105142
114. Zong J, Zhu Y, Yang X, Shen J, Li C (2011) Synthesis of photoluminescent carbogenic dots using mesoporous silica spheres as nanoreactors. Chem Comm (Camb) 47:764–766. https://doi.org/10.1039/C0CC03092A

Supramolecular Ion-Exchange Resins Based on Calixarene Derivatives for Pollutant Removal from Aquatic Environmental Samples

Jumina and Yehezkiel Steven Kurniawan

Abstract Water pollutants, i.e., heavy metal ions, pesticides, dyes, and pigments, are contaminating our aquatic environment and generating serious health damage for the human body. Adsorption using ion-exchange resin materials is one of the versatile techniques for water treatment due to its simple and low-cost process. However, when the separation and pre-concentration of the water pollutants only rely on the ion-exchange mechanism, selective removal of water pollutants cannot be easily achieved. In contrast, a supramolecular organic compound named calixarene offers a promising technology for selective ion and molecular discrimination due to strict host–guest interactions. We reviewed the up-to-date research of supramolecular ion-exchange resins based on calixarene derivatives for pollutant removal from aquatic environmental samples. There are three techniques to prepare the supramolecular ion-exchange resins based on calixarene derivatives: (1) impregnation of calixarene on the commercially available resins, (2) polymerization of calixarene, and (3) crosslink-reaction of the calixarene derivatives. The description, advantages, and disadvantages of each technique are discussed. By using these techniques, hundreds of ion-exchange resin materials have been successfully prepared and they showed a remarkable capability and selectivity for heavy metal ions, pesticides, dyes, and pigments removal.

Keywords Supramolecular · Ion-exchange · Resin · Calixarene · Water pollutant · Heavy metal ions · Dyes · Pigments · Pesticides · Adsorption

---

Jumina (✉) · Y. S. Kurniawan
Department of Chemistry, Faculty of Mathematics and Natural Sciences, Universitas Gadjah Mada, Sekip Utara, Yogyakarta 55281, Indonesia
e-mail: jumina@ugm.ac.id

Y. S. Kurniawan
e-mail: yehezkiel.steven.k@mail.ugm.ac.id

Y. S. Kurniawan
Ma Chung Research Center for Photosynthetic Pigments, Universitas Ma Chung, Villa Puncak Tidar N-01, Malang 65151, Indonesia

E. Lichtfouse et al. (eds.), *Inorganic-Organic Composites for Water and Wastewater Treatment*, Environmental Footprints and Eco-design of Products and Processes,
https://doi.org/10.1007/978-981-16-5928-7_5

Abbreviations

α and β	Elovich coefficients (mg g^{-1} min^{-1})
b_T	Adsorption heat constant (J mol^{-1})
C_{ads}	Concentration of adsorbed pollutants on the stationary phase (g L^{-1})
C_e	Adsorbate concentration at the equilibrium condition (mg L^{-1})
C_i	Initial concentration of pollutants (g L^{-1})
C_t	Concentration of pollutants after adsorption for t min (g L^{-1})
E_s	Mean free energy of adsorption (J mol^{-1})
k	Adsorption rate constant
k_{BA}	Bohart-Adams adsorption rate constant (L g^{-1})
k_{TH}	Thomas adsorption rate constant (L g^{-1} min^{-1})
k_{YN}	Yoon and Nelson adsorption rate constant (min^{-1})
K	Equilibrium constant
K_{DR}	Dubinin-Radushkevich constant (mol^2 J^{-2})
K_F	Freundlich constant (mg g^{-1})
K_{IPD}	Intraparticle diffusion constant (mg g^{-1})
K_L	Langmuir constant (L mg^{-1})
K_T	Temkin constant (L g^{-1})
m	Mass of the used adsorbent in the fixed-bed adsorption (g)
m_t	Total amount of pollutants injected to the stationary phase (g)
n	The heterogeneity factor
Q	Flow rate of the aqueous phase (L min^{-1})
q_e	Equilibrium adsorption capacity of adsorbate (mg g^{-1})
q_{max}	Maximum adsorption capacity (mg g^{-1})
q_o	Initial adsorption capacity of adsorbate (mg g^{-1})
q_t	Adsorption capacity of adsorbate at a certain time (mg g^{-1})
q_{total}	Total adsorbed pollutants on the stationary phase (g)
R	Ideal gas constant (8.314 J mol^{-1} K^{-1})
t	Adsorption time (min)
t"	Breakthrough time or required time to reach 50% of adsorbate adsorption (min)
T	Adsorption temperature (K)
Z	Column height (m)
ΔG^o	Change in Gibbs energy (J mol^{-1})
ΔH^o	Change in enthalpy energy (J mol^{-1})
ΔS^o	Change in entropy energy (J mol^{-1} K^{-1})

1 Introduction

Nowadays, water pollution has reached alarming levels, especially in developing countries [66, 76, 9]. In Indonesia, dangerous water pollutants which are environmentally hazardous and poisonous to humans have been reported in Jakarta and Surabaya cities [6, 73]. However, in these locations, the upstream river waters are still being used for drinking water sources and as a result, cancer, diarrhea, and other digestive problems happen [5, 37]. Even though several efforts have been made to overcome these serious problems, the handling of water pollution is sometimes not easy due to the wide area of the involved aquatic environment. Furthermore, the water pollutants may exist as colorless compounds and are not easily detected by the naked eye [28].

The water pollutants are hazardous chemicals such as heavy metal ions, pesticides, dyes, and pigments, which are found in higher amounts than the allowed concentrations [122]. Heavy metal ions are cation or anion species containing high-density metal (more than 5 g mL^{-1}) as well as high atomic weight (higher than 50 g mol^{-1}), such as copper (Cu(II)), cobalt (Co(II)), cadmium (Cd(II)), lead (Pb(II)), mercury (Hg(II)), chromium (Cr(III) and Cr(VI)), and arsenic (As(III) and As(V)) [16, 27]. These metal ions are highly toxic for human health even in trace concentrations [32, 75]. Cu(II) and Co(II) ions cause insomnia, anxiety, anemia, chronic headaches, endocrine and congenital disruption [16]. The Cd(II) ion causes nervous, respiratory, cardiovascular, and reproduction degenerative diseases while Pb(II) and Hg(II) ions generate fatigue, sperm dysfunction, nervous disorders, and cognitive deficits [124]. Additionally, Cr(III), As(III), and As(V) ions are carcinogenic for the skin, sinuses, and lungs [27]. Cr(VI) in the form of $Cr_2O_7^{2-}$ anions have been widely used for electroplating, wood preserving, and textile dyeing in many industrial processes, however, Cr(VI) ions generate serious health problems to the lungs, liver, and kidneys as well as degrading the biodiversity of aquatic environments [48].

Pesticides are chemicals used in farmlands to preserve the crops from undesired insects or other organisms [151, 137]. Unfortunately, up to 90% of pesticide compounds are accumulated in the soil and water environment [126]. Commonly used pesticides are hexachlorocyclohexane, endosulfan, paraquat, hexaconazole, and chlorpyrifos [41, 114]. The chemical structures of each pesticide are shown in Fig. 1. The hexachlorocyclohexane, endosulfan, paraquat, hexaconazole, and chlorpyrifos pesticides are categorized as highly toxic pollutants according to the United Nations Environmental Protection Agency [95]. The presence of pesticides in the aquatic ecology generates serious health disorders such as paresthesia, nausea, muscular fasciculations, pulmonary edema, endocrine disruption, and cardiac arrhythmias [114].

Meanwhile, dyes and pigments have been extensively used in our life especially for food colorants, textiles, paint, cosmetics, and other daily household items [53, 20]. Butyl rhodamine B, methyl red, methyl orange, methylene blue, methyl green, methyl violet, eosin, acid orange 5, reactive black-5, reactive black-45, and congo red are common dyes and pigments in the industrial fields [19]. The chemical structures

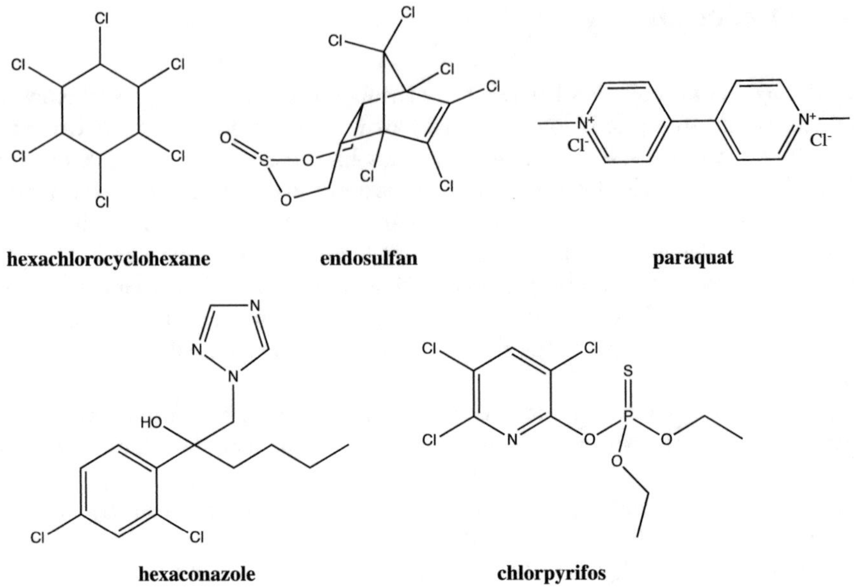

Fig. 1 The chemical structures of the commonly used pesticides, i.e., hexachlorocyclohexane, endosulfan, paraquat, hexaconazole, and chlorpyrifos

of common dyes and pigments are shown in Fig. 2. Because of their popular use around the world, the release of dyes and pigments into our environment is inevitable thus generating serious health and environmental problems. The condition becomes worse when the dyes and pigments are very stable, highly soluble in the water, slow to degrade, and highly toxic for human health and aquatic organisms [86].

Due to the serious effects on human health and environment, these water pollutants should be removed or at least reduced to an acceptable concentration level. Several techniques for water treatment have been established such as adsorption, bioremediation, photocatalysis, membrane filtration, coagulation, and precipitation [127, 23, 51, 64, 72, 8, 49]. Among the water treatment techniques, adsorption is the most favorable process with its simple and facile operation conditions which are very suitable for commercial purposes [1, 24]. In contrast, bioremediation is a complicated process since it is very sensitive for pH, oxygen level, temperature, and light intensity and thus not favorable for commercial application [8]. The photocatalysis process is a simple process, however, it requires a long time to degrade the pollutants with strong light intensity which is not cost-effective nor feasible [64]. Membrane filtration is prone to fouling effects and involves an expensive regeneration process [49]. On the other hand, coagulation and precipitation processes are not a selective process and thus less effective while the processes of separation and preconcentration are difficult to be achieved [72, 127].

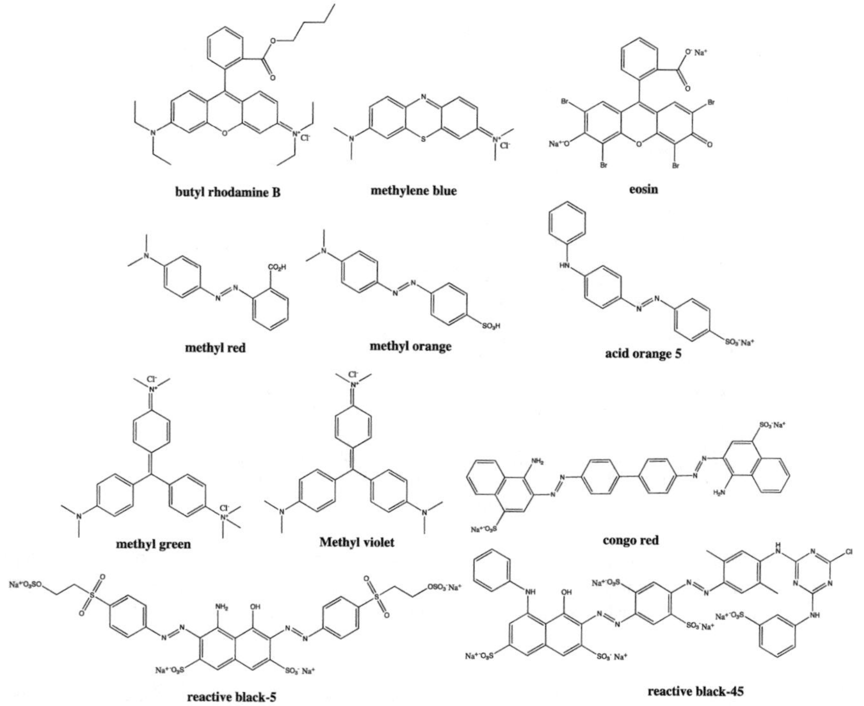

Fig. 2 The chemical structures of the commonly used dyes and pigments, i.e., methyl red, methyl orange, acid orange 5, methyl green, methyl violet, congo red, reactive black-5, and reactive black-45

2 Adsorption Process Using Ion-Exchange Resin Materials

2.1 *Adsorption Process*

Adsorption is an adhesion process of chemicals (adsorbates) in the form of cation, anion, gas, or liquid onto the surface of heterogeneous material due to physical and/or chemical interactions (see Fig. 3) [69]. The most common adsorption process is the adsorption of ions and small molecules on the solid materials' surface [89]. The excellent adsorption process is composed of high maximum adsorption capacity and high selectivity thus the adsorption process could selectively eliminate as much as possible of pollutants among a mixture in the real samples [51]. Moreover, high regenerability of the adsorbent material is also important thus it could be used several times without losing the adsorption capability [107].

Several factors such as adsorbent dosage, shaking time, shaking speed, pollutant concentration, pH, and temperature influence the adsorption process. Each factor should be evaluated independently to find the optimum adsorption conditions while the other factors should be kept as constant [92]. By increasing the adsorbent dosage, the adsorbent percentage will increase until reaching a plateau. The same

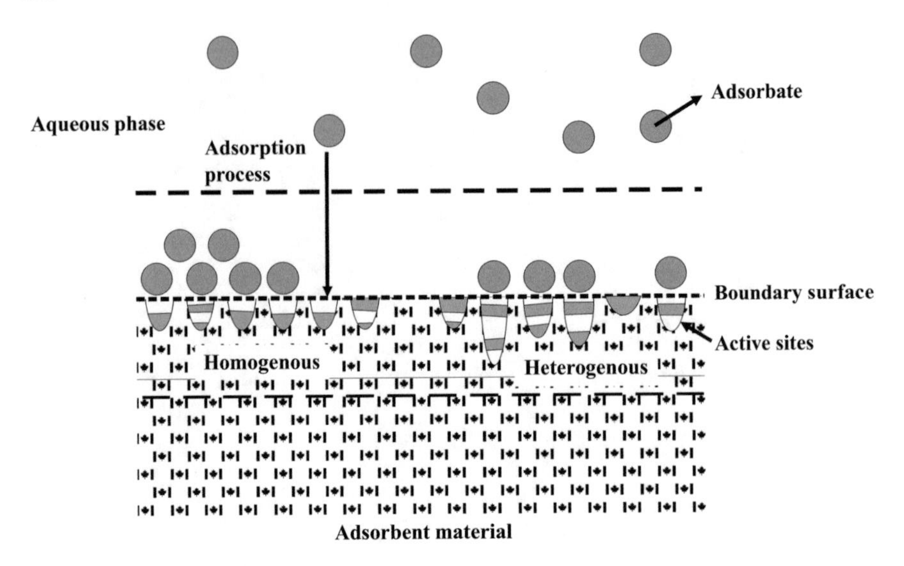

Fig. 3 Schematic representation of adsorption process. Adsorbates are adsorbed through an adsorption process from the aqueous phase onto the active sites of the adsorbent materials. In general, adsorbent material contains homogenous and heterogenous active sites on the boundary surface

phenomenon will be observed for shaking time and shaking speed. When the shaking time and shaking speed are increased, the adsorbent percentage will increase until reaching the maximum value. In contrast, increasing pollutant concentration will decrease the adsorption percentage [108].

On the other hand, the trend of pH and temperature could not be predicted as easily as for the other factors [103]. For example, cationic resin materials are not effective at high pH values because the positive charges of the active functional groups are neutralized by the alkaline media. The cationic resin needs a lower pH value to adsorb anionic chemicals. However, acidic condition (very low pH value) is not always suitable because the adsorbates could become protonated and thus the adsorption percentages are significantly decreased [51]. Similar to the pH value, temperature affects the adsorption process according to the enthalpy value of the reaction. When the adsorption reaction is exothermic (owing to the negative value of enthalpy), the lower temperature is desirable for the adsorption process. However, low temperature decreases the adsorption rate making the complete adsorption process time-consuming [112]. Because of that, it is crucial to adjust the adsorption conditions to reach an efficient process for the most effective removal of water pollutants.

2.2 *Kinetic and Thermodynamic Aspects of the Adsorption Process*

In general, the adsorption process shall be investigated in both kinetic and thermodynamic studies [84, 98]. The kinetic study revealed how fast an adsorption process occurs at a certain operation condition [140]. Several kinetic models have been developed such as zeroth-order, first-order, pseudo-first-order (Lagergren), second-order, pseudo-second-order (Ho and McKay), Elovich, intraparticle diffusion, and liquid film diffusion [51, 70, 46, 68]. Each mathematical equation of these models is given as the following Eqs. (1)–(8), respectively.

$$q_t = -kt + q_o \tag{1}$$

$$ln(q_t) = -kt + \ln(q_o) \tag{2}$$

$$ln(q_e - q_t) = \ln(q_e) - kt \tag{3}$$

$$\frac{1}{q_t} = kt + \frac{1}{q_o} \tag{4}$$

$$\frac{t}{q_t} = \frac{1}{kq_e^2} + \frac{t}{q_e} \tag{5}$$

$$q_t = \frac{t\ln(\alpha\beta)}{\beta} + \frac{\ln(t)}{\beta} \tag{6}$$

$$q_t = kt^{1/2} + K_{IPD} \tag{7}$$

$$ln\left(1 - \frac{q_t}{q_e}\right) = -kt \tag{8}$$

where q_o, q_t, and q_e denote initial adsorption capacity of adsorbate (mg g^{-1}), the adsorption capacity of adsorbate at a certain time (mg g^{-1}), and equilibrium adsorption capacity of adsorbate (mg g^{-1}), respectively. The k, t, K_{IPD}, α, and β denote the adsorption rate constant, adsorption time (min), intraparticle diffusion constant (mg g^{-1}), Elovich coefficients (mg g^{-1} min^{-1}), respectively.

Additionally, adsorption isotherm is defined as the mathematic relation of the equilibrium adsorption capacity and concentration of the adsorbate on the solid surface at a certain temperature [10]. So far, there are four common adsorption isotherm models, i.e., Langmuir, Freundlich, Temkin, and Dubinin-Radushkevich [133, 36, 30, 69]. The mathematical equations of the Langmuir, Freundlich, Temkin, and Dubinin-Radushkevich models are given as Eqs. (9)–(12), respectively.

$$\frac{C_e}{q_e} = \frac{1}{K_L q_{max}} + \frac{C_e}{q_{max}} \quad (9)$$

$$ln(q_e) = \ln(K_F) + \frac{ln(C_e)}{n} \quad (10)$$

$$q_e = \frac{RTln(C_e)}{b_T} + \frac{RTln(K_T)}{b_T} \quad (11)$$

$$ln(q_e) = \ln(q_{max}) - K_{DR}\left[RT\ln\left(1 + \frac{1}{C_e}\right)\right]^2 \quad (12)$$

$$Es = (2K_{DR})^{-2} \quad (13)$$

where C_e, q_{max}, n, R, b_T, and T represent adsorbate concentration at the equilibrium condition (mol L^{-1} or mg L^{-1}), maximum adsorption capacity (mol g^{-1} or mg g^{-1}), the heterogeneity factor, ideal gas constant (8.314 J mol^{-1} K^{-1}), adsorption heat constant (J mol^{-1}), and temperature (K), respectively. While K_L, K_F, K_T, K_{DR}, and *Es* represent Langmuir constant (L mg^{-1}), Freundlich constant (mg g^{-1}), Temkin constant (L g^{-1}), Dubinin-Radushkevich constant (mol^2 J^{-2}), and mean free energy of adsorption (J mol^{-1}), respectively.

The Langmuir model describes a well-ordered monolayer adsorption process of the adsorbate onto the homogeneous surface of the adsorbent material. The adsorbates are adsorbed in the limited number of identical adsorption sites on the surface of the adsorbent material. When the isotherm adsorption process of water pollutants follows the Langmuir model, the adsorption process mainly occurs through chemisorption reaction and the adjacent adsorbates do not interact with each other [69]. Meanwhile, the Freundlich model describes a multilayer adsorption process of the adsorbate onto the heterogeneous surface of the adsorbent material. The heterogeneity factor (n) describes the surface heterogeneity degree, as well as the distribution factor of the adsorbate molecules on the surface of the adsorbent material. The higher heterogeneity factor gives higher adsorption density on a certain area of the adsorbent material. A favorable adsorption process is reflected from the value of the heterogeneity factor higher than 1.0. When the isotherm adsorption process of water pollutants follows the Freundlich model, the adsorption process occurs through physisorption reaction and the adjacent adsorbates do interact with each other [36].

The Temkin model describes the decrement of adsorption heat that is caused by the increment of the number of adsorbate molecules on the surface of the adsorbent [133]. Meanwhile, the Dubinin–Radushkevich model describes an adsorption process of the adsorbate onto the micropores of the adsorbent material. When the isotherm adsorption process of water pollutants fits well with the Dubinin–Radushkevich model, the adsorbates may be adsorbed in monolayer or multilayer formation. The Dubinin–Radushkevich constant could be converted to the mean free energy of adsorption through an Eq. (13). When the value of mean free energy (*Es*) of adsorption is less than 16 kJ mol^{-1}, the adsorption process occurs through chemisorption or ion-exchange reaction [30]. Further investigation on the thermodynamic parameters of the adsorption process could be calculated by using the following Eqs. (14)–(16):

$$K = \frac{C_s}{C_e} \tag{14}$$

$$\Delta G^o = -RTln(K) \tag{15}$$

$$ln(K) = \frac{\Delta S^o}{R} - \frac{\Delta H^o}{RT} \tag{16}$$

where K, ΔG^o, ΔS^o, and ΔH^o represent equilibrium constant, change in Gibbs (J mol^{-1}), entropy (J mol^{-1} K^{-1}), and enthalpy (J mol^{-1}) energy, respectively.

A negative change in Gibbs energy denotes a favorable adsorption process at a certain temperature. A positive value of change in enthalpy reflects an endothermic process while a negative value of change in enthalpy reflects an exothermic process [84]. The physisorption process is indicated by the change in enthalpy value less than 20 kJ mol^{-1} while the chemisorption process is indicated by the change in enthalpy value higher than 20 kJ mol^{-1}. A positive value of the change in entropy reflects the increase of disorder (degree of freedom) while a negative value of the change in entropy reflects the decrease of disorder of the adsorption reaction equation [98].

2.3 *Batch and Fixed-Bed Adsorption Process*

In the experimental investigation, the adsorption process could be carried out through two procedures, i.e. batch and fixed-bed adsorption [82] (see Fig. 4). Batch adsorption is a non-continuous system that is carried out in a closed reactor [1]. The main disadvantage of batch adsorption is the re-conditioning process after the adsorption reaction takes a long time. The adsorbent material should be filtered from the closed reactor and then the adsorbent material should be regenerated and readded into another batch reactor making the process not simple nor convenient [112].

On the other hand, fixed-bed adsorption (also known as column adsorption) is a continuous system in which the adsorbent material is placed in the tube reactor as a stationary phase and then in the aqueous phase flows through a column reactor [40]. The main disadvantage of fixed-bed adsorption is the maintenance of adsorption conditions is not as simple as batch adsorption. However, the fixed-bed adsorption process could be continuously operated [130]. The elution of the adsorbates from the adsorbent material could be easily conducted using an elution or stripping reagent. The common elution agent for ion-exchange resin material for heavy metal adsorption is an acidic solution (HCl or HNO_3 or H_2SO_4) [141].

From the batch adsorption, the obtained data involve adsorption percentages (%Adsorption) only (see Eq. 17) [1]. In contrast, the fixed-bed adsorption process

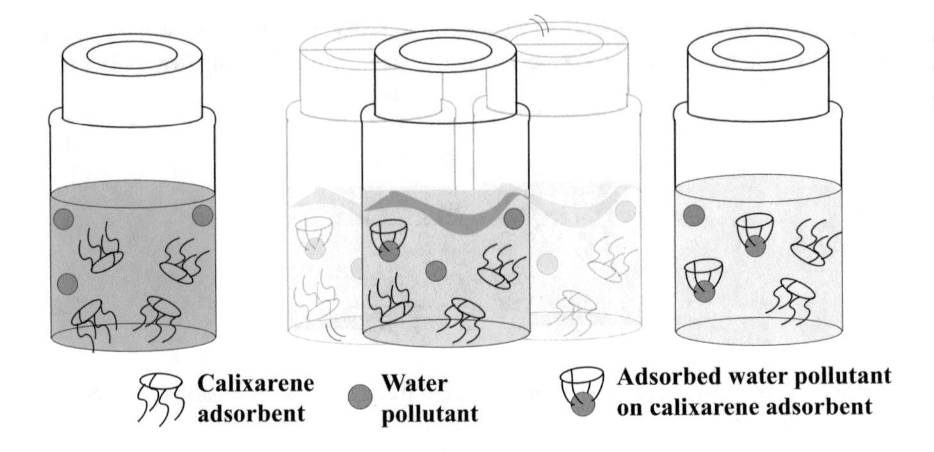

Fixed-bed adsorption process

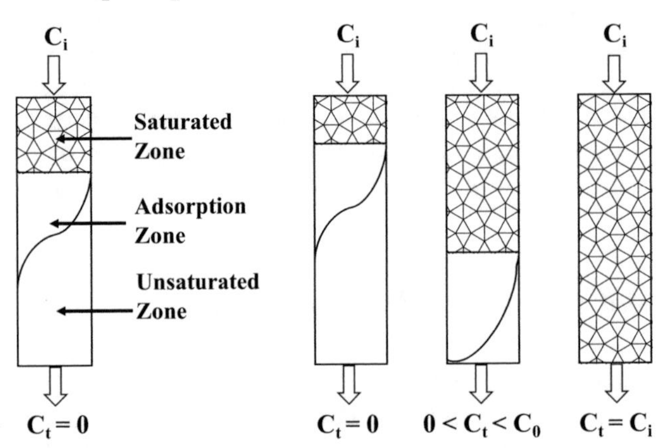

Fig. 4 Schematic representation of batch and fixed-bed adsorption experiment. In the batch adsorption process, the water pollutants were adsorbed on the active sites of calixarene adsorbent which is homogeneously distributed in the aqueous phase. In the fixed-bed adsorption process, the water pollutants flow from the inlet (upper part) to the outlet (lower part) of a column reactor. During the adsorption process, the unsaturated zone ($C_t = 0$) gradually changes to the saturated zone ($0 < C_t < C_o$) to reach the equilibrium state ($C_t = C_i$). Whereas C_i = initial concentration of pollutants (g L^{-1}). C_t = concentration of pollutants after adsorption for t min (g L^{-1})

could be presented as a breakthrough curve using an Eq. (18). Then, the removal percentage (%Removal) of water pollutants using a fixed-bed adsorption process could be calculated using Eq. (19) [82].

$$\%Adsorption = \frac{(C_i - C_t)}{C_i} \times 100\% \tag{17}$$

$$q_{total} = \frac{Q}{1000}\int_0^t C_{ads}dt \tag{18}$$

$$\%Removal = \frac{q_t}{m_t} \times 100\% \tag{19}$$

where C_i, C_t, q_{total}, Q, C_{ads}, and m_t are initial concentration of pollutants (g L^{-1} or mol L^{-1}), concentration of pollutants after adsorption for t min (g L^{-1} or mol L^{-1}), total adsorbed pollutants on the stationary phase (g), flow rate of the aqueous phase (L min^{-1}), concentration of adsorbed pollutants on the stationary phase (g L^{-1} or mol L^{-1}), and total amount of pollutants injected to the stationary phase (g).

The breakthrough curve of the fixed-bed adsorption could be modeled through three models, i.e. Thomas, Bohart-Adams, and Yoon and Nelson models [120, 22]. The Thomas model predicts plug flow of adsorbate through the stationary phase thus no axial dispersion happens [135]. Meanwhile, the Bohart-Adams model describes that the adsorption rate is correlated to the adsorption capacity and the initial concentration of the adsorbate in the aqueous phase. The Bohart-Adams model assumes that the equilibrium state is reached by a stepwise occupation of the adsorption sites on the surface of the adsorbent materials. However, the Bohart-Adams model is suitable for only the first 50% of the breakthrough curve [15]. On the other hand, Yoon and Nelson's model describes that the adsorption rate is decreased by increasing the sorption capacity and the breakthrough of the adsorbate in the aqueous phase. Yoon and Nelson's model is the simplest model due to the least required experimental data compared to the other models [148]. The mathematical equations of Thomas, Bohart-Adams, and Yoon and Nelson models are given as the following Eqs. (20)–(22):

$$ln\left(\frac{C_0}{C_t} - 1\right) = k_{TH}C_o + k_{TH}\frac{q_{max}m}{Q} \tag{20}$$

$$ln\left(\frac{C_t}{C_0}\right) = k_{BA}C_o t - k_{BA}q_{max}\frac{Z}{Q} \tag{21}$$

$$t = t'' + \frac{1}{k_{YN}}ln\left(\frac{C_0}{(C_0 - C_e)}\right) \tag{23}$$

where m, Z, t'', k_{TH}, k_{BA}, and k_{YN} represent mass of the used adsorbent in the fixed-bed adsorption (g), column height (m), breakthrough time (required time to reach 50% of adsorbate adsorption, min), Thomas adsorption rate constant (L g^{-1} min^{-1}), Bohart-Adams adsorption rate constant (L g^{-1}), and Yoon and Nelson adsorption rate constant (min^{-1}).

Pollutants removal from the aquatic media is being evaluated over the past several years through an adsorption process employing either organic or inorganic materials. These materials are activated carbon, clay, chitosan, zeolite, metal oxide, and ion-exchange resins [93, 144, 97, 54, 74, 14, 43, 109, 105, 20, 55]. Among these materials, ion-exchange resins are the most special ones because they possess a strong binding affinity with water pollutants [26]. Furthermore, ion-exchange resins are easily regenerated by adjustment of the pH value thus the usage lifetime is longer than the other adsorbent materials [115].

2.4 *Ion-Exchange Resin Material*

A resin material is defined as a macroporous solid material that is generated from the polymerization of organic compounds [17]. In general, phenolic, silicon, and epoxide compounds provide thermally and chemically stable resin materials [88]. Moreover, these resin materials exhibit high adsorption capability, in addition to the difficult-free regeneration process, thus the adsorption process using resin materials is simple, efficient, and low-cost to be employed in commercial applications [38]. Several resin materials have been commercially available such as Amberlite XAD-4, Amberlite XAD-16, Merrifield, cellulose, silica, and so on [18]. The chemical structures of these resin materials are shown in Fig. 5. However, when the separation and preconcentration of the water pollutants only rely on the ion-exchange mechanism, selective removal of water pollutants is not possible to be established [58]. Therefore,

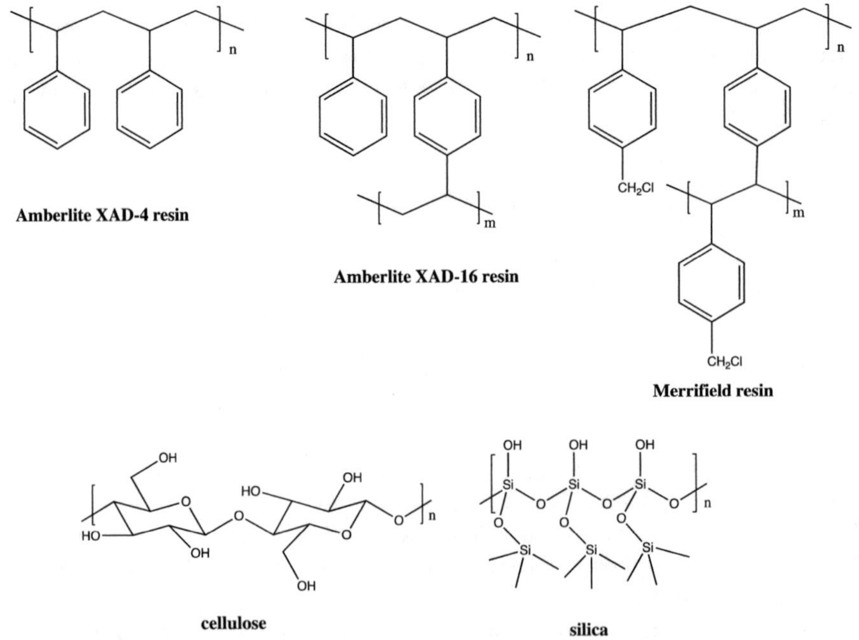

Fig. 5 The chemical structures of the commonly used resin materials, i.e., Amberlite XAD-4 resin, Amberlite XAD-16 resin, Merrifield resin, cellulose, and silica

further modification is still required to optimize the removal of water pollutants using an ion-exchange resin material.

3 Calixarene

Calixarene is one of the supramolecular host compounds with remarkable discriminative effects due to excellent non-covalent and size-exclusion interactions [47, 118]. Calixarene is a cyclic oligomer consisting of aldehyde and *p*-alkylphenol derivatives. Calixarene compounds were introduced by Gutsche in 1978 and afterward, the calixarene field attracted the world's attention due to its high stability, rigid structure, large-scale synthesis, and ease in modifications [117]. By using the ion template on the synthesis process of calixarene, the cavity size of calixarene could be controlled to form calix[4]arene, calix[5]arene, calix[6]arene, and calix[8]arene [65]. The chemical structures of calix[n]arene are shown in Fig. 6. Besides, the conformation of calixarene could be adjusted in several modes [59]. As an example, calix[4]arene exists in cone, partial cone, 1,2-alternate, and 1,3-alternate conformations. Other families of calix[4]arene such as calix[4]resorcinarene, calix[4]pyrogallolarene, calix[4]pyrrole, and oxacalix[4]arene (see Fig. 7) have been introduced and thoroughly investigated for several applications [33, 56, 77, 52, 61, 34, 111]. The chemical modification of calixarene on either the lower or upper rim could be easily conducted while maintaining the conformation of the calixarene [71, 65]. Accordingly, researchers are giving high interest to the calixarene field [60, 62, 63, 66].

These macrocyclic calixarene derivatives exhibit outstanding separation and preconcentration of pollutants due to favorable chelates, size discrimination, and strong electrostatic interactions [103]. Furthermore, these calixarene compounds could be easily regenerated by adjustment of the pH value. By adjusting the pH value, the functional groups will have different charges thus the conformation of calixarene is influenced. The flexibility of calixarene conformation is pivotal to establish a highly efficient adsorption process [123]. However, the main drawback of employing

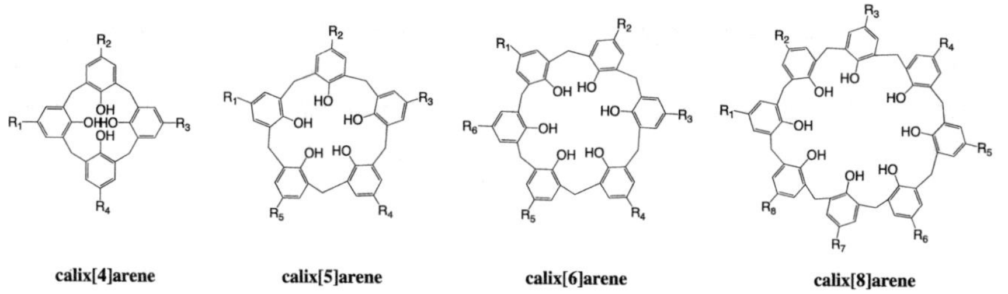

calix[4]arene calix[5]arene calix[6]arene calix[8]arene

Fig. 6 The chemical structures of calix[n]arene including calix[4]arene, calix[5]arene, calix[6]arene, and calix[8]arene. The "n" represents the number of the phenolic aromatic ring in the cyclic structure

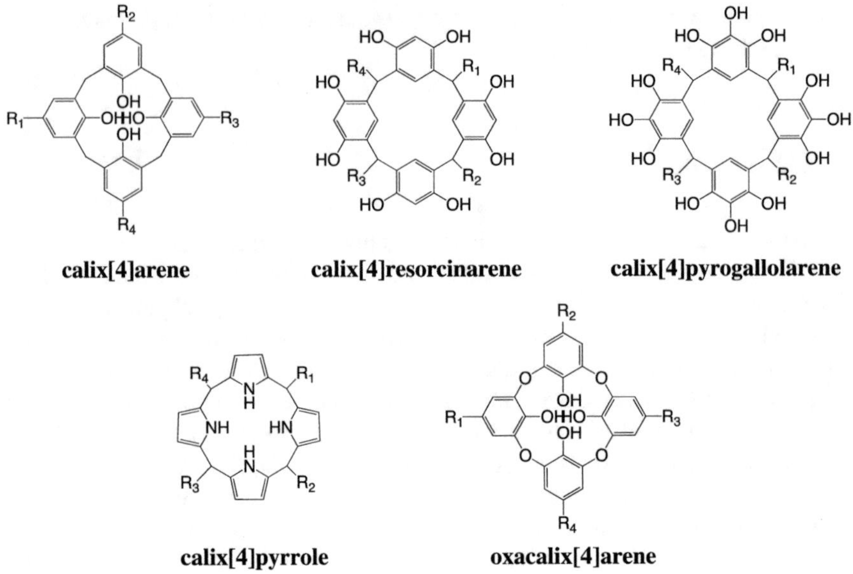

calix[4]arene **calix[4]resorcinarene** **calix[4]pyrogallolarene**

calix[4]pyrrole **oxacalix[4]arene**

Fig. 7 The chemical structures of calix[4]arene, calix[4]resorcinarene, calix[4]pyrogallolarene, calix[4]pyrrole, and oxacalix[4]arene

calixarene for the removal of pollutants is their solubility in the aqueous media. The phenolic functional groups of calixarene easily form hydrogen bonding with water molecules thus the adsorbent loss is inevitable [51]. Because of that, impregnation and polymerization of calixarene derivatives are important to prevent the adsorbent loss for real applications [104].

So far, there are three common techniques to prepare supramolecular ion-exchange resins based on calixarene derivatives. They are impregnation of calixarene on the commercially available resin materials, polymerization of calixarene, and crosslinking-reaction of the calixarene derivatives. A schematic process of these three techniques is shown in Fig. 8. By using these techniques, hundreds of ion-exchange resin materials have been successfully prepared and they showed a remarkable capability for heavy metal ions, pesticides, dyes, and pigments removal. The present book chapter discusses the up-to-date research of supramolecular ion-exchange resins based on calixarene derivatives for pollutant removal from aquatic environmental samples.

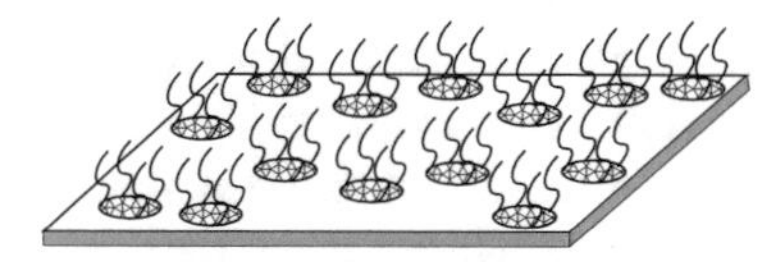

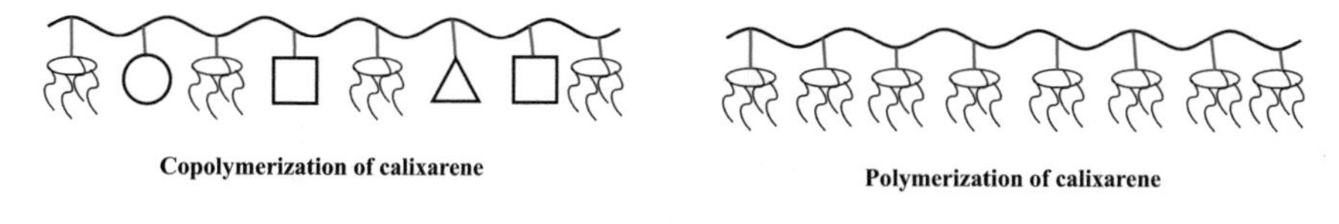

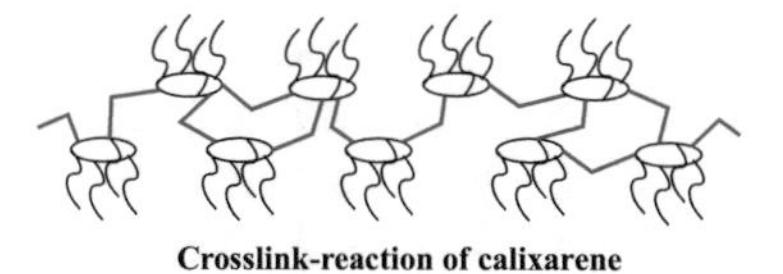

Fig. 8 A schematic process of impregnation of calixarene on the commercially available resin materials, copolymerization of calixarene, polymerization of calixarene, and crosslink-reaction of calixarene derivatives

4 Removal of the Pollutants from Aquatic Environmental Samples Using Supramolecular Ion-Exchange Resins

4.1 Impregnated Calixarene on the Commercially Available Resin Materials

Impregnation of calixarene on the commercially available resins is the simplest technique to prepare supramolecular ion-exchange resin materials. Early study of the impregnation of calixarene derivatives on Merrifield's resin has been reported in 2004 [13]. Three types of calix[4]arene have been used, i.e., calix[4]arene **1**, calix[4]arene **2**, and calix[4]arene **3** (see Fig. 9). Either calix[4]arene **1** or calix[4]arene **3** represent hydrophilic calix[4]arene while calix[4]arene **2** represents hydrophobic calix[4]arene. The impregnation process was conducted through a chemical bond formation between free calix-OH and Cl-CH_2-resin functional groups under alkaline conditions (NaH in tetrahydrofuran). By using a stirring method at room temperature, the calix[4]arene-resin materials were obtained in 70–86% yield with 0.38–0.78 mmol calix[4]arene amount per 1 g of resin material [13]. The study demonstrated that the calixarene compound was easily and efficiently impregnated onto the surface of the polymer material through a simple stirring procedure, which is remarkable.

Modification of polystyrene resin material with calix[4]arene derivative **4** has been prepared and applied for metal ions adsorption application [4]. The polystyrene material was prepared from vinylbenzyl chloride monomer through suspension polymerization with the addition of divinylbenzene as the cross-linker agent. The impregnation of calix[4]arene derivative was done by reacting the calix[4]arene with NaH thus the phenolate ion attacked the benzylic carbon forming an ether bond. The resin material was able to adsorb 97% of Cs(I) ion while a small amount of Cu(II) (14%) and Pb(II) (7.4%) ions were also adsorbed. When calix[4]arene **5** was employed for the impregnation of polystyrene material, the selectivity was reversed. The polystyrene-calix[4]arene **5** resin material could adsorb Cu(II) (67%) and Pb(II) (100%) ions in high percentages, however, the Cs(I) adsorption dropped to 44% [4]. It can be noted that the metal ions' adsorption ability depends on the calix[4]arene substituents.

A study on the heavy metal adsorption process using grafted-calix[4]arene on the cellulose has also been performed [129]. The evaluated heavy metals were Cu(II), Cd(II), Hg(II), Pb(II), and Cr(VI) ions. The resin material was prepared from high-purity cellulose and calix[4]arene **1** using 3-glycidoxypropyl triethoxysilane and 3-aminopropyl triethoxysilane linker agents under alkaline conditions. From the elemental analysis, it was found that as much as 0.64 mmol of calix[4]arene **1** was successfully grafted on 1 g of cellulose material. The thermogravimetric (TGA) analysis showed that the resin material was stable up to 570 K. The cellulose-calix[4]arene **1** resin material gave similar q_{max} values in the range of 0.076–0.095 mmol g^{-1} (equivalent to 1.27–3.56 mg g^{-1}) for Cu(II), Cd(II), Hg(II), and Pb(II) metal ions due to similar charges of divalent metal ions. On the other hand, since the Cr(VI) ion was found as $Cr_2O_7^{2-}$ anion thus lower pH is required for Cr(VI) adsorption. It was found that the Cr(VI) adsorption percentage was achieved in 2.1, 34, 67, and 91% for pH of 4.5, 3.5, 2.5, and 1.5, respectively. Lower pH value caused protonation of calix[4]arene **1** thus the –OH functional groups could interact with $Cr_2O_7^{2-}$ anion through dipole–dipole interactions [129].

Simultaneous adsorption of Co(II), Ni(II), Cu(II), and Cd(II) metal ions was investigated employing modified Amberlite XAD-16 resin using calix[4]resorcinarene **6** [40]. These metal ions were preconcentrated from the real samples such as spinach, tobacco, black tea, mushroom, wheat, and commercial juices from Iran through an adsorption process. The preparation of Amberlite XAD-16-calix[4]resorcinarene **6** resin material was done by impregnating the calix[4]resorcinarene **6** solution in water through a fixed-bed column of Amberlite XAD-16 resin. Since the ion-exchange resin was employed as the adsorbent material, effort on pH optimization is required. The adsorption percentages of Co(II), Ni(II), Cu(II), and Cd(II) ions reached the highest value at pH 6.0. Below pH 6.0, these metal ions' adsorption was low due to protonation of phenolic groups of calix[4]resorcinarene **6** thus the ion-exchange reaction did not happen. On the other hand, above pH 6.0, the metal ions form hydroxo-complexes with hydroxide ions thus lowering the effective charge of metal ions and weakening the ion-exchange interaction. At pH 6.0, the adsorption percentages of Co(II), Ni(II), Cu(II), and Cd(II) metal ions were 96, 97, 97, and 96%, which is remarkable. The adsorption percentage was similar due to the similar charges of these divalent metal ions. The Amberlite XAD-16-calix[4]resorcinarene **6** resin was regenerated using

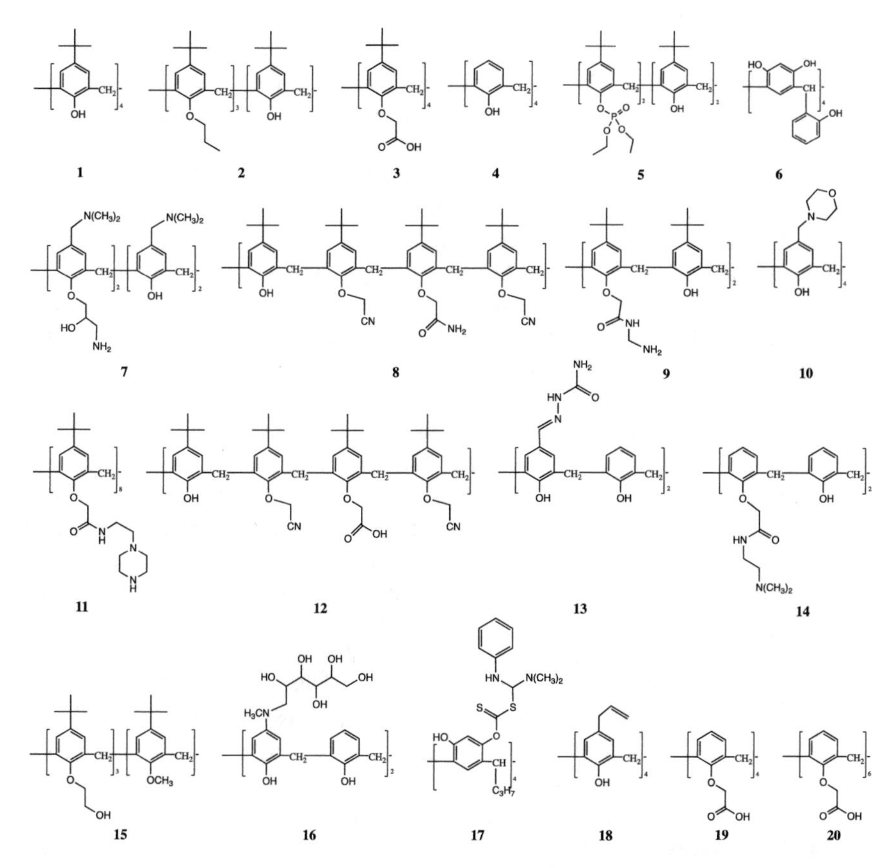

Fig. 9 Functionalized calixarene and calixresorcinarene derivatives as the active site on the supramolecular ion-exchange resin materials. **1**: Unmodified calix[4]arene. **2**: Alkoxy calix[4]arene. **3**, **19**, **20**: Carboxylic acid calixarene. **4**: Dealkylated calix[4]arene. **5**: Phosphonate calix[4]arene. **6**: Calix[4]resorcinarene. **7**–**14**: Nitrogenated calixarene. **15**, **16**: Hydroxylated calix[4]arene. **17**: Nitrogenated calix[4]resorcinarene. **18**: Allylated calix[4]arene

3.0 M HCl by protonating the phenolic groups of calix[4]resorcinarene thus the adsorbed metal ions were released. In the real sample analysis, it was found that the recovery percentages of Co(II), Ni(II), Cu(II), and Cd(II) ions were found in the range of 96–105% with the preconcentration factor up to 208, which is again remarkable [40].

The modified silica material using calix[4]arene **7** containing dimethylamino-functional group has been prepared and evaluated for Co(II), Cu(II), Zn(II), Ni(II), and Cr(III) ions adsorption [96]. The modification of silica material was done using 1-chloro-2,3-epoxypropane as the linker agent in dry toluene solvent. From the elemental analysis, it was found that 0.49 mmol calix[4]arene **7** was successfully impregnated on 1 g of silica material. The resin material was stable up to 590 K as shown from the TGA analysis. A quick adsorption process of Co(II), Cu(II), Zn(II),

Ni(II), and Cr(III) metal ions was achieved in less than 20 min. From the pH study, the adsorption percentages of Co(II), Cu(II), Zn(II), Ni(II), and Cr(III) metal ions were increased at a higher pH value. A quantitative adsorption percentage of Co(II), Cu(II), Zn(II), Ni(II), and Cr(III) metal ions was achieved at pH 4.0. The highest adsorption ability was achieved at pH 4.0 giving the q_{max} values of Co(II), Cu(II), Zn(II), Ni(II), and Cr(III) metal ions of 30, 34, 36, 53, and 47 mg g^{-1}, respectively. The resin material could be easily recovered employing 0.5 M HCl. From the reusability study, the silica-calix[4]arene **7** resin did not lose the adsorption capability even after twelve repeated cycles. The silica-calix[4]arene **7** resin material has been evaluated also for the real water samples, i.e. tap water and Yellow River water. It was found that the recovery percentage was 97–100 and 96–101% for tap water and Yellow River water samples, respectively [96].

The adsorption studies of Cu(II) ions employing a modified silica gel-calix[4]arene **8** have been investigated [130]. The modification of silica gel was conducted using a 3-aminopropyl triethoxysilane linker agent to form an amide bond with the calix[4]arene **8** [39]. By using silica gel-calix[4]arene **8** resin material, as much as 80% of Cu(II) ions were removed. The adsorption isotherm of Cu(II) ions fit the Langmuir model yielding the q_{max} value of 0.079 mmol g^{-1} at pH 6.0. The observed q_{max} value (0.079 mmol g^{-1}) was bigger than other materials such as fly ash (0.022 mmol g^{-1}), sawdust (0.028 mmol g^{-1}), zeolite (0.054 mmol g^{-1}), and activated carbon (0.070 mmol g^{-1}), which is remarkable [139, 7, 149, 106]. On the other side, the fixed-bed column study followed Yoon and Nelson model ($R^2 = 0.9924$) rather than the Thomas model ($R^2 = 0.9896$). From the Yoon and Nelson model, the k_{YN} of Cu(II) ions was 0.048 min^{-1} and the t" was 22.4 min. The reusability study of the resin material for the three-cycle process showed that the resin material exhibited similar adsorption capability for Cu(II) ions with negligible decrement (about 2% deviation) [130].

Modification of Amberlite XAD-4 resin with calix[4]arene **9** for As(III) removal from real samples has been evaluated [84]. The calix[4]arene **9** was immobilized on the Amberlite XAD-4 by the stirring method for 120 h in ethanol solvent at room temperature. Through the stirring method, it was found that as much as 0.63 mmol calix[4]arene **9** was successfully impregnated on 1 g of the Amberlite XAD-4 resin material. The resin material was stable up to 760 K from the TGA analysis. The adsorption process showed that pH 4.5 for 45 min was the best condition for As(III) adsorption. The reason is As(III) exists as AsO_3^{3-} anion thus the acidic media was favorable for As(III) adsorption using Amberlite XAD-4-calix[4]arene **9** resin material. At pH higher than 4.5, the amino-functional group was deprotonated thus the electron repulsion happened between lone pairs of electrons of amine and AsO_3^{3-} anion thus the adsorption percentage was quenched. In contrast, at pH lower than 4.5, the AsO_3^{3-} anion was protonated to form $HAsO_3^{2-}$, $H_2AsO_3^{-}$, and H_3AsO_3 thus weakening the electrostatic interaction with the impregnated calix[4]arene **9** on the resin material.

The adsorption kinetics of As(III) using Amberlite XAD-4-calix[4]arene **9** resin material followed with the pseudo-second-order kinetic model while the isotherm adsorption of As(III) followed the Langmuir model. It was found that the q_{max} value reached 12 mg g^{-1} which is better than other adsorbent materials such as pine wood char (0.001 mg g^{-1}), magnetic sand (0.136 mg g^{-1}), biomass (0.164 mg g^{-1}), and Fe-coated mesoporous carbon (5.96 mg g^{-1}) [134, 42, 90, 21]. The favorable adsorption was reflected from the negative ΔG^o energy (−9.85 kJ mol^{-1}) demonstrating that the chemisorption occurred through electrostatic interactions. When the Amberlite XAD-4-calix[4]arene **9** material was employed for a real application, as much as 76% of As(III) could be adsorbed from the Iranian wastewater sample, which is remarkable [84]. On the other hand, the evaluation of As(III) and As(V) adsorption using Merrifield-calix[4]arene **10** resin material has been conducted [113]. The Merrifield-calix[4]arene **10** resin material was prepared by a reflux method for 24 h under a nitrogen atmosphere. The resin material contained 0.34 mmol calix[4]arene **10** in 1 g of Merrifield support. By employing the Merrifield- calix[4]arene **10** resin material, as much as 97% and 89% of As(III) and As(V) ions were successfully adsorbed at pH 1. Since the calix[4]arene contained morpholine moieties thus the adsorbed As(III) and As(V) ions could be stripped adopting 0.1 M NaOH to give the regenerated resin material [113].

Investigation of the Cr(VI) adsorption using modified Amberlite XAD-4-calix[8]arene **11** has been reported [112]. The resin material was prepared through a simple stirring method for 60 h in water:ethanol 1:1 v/v media. Through the stirring method, it was found that as much as 0.1 mmol of calix[8]arene **11** was successfully impregnated in 1 g of the Amberlite XAD-4 resin. The adsorption process was conducted using a fixed-bed method employing 1 g of modified resin with a volumetric rate of 2 mL min^{-1}. The pH 3.0 media was reported as the best media for the adsorption of Cr(VI). The acidic condition is required to protonate the tertiary-amine functional groups thus the calix[8]arene has a positive charge to interact with the Cr(VI) ion in the form of $Cr_2O_7^{2-}$ anion. The batch adsorption isotherm of Cr(VI) ions onto Amberlite XAD-4-calix[8]arene **11** resin followed the Langmuir model with q_{max} value of 88 mg g^{-1}. Meanwhile, the fixed-bed column investigation followed the Thomas model with k_{TH} of 9 mL mg^{-1} min^{-1}. The resin material was easily regenerated using 4.0 M HCl to elute the Cr(VI) thus the resin material could be reused [112].

Modification of chitosan resin material with calix[4]arene **12** has been studied [131]. The calix[4]arene **12** and chitosan resin material were combined by using *N*,*N*'-diisopropylcarbodiimide as the coupling reagent through a stirring method under nitrogen atmosphere for 72 h. The chitosan-calix[4]arene **12** resin material was obtained in 61% yield owing to 0.72 mmol calix[4]arene **12** content in 1 g of chitosan material. The chitosan-calix[4]arene **12** material was stable up to 470 K as reflected from the TGA analysis. The chitosan-calix[4]arene **12** material gave a three times higher adsorption percentage (89%) than the unmodified chitosan material (29%) for Cr(VI) adsorption at the same pH value (4.5). Furthermore, the chitosan-calix[4]arene **12** material reached a quantitative adsorption percentage (100%) for Cr(VI) adsorption at a lower pH value (1.5), which is remarkable. The calix[4]arene-chitosan resin material also gave high adsorption percentages for other heavy metal ions such as Cu(II), Cd(II), Pb(II), and Hg(II) with 84, 87, 90, and 93%, respectively. These adsorption percentages were much higher than the adsorption percentage of

Cu(II) (27%), Cd(II) (34%), Pb(II) (31%), and Hg(II) (30%) ions with the unmodified chitosan material, which is again remarkable [131].

Modified Merrifield resin material with calix[4]arene **13** has been used to preconcentrate the radioactive metal ions [50]. The Merrifield's resin material was refluxed in the presence of calix[4]arene **13** in dimethylformamide:tetrahydrofuran 1:5 v/v solvent for 24 h under argon atmosphere. The amount of loaded calix[4]arene **13** on Merrifield's resin was around 8.9% w/w. The optimum pH values for 100% adsorption of Ce(III), La(III), Th(IV), and U(VI) were 6.5–8.5; 6.5–8.5; 2.5–4.5; and 5.5–7.0, respectively. These metal ions were easily eluted using HCl solution to give the preconcentration factors of 130, 125, 102, and 108 for Ce(III), La(III), Th(IV), and U(VI) ions, respectively. The reusability of the resin material was evaluated for 12 repeated cycles with only a 2% decrement in the adsorption capability, which is remarkable. The preconcentration of the radioactive metal ions from real samples, i.e., Monazite sand and other geological solids, was also investigated demonstrating that the Merrifield-calix[4]arene **13** resin material has excellent potential to be used in a commercial application [50].

Impregnation of calix[4]arene **14** on the Amberlite XAD-4 resin material was found to be effective for ClO_4^- (perchlorate) anion removal from the aqueous phase [83]. The removal of perchlorate anion is crucial due to its harmful effects such as physical growth and mental disorders, and carcinogenic effect on human organs. The Amberlite XAD-4-calix[4]arene **14** resin material was prepared by several reaction steps, such as nitration, reduction, diazotization, and coupling reactions to obtain the Amberlite XAD-4 resin with a connection with aminocalix[4]arene through an azo bond (see Fig. 10). The Amberlite XAD-4-calix[4]arene **14** resin material was able to adsorb 94% of perchlorate anion at pH 4.5. The isotherm of perchlorate anion adsorption followed the Langmuir model with the q_{max} value of 139 mmol g^{-1}.

The q_{max} value (139 mmol g^{-1}) of perchlorate adsorption using Amberlite XAD-4-calix[4]arene **14** resin material was higher than the other adsorbent materials such as oxidized carbon nanotube (0.068 mmol g^{-1}), iron hydroxide-doped granular activated carbon nanomaterial (0.16 mmol g^{-1}), granular activated carbon (0.36 mmol g^{-1}), MIEX resin (0.96 mmol g^{-1}), quaternary amine-modified reed (3.25 mmol g^{-1}), and calcined Zn-Al hydroxides (5.49 mmol g^{-1}), which is remarkable [35, 142, 145, 12, 132, 146]. Meanwhile, the adsorption kinetics fit well with the pseudo-second-order kinetic model ($R^2 = 0.999$) with the k of 162 g mol^{-1} min^{-1}. The other anions, such as NO_3^- (nitrate), NO_2^- (nitrite), Br^- (bromide), Cl^- (chloride), F^- (fluoride), SO_4^{2-} (sulfate), and PO_4^{3-} (phosphate), did not significantly influence the perchlorate ion adsorption percentage. The resin material selectively removed the perchlorate anion from the aqueous solution. The adsorption of the perchlorate anion was favorable due to the chemisorption process as shown from the negative ΔG^o energy (−5 kJ mol^{-1}). Furthermore, the Amberlite XAD-4-calix[4]arene **14** resin material was easily regenerated using HCl solution [83].

Impregnated calix[4]arene **15** on Amberlite XAD-4 resin material has been employed for the removal of several dyes, i.e. methyl violet, methyl green, methyl red, methyl orange, eosin, and methylene blue [86]. The resin material was prepared

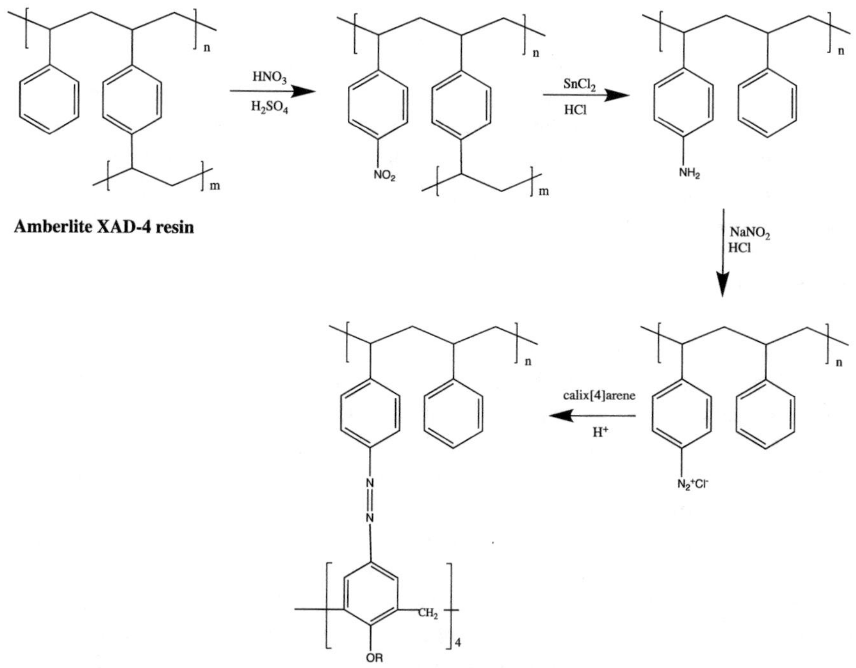

Fig. 10 Preparation of calix[4]arene-Amberlite XAD-4 resin material through a chemical impregnation process. First step is the nitration of aromatic rings of Amberlite XAD-4 resin. Second step is the reduction of nitro to amine functional group. Third step is the diazotization reaction of amino to the diazonium functional group. Last step is coupling reaction of diazo-Amberlite XAD-4 resin with calix[4]arene derivatives

through a stirring method thus as much as 0.052 mmol calix[4]arene **15** was immobilized in 1 g of the Amberlite XAD-4 resin material. Through a batch study, it was found that the resin material was selective for methylene blue, methyl green, and methyl violet dyes in around 90% adsorption percentage. Meanwhile, the other dyes, i.e., methyl red, methyl orange, and eosin were adsorbed in lower percentages (30–40%). The optimum reaction condition was 5 mg of adsorbent dose for 30 min adsorption process at pH 6.0 at 293 K. The isotherm adsorption followed the Freundlich adsorption model with *n* values in the range of 2.0–2.1 showing that a favorable adsorption process was observed [86].

The q_{max} value for methylene blue, methyl green, and methyl violet using impregnated calix[4]arene **15** on Amberlite XAD-4 resin material were 0.89, 1.04, and 1.25 mol g^{-1} (or mg g^{-1}), respectively [86]. The q_{max} value of methyl violet adsorption using calix[4]arene-resin material (1.25 mol g^{-1}) was bigger than the other adsorbents such as activated carbon (3.7×10^{-6} mol g^{-1}), carbon nanotubes (1.1×10^{-4} mol g^{-1}), and wood sawdust (1.2×10^{-5} mol g^{-1}) [99, 91]. The q_{max} value for methyl green adsorption using calix[4]arene-resin material (1.04 mol g^{-1}) was higher than activated carbon (1.0×10^{-5} mol g^{-1}), and $NiFe_2O_4$-carbon nanotubes (2.4×10^{-6} mol g^{-1}) [45, 11]. Meanwhile, the q_{max} value of methylene blue adsorption using calix[4]arene-resin material (0.89 mol g^{-1}) was higher than activated carbon (3.7×10^{-6} mol g^{-1}), and the unmodified resin adsorbent (6.0×10^{-6} mol g^{-1}) [150, 116]. These results demonstrated that the calix[4]arene-resin material exhibits much better adsorption capability for organic dyes than the other adsorbent materials, which is remarkable. It is pivotal to be noted that the impregnated calix[4]arene **15** on Amberlite XAD-4 resin material exhibited a 99% adsorption percentage of methylene blue, methyl green, and methyl violate dyes from the real wastewater samples. The favorable adsorption was generated from the negative ΔG^o energy, i.e. −16.1, −12.9, and −13.8 kJ mol^{-1} for methylene blue, methyl green, and methyl violate, respectively. Furthermore, the Amberlite XAD-4-calix[4]arene **15** resin material was easily regenerated through elution of 0.6 M HCl solution and the resin material gave around 90% of organic dyes adsorption after the five-cycles process [86].

Grafted calix[n]arene (n = 4, 6, and 8) on the starch biopolymer gave high adsorption capability toward butyl rhodamine B in the aqueous solution [20]. The starch-calix[n]arene resin material was prepared by grafting the starch resin material with calix[n]arene derivative using epichlorohydrin as the linker agent under alkaline conditions. It was found that the adsorbent was stable up to 480 K. The adsorption process reached the best performance at pH 9.0. Through the grafting process, the q_{max} value of the starch-calix[n]arene resin material was remarkably enhanced up to 17 times higher than the unmodified starch material (0.58 mg g^{-1}). The kinetic of butyl rhodamine B adsorption on the grafted-polymer fit well with the second-order kinetic model. Meanwhile, the adsorption isotherm of the butyl rhodamine B fit the Langmuir model giving the q_{max} value of 13 mg g^{-1}. Furthermore, the adsorbed butyl rhodamine B on the calix[8]arene-starch material was easily desorbed using ethanol:water 4:1 v/v thus regenerating the free adsorbent material. The starch-calix[8]arene material could be used for ten cycles of butyl rhodamine B adsorption process without a significant decrement of the adsorption capability [20].

The Amberlite XAD-4-calix[4]arene **4** resin material has been applied for azo dyes adsorption from the aqueous solution [53]. The congo red, reactive black-5, and reactive black-45 were evaluated as the azo model compounds. To modify the Amberlite XAD-4 resin, at first, the resin was nitrated using concentrated nitric acid/sulfuric acid reagents. Then the nitro functional group was reduced to the amine group followed by Sandmeyer reaction to form diazonium chloride salt. The diazonium chloride salt of Amberlite XAD-4 resin was reacted with calix[4]arene **4** thus the Amberlite XAD-4 resin and calix[4]arene **4** was connected through –N = N– bonds. It was found that the optimum adsorption condition was achieved at 100 mg dose of adsorbent, 1 h contact time, and 0.2 M NaCl media as the aqueous phase. Since the adsorption of azo dyes occurred through the electrostatic interactions, it is reasonable that each azo dye required different pH conditions. Congo red reached the maximum adsorption percentage (72%) at pH 6 while reactive black-5 reached the maximum adsorption percentage (82%) at higher pH (11). The reactive black-45 compound requires pH 3.0 to reach 60% adsorption percentage using Amberlite XAD-4-calix[4]arene **4** resin material [53].

Modified silica material with calix[4]arene **1** has been prepared using silicon tetrachloride as the linker agent under alkaline conditions for 50 h [82]. By optimizing several adsorption parameters and conditions such as adsorbent dose, contact time, pH, and shaking speed, it was found that quantitative endosulfan adsorption (98–99%) was reached employing 50 mg of resin material at pH 2.0 with a shaking speed of 125 rpm for 60 min. The isotherm of endosulfan adsorption followed the Freundlich model ($R^2 = 0.992$). From the fixed-bed adsorption study, it was found that 0.002 mmol endosulfan could be adsorbed after 30 min process with a volumetric rate of 6 mL min^{-1} according to the Thomas model. The adsorbent regeneration was easily achieved using 5 mL of a binary mixture of *n*-hexane and ethyl acetate giving around 90% recovery percentage. The adsorption ability of the silica-calix[4]arene **1** resin material did not significantly change after the five-cycles adsorption process at the optimum condition. Furthermore, the silica-calix[4]arene **1** resin material was able to remove 90–94% of endosulfan from real polluted-water samples, which is promising for environmental applications [82].

Calix[4]arene-composite resin has been employed for hexaconazole and chlorpyrifos pesticides adsorption from the aqueous solution [98]. The composite resin was prepared from a magnetic nanoparticle, graphene oxide, and glucamine-calix[4]arene **16** through a crosslinking reaction using 3-glycidyloxypropyl trimethoxysilane for 24 h. The composite resin exhibited 90–100% adsorption percentages of hexaconazole and chlorpyrifos pesticides at a pH range of 5.0–7.5 for 30 min batch process. The isotherm adsorption of these pesticides followed the Langmuir model with q_{max} values of 79 and 94 mg g^{-1}, respectively. The high maximum adsorption capacity of hexaconazole and chlorpyrifos pesticides was achieved due to the favorable adsorption process as reflected from the ΔG^o energy, i.e., −33.97 and −36.39 kJ mol^{-1}, respectively. When the composite resin material was applied for real samples, around 80–95% of the pesticides were adsorbed from the tap water, river water, and wastewater. The resin nanomaterial was easily regenerated using acetone solvent for 5 min shaking process. After the twenty-cycles process, the resin nanomaterials still gave 80–84% adsorption, which is remarkable [98].

Modified silica material with *p*-tert-butylcalix[8]arene was found as a useful technique for hexachlorocyclohexane removal from the aqueous media [85]. The silica material was modified using silicon tetrachloride and *p*-tert-butylcalix[8]arene in dry dichloromethane solvent for 60 h. The optimum condition was 20 mg adsorbent dose at pH 8.0 for 60 min adsorption process. The resin material gave 103 mg g^{-1} as the q_{max} value according to the Langmuir model. The observed q_{max} value (103 mg g^{-1}) was much bigger than bagasse (0.251 mg g^{-1}) and clinoptilolite (0.244 mg g^{-1}) adsorbent materials [44, 125]. From the real sample study, the resin material exhibited 79–86% removal of hexachlorocyclohexane from the wastewater samples, which is remarkable [85].

The impregnation technique is low-priced because only a small amount of calixarene is required [3]. The impregnation process could be conducted in two ways, i.e., physical and chemical methods [78, 80, 81]. Physical impregnation is generally done by stirring of commercially available resin material and calixarene derivatives for a long time (24–120 h). The calixarene derivatives will be adsorbed in the resin material thus increasing the adsorption capability of the resin material [79]. The main drawback of physical impregnation is the long time required in the preparation of composite material and slow leaching of the calixarene derivative thus the adsorption capability of resin material is decreased over time [83]. In contrast, chemical impregnation is generally done by stirring of commercially available resin material and calixarene derivatives in the presence of the linker agent in a shorter time [94]. The calixarene derivatives will be connected with the resin material through covalent bondings thus the leaching of calixarene is suppressed. However, the main drawback of chemical impregnation is the higher preparation cost of composite material [3].

4.2 Polymerization of Calixarenes

Polymerization of calixarene could be conducted as copolymerization or total-polymerization reactions. The copolymerization of calixarene with the other monomers has been evaluated due to the low-cost process [87]. The copolymerization technique is conducted by mixing the commercially available monomer such as bisphenol A, vinyl acetate, diisocyanate, styrene, norbornene, and tetraphthaloyl dichloride with functionalized calixarene [57, 138]. Meanwhile, the total-polymerization reaction of calixarene has been performed by adding the allyl functional groups on the calixarene framework. The polymerization of calixarene could be conducted through free-radical, cationic, or anionic reactions [147].

Calix[4]arene-based polyurethane and copolyether materials have been prepared through a copolymerization reaction [29]. The calix[4]arene-based polyurethane material was prepared by mixing calix[4]arene derivative, 2,4-tolylendiisocianate, and dibutyltin dilaurate in dry toluene at 350 K for 8 h. Meanwhile, the calix[4]arene-based copolyether material was prepared by mixing calix[4]arene dialcohol derivative, bisphenol-A, and NaH in dry tetrahydrofuran at 350 K for 12 h. By using various calix[4]arene derivatives, the calix[4]arene-based polyurethane material was obtained in 35–55% yield with an index of polydispersity (*IPD*) of 1.70–2.10. The calix[4]arene-based polyurethane material had a glass transition temperature (*Tg*) in a range of 370–400 K with the molecular weight of 12.3–19.5 kg mol^{-1}. The calix[4]arene-based copolyether material was obtained in 47–58% yield with an *IPD* of 2.42–2.68. The calix[4]arene-based copolyether material had *Tg* in a range of 348–372 K with the molecular weight of 11.1–11.6 kg mol^{-1}. The calix[4]arene-based polyurethane material exhibited high adsorption percentage (85–97%) for Ag(I) and Cs(I) ions over the other metal ions (less than 15%), which is remarkable. Meanwhile, the calix[4]arene-based copolyether material gave a low adsorption percentage (up to 18%) toward Cs(I) ions only [29].

Copolymer material of polyaniline, polyacrylic acid, and calix[4]resorcinarene **17** was reported as a promising adsorbent for paraquat pesticide from the aquatic media [31]. The copolymerization reaction was conducted through a free-radical reaction using ammonium persulfate as an initiator in carbon disulfide and dimethylformamide as solvents. The optimum adsorption condition was 50 mg of adsorbent material at pH 8 for 5 min process to give 93% of adsorption percentage of paraquat. The isotherm adsorption process followed the Freundlich model with q_{max} value of 0.4 mg g^{-1} [31].

Our research group conducted total-polymerization of calixarene derivatives by the addition of allyl functional group [110]. The addition of allyl functional group would lead to cationic polymerization under acidic conditions thus the calixarene derivatives are connected to each other to form an ion-exchange resin material. Calix[6]arene-polymer was successfully prepared from the monoallyloxy-calix[6]arene. The polymerization reaction was conducted for 8 h in chloroform catalyzed by sulfuric acid. The polymerization process was terminated with the addition of a limited amount of methanol to form calix[6]arene-polymer. The successful polymerization reaction was indicated by the disappearance of $-CH=CH_2$ functional group from the nuclear magnetic resonance and Fourier transform infrared spectra. The polymer material was produced as a brown solid with a melting point of 472–474 K. The molecular weight of calix[6]arene-polymer was 24.6–30.2 kg mol^{-1} [110].

The synthesized calix[4]arene-polymer has also been synthesized from calix[4]arene **18** monomer under acidic conditions [67]. The monomer was dissolved in chloroform and then polymerized in the presence of sulfuric acid solution for 7 h. The polymerization reaction was stopped by the addition of a small amount of methanol to form calix[4]arene-polymer. The calix[4]arene-polymer was obtained as a brown solid in 80% yield. The melting point of the calix[4]arene-polymer is somewhat above 650 K. The surface area of the polymer was 46.6 m^2 g^{-1} with a pore diameter of 300 nm. When the calix[4]arene-polymer material was employed for Cd(II) adsorption from the aqueous media, the best adsorption condition for Cd(II) was at pH 6.0 for 50 min shaking. The adsorption isotherm of Cd(II) ions followed the Langmuir isotherm model with 96 mg g^{-1} as the q_{max} value. The ΔG^o adsorption energy value was -32.65 kJ mol^{-1} demonstrating that chemisorption of Cd(II) ions onto the calix[4]arene polymer is highly favorable [67].

On the other hand, the calix[4]arene-, calix[6]arene-, and calix[8]arene-polymer also gave remarkable adsorption capability for paraquat from the aqueous phase [119]. The calix[n]arene-polymer was prepared through Sonogashira-Hagihara cross-coupling reaction of calix[n]arene with tetracetylene pyrene in tetrahydrofuran at 340 K. The calix[n]arene-polymer gave 69, 70, and 100% adsorption percentages of paraquat for $n = 4$, 6, and 8, respectively. The calix[8]arene-polymer was found as the best adsorbent material with q_{max} value of 419 mg g^{-1}. Meanwhile, the calix[4]arene- and calix[6]arene-polymer gave the same q_{max} value of paraquat (411 mg g^{-1}). It was reported that the calix[4]arene-, calix[6]arene-, and calix[8]arene-polymer have a surface area of 759, 725, and 635 m^2 g^{-1}, respectively. Therefore, the adsorption of paraquat was influenced by the larger ring size of calix[n]arene rather than the larger surface area. The calix[n]arene-polymer was easily regenerated by washing for 15 min with methanol to elute the adsorbed paraquat compound. After the three-recycles process, the adsorption percentage of paraquat using calix[n]arene-polymer did not significantly change demonstrating that they are potential polymers to be used in the real process [119].

Compared to the impregnation process, the polymerization process gave higher adsorption capability and faster reaction time in material preparation. Higher adsorption capability is caused by a higher amount of calixarene in the resin material. Meanwhile, faster reaction time is caused by more elevated reactivity of the monomer species in a comparison with the inert resin material. The main disadvantage of employing calixarene polymer is higher production cost because a higher amount of calixarene derivative is used in the preparation of resin material [128].

4.3 Crosslinking Reaction of the Calixarenes

Instead of impregnation and polymerization of calixarene, utilizing the unmodified calixarene to prepare resin material is also possible through crosslinking reactions [102]. Since calixarenes have phenol moiety thus they could be linked to each other through a Friedel-Craft reaction. Compared to the impregnation and polymerization process, the crosslinking reaction gave much higher adsorption capability [1]. The higher adsorption capability is caused by a higher amount of calixarene in the resin material. However, the preparation cost of crosslinked calixarene resin material is the most expensive compared to the other techniques.

Crosslinking reaction of calixarene has been thoroughly studied by the Ohto group in Japan. The calix[4]arene-resin material was synthesized from a crosslinking of calix[4]arene **19** derivatives using *s*-trioxane as the crosslinker agent [121]. The calix[4]arene **19** and crosslinker agent were reacted in acetic acid media at 380 K for 8 h. The resin material is insoluble in water, chloroform, methanol, ethanol, acetone, and dimethylsulfoxide. Furthermore, the resin material is also insoluble in NaOH, NH_3, HNO_3, and HCl solution media thus calix[4]arene-resin material could be used in almost all media for the adsorption of heavy metal ions [102].

The calix[4]arene-resin material was selective for Pb(II) adsorption over Ni(II), Co(II), Zn(II), and Cu(II) ions at pH 3.5 or lower. The Pb(II) adsorption fit the Langmuir isotherm model with the q_{max} value of 278 mg g^{-1}. The observed q_{max} value (278 mg g^{-1}) is 2.2 times higher than the free calix[4]arene **19** (128 mg g^{-1}) confirming that the crosslinking reaction is favorable for higher adsorption capability. Preconcentration of Pb(II) from a model mixture of Zn(II) and Pb(II) solution (initial concentration of each ion was 100 mg L^{-1}) was achieved through a fixed-bed process. The resin material was easily regenerated by flowing 0.1 M HCl to obtain 3,000 mg L^{-1} Pb(II) solution yielding the preconcentration factor equals to 30 [121].

In our previous work, we also prepared calix[6]arene-resin material from a crosslinking reaction of calix[6]arene **20** using *s*-trioxane as the crosslinker agent [1]. The crosslinking reaction conditions were similar to the preparation of calix[4]arene-resin material. The particle size of the calix[6]arene-resin material was less than 0.15 mm thus providing a relatively large surface area. The q_{max} value for Pb(II) ions was 269 mg g^{-1} which is slightly lower than calix[4]arene-resin material. We expected that the larger number of the carboxylic acid functional group would

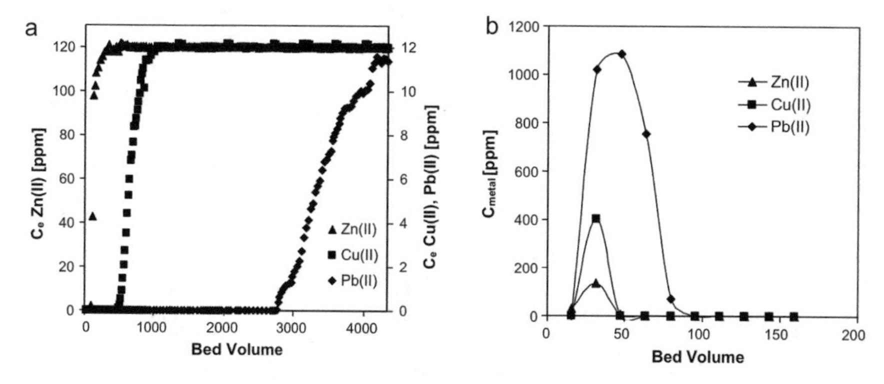

Fig. 11 Profile of **a** breakthrough and **b** elution of Pb(II), Zn(II), and Cu(II) metal ions employing calix[6]arene-resin material. The calix[6]arene-resin material selectively adsorbed Pb(II) with a high preconcentration factor. Adsorbent mass = 150 mg. Concentration of Pb(II) = concentration of Cu(II) = 12 mg L^{-1}. Concentration of Zn(II) = 120 mg L^{-1}. pH = 4.1. Elution agent = 2.0 M HCl solution [1]

improve the maximum adsorption capacity for Pb(II) ions because calix[6]arene-resin has a higher density of carboxylic acid group (6-COOH/monomer) than the calix[4]arene-resin (4-COOH/monomer). In contrast, upgrading the calix[4]arene-resin to calix[6]arene-resin material also leads to a larger cavity size of calixarene thus weakening the chelating ability of calixarene itself. Both factors seem to influence the maximum adsorption capacity value for Pb(II) ions thus both materials gave similar q_{max} value (278 mg g^{-1} for calix[4]arene-resin vs 269 mg g^{-1} for calix[6]arene-resin) [1].

When the Pb(II) adsorption capability of calix[n]arene-resin materials was compared to the other adsorbent materials, the calix[n]arene-resin materials demonstrated a much higher maximum adsorption capacity value, which is remarkable. The calix[n]arene-resin materials (269–278 mg g^{-1}) gave several times enhancement on the maximum adsorption capacity value than carbon nanotubes (10 mg g^{-1}), imprinted polymer (20 mg g^{-1}), silica gel-ofloxacin (49 mg g^{-1}), iminodiacetate-type cellulose (178 mg g^{-1}), peanut shell-phosphonic acid (117 mg g^{-1}) [2, 25, 136, 143, 152]. Furthermore, the crosslinked calix[n]arene-resin materials also exhibited a higher q_{max} value than the impregnated resin of calix[4]arene (68 mg g^{-1}) and calix[6]arene (31 mg g^{-1}), as well as calix[4]arene polymer-supported resin material (128 mg g^{-1}) [100, 101]. These results demonstrate that crosslinked types of calixarene resin are promising adsorbent material for heavy metal removal than calixarene-impregnated and -supported resin materials. The crosslinked calix[6]arene resin material was easily regenerated with elution of 1.0 M HCl and stable enough for the five-cycles adsorption process. The fixed-bed column experiment revealed that the crosslinked calix[6]arene resin material was selectively preconcentrated Pb(II) ions up to 10 times higher than the initial concentration over Cu(II) and Zn(II) metal ions from lead-bearing effluents (see

Table 1 Summary of adsorption condition and maximum adsorption capacity value of water pollutants using supramolecular ion-exchange resin based on calixarene derivatives

No	Material	Analyte	pH	q_{max} (mg g^{-1})	References
		Heavy metal ions			
1	Cellulose-calix[4]arene **1** (25 mg)	Cu(II), Cd(II), Hg(II), Pb(II) (10 mL, 0.02 mM) Cr(VI) (10 mL, 0.10 mM)	1.5	Cu(II): 1.3 Cd(II): 2.0 Hg(II): 3.4 Pb(II): 3.6 Cr(VI): 19	[129]
2	Amberlite XAD-16-calix[4]resorcinarene **6** (500 mg)	Ni(II), Cu(II), Co(II), Cd(II) (250 mL, 0.20 mg L^{-1})	6.0	Ni(II): 97 Cu(II): 97 Co(II): 96 Cd(II): 96	[40]
3	Silica-calix[4]arene **7** (30 mg)	Co(II), Ni(II), Zn(II), Cu(II), Cr(III) (25 mL, 10.0 mg L^{-1})	4.0	Co(II): 30 Ni(II): 53 Zn(II): 36 Cu(II): 34 Cr(III): 47	[96]
4	Silica-calix[4]arene **8** (25 mg)	Cu(II) (10 mL, 6.35 mg L^{-1})	6.0	5.1	[130]
5	Amberlite XAD-4-calix[4]arene **9** (50 mg)	As(III) (10 mL, 30.0 mg L^{-1})	4.5	12	[84]
6	Merrifield-calix[4]arene **10** (50 mg)	As(III), As(V) (10 mL, 40.0 mg L^{-1})	1.0	As(III): 456 As(V): 584	[113]
7	Amberlite XAD-4-calix[4]arene **11** (100 mg)	Cr(VI) (10 mL, 0.10 mM)	3.0	88	[112]

(continued)

Table 1 (continued)

No	Material	Analyte	pH	q_{max} (mg g^{-1})	References
8	Chitosan-calix[4]arene **12** (25 mg)	Cu(II), Cd(II), Hg(II), Pb(II) (10 mL, 0.02 mM) Cr(VI) (10 mL, 0.10 mM)	1.5	Cu(II): 4.3 Cd(II): 7.9 Hg(II): 15 Pb(II): 15 Cr(VI): 21	[131]
9	Amberlite XAD-4-calix[4]arene **19** (20 mg)	Pb(II) (10 mL, 0.10 mM)	2.5	68	[100]
10	Amberlite XAD-4-calix[6]arene **20** (20 mg)	Pb(II) (10 mL, 0.10 mM)	2.5	31	[100]
11	Calix[4]arene-polymer (4 mg)	Cd(II) (5 mL, 10.0 mg L^{-1})	6.0	96	[67]
11	Crosslinked-calix[4]arene-resin (20 mg)	Pb(II) (10 mL, 0.10 mM)	2.5	278	[121]
13	Crosslinked-calix[6]arene-resin (60 mg)	Pb(II) (50 mL, 2.00 mM)	4.1	269	[1]
		Toxic anion			
14	Amberlite XAD-4-calix[4]arene **14** (100 mg)	ClO_4^- (10 mL, 0.10 mM)	4.5	139	[83]
		Pesticides			
18	Silica-calix[8]arene (20 mg)	Hexachlorocyclohexane (10 mL, 1.0 mg L^{-1})	8.0	103	[85]
19	Silica-calix[4]arene **1** (50 mg)	Endosulfan (10 mL, 0.05 mg L^{-1})	2.0	1.1	[82]

(continued)

Table 1 (continued)

No	Material	Analyte	pH	q_{max} (mg g^{-1})	References
20	Fe_3O_4-graphene oxide-calix[4]arene **16** (50 mg)	Chlorpyrifos, hexaconazole (50 mL, 100 mg L^{-1})	Chlorpyrifos: 6.0 Hexaconazole: 5.0	Chlorpyrifos: 79 Hexaconazole: 94	[98]
21	Polyaniline-polyacrylic acid-calix[4]resorcinarene **17** (50 mg)	Paraquat (20 mL, 20.0 mg L^{-1})	8.0	0.4	[31]
22	Calix[8]arene polymer (2.5 mg)	Paraquat (5 mL, 2.00 mM)	7.0	419	[119]
		Dyes and pigments			
15	Starch-calix[8]arene (200 mg)	Butyl rhodamine B (50 mL, 20 mg L^{-1})	9.0	13	[20]
16	Amberlite XAD-4-calix[4]arene **4** (100 mg)	Reactive black-5, reactive black-45, congo red (10 mL, 0.02 mM)	Reactive black-5: 9.0 Reactive black-45: 5.0 Congo red: 6.0	Reactive black-5: 16 Reactive black-45: 14 Congo red: 11	[53]
17	Amberlite XAD-4-calix[4]arene **15** (5 mg)	Methyl violate, methyl green, methylene blue (25 mL, 0.03 mM)	6.0	Methyl violate: 492 Methyl green: 679 Methylene blue: 285	[86]

q_{max}: maximum adsorption capacity

Fig. 11), which is remarkable [1]. Table 1 shows the summary of adsorption condition and maximum adsorption capacity value of water pollutans removal using supramolecular ion-exchange resin based on calixarene derivatives.

5 Conclusions

Supramolecular ion-exchange resins based on calixarenes are found to be an efficient adsorbent material for water pollutant adsorption and removal from the aquatic media. The supramolecular ion-exchange resins based on calixarenes could be prepared through three techniques: (1) impregnation of calixarene on the commercially available resins, (2) polymerization of calixarene, and (3) crosslink-reaction of the calixarene derivatives. Impregnation of calixarene on the commercially available resins is the simplest technique, however, the material preparation is time-consuming and had low adsorption capacity. Furthermore, the impregnated calixarene is slowly leached thus the adsorption capability of the resin material is decreased over time. Polymerization gave faster material preparation due to a higher reactivity of the monomers. Meanwhile, the crosslink reaction gave the adsorbent material with the highest adsorption capacity among the other techniques. The supramolecular ion-exchange resin materials possess a strong binding affinity with the water pollutants (heavy metal ions, pesticides, dyes, and pigments). Furthermore, the supramolecular ion-exchange resins are easily regenerated by adjustment of the pH value thus the usage lifetime is longer for commercial applications.

Acknowledgements Financial support from KEMENDIKBUD-RISTEK, The Republic of Indonesia, through the PTUPT Scheme for the budget year 2020–2022 is greatly acknowledged.

References

1. Adhikari BB, Gurung M, Kawakita H, Jumina J, Ohto K (2011) Methylene crosslinked calix[6]arene hexacarboxylic acid resin: a highly efficient solid phase extractant for decontamination of lead bearing effluents. J Hazard Mater 193:200–208. https://doi.org/10.1016/j.jhazmat.2011.07.051
2. Akama Y, Yamada K, Itoh O (2003) Solid phase extraction of lead by chelest fiber iry (aminopolycarboxylic acid-type cellulose). Anal Chim Acta 485:19–24. https://doi.org/10.1016/S0003-2670(03)00399-4
3. Akkus GU, Memon S, Gurkas DE, Aslan S, Yilmaz M (2008) The synthesis and metal cation extraction studies of novel polymer-bound calix(aza)crowns. React Funct Polym 68:125–133. https://doi.org/10.1016/j.reactfunctpolym.2007.10.005
4. Alexandratos SD, Natesan S (2001) Synthesis and ion-binding affinities of calix[4]arenes immobilized on cross-linked polystyrene. Macromolecules 34:206–210. https://doi.org/10.1021/ma0012550

5. Amin MT, Alazba AA, Manzoor U (2014) A review of removal of pollutants from water/wastewater using different types of nanomaterials. Adv Mater Sci Eng 2014:825910. https://doi.org/10.1155/2014/825910
6. Amira S, Astono W, Hendrawan D (2018) Study of pollution effect on water quality of Grogol river, DKI Jakarta. IOP Conf Ser Earth Environ Sci 106:012023. https://doi.org/10.1088/1755-1315/106/012023
7. An KH, Park YB, Kim SD (2001) Crab shell for the removal of heavy meals from aqueous solution. Water Res 35:3551–3556. https://doi.org/10.1016/S0043-1354(01)00099-9
8. Ariyanti D, Iswantini D, Sugita P, Nurhidayat N, Effendi H, Ghozali AA, Kurniawan YS (2020) Highly sensitive phenol biosensor utilizing selected Bacillus biofilm through an electrochemical method. Makara J Sci 24:24–30. https://doi.org/10.7454/mss.v24i1.11726
9. Aryal N, Wood J, Rijal I, Deng D, Jha MK, Boadu AO (2020) Fate of environmental pollutants: a review. Water Environ Res 92:1587–1594. https://doi.org/10.1002/wer.1404
10. Ayawei N (2017) Modelling and interpretation of adsorption isotherms. J Chem 2017:3039817. https://doi.org/10.1155/2017/3039817
11. Bahgat M, Farghali AA, Rouby WE, Khedr M, Ahmed MYM (2013) Adsorption of methyl green dye onto multi-walled carbon nanotubes decorated with Ni nanoferrite. Appl Nanosci 3:251–261. https://doi.org/10.1007/s13204-012-0127-3
12. Baidas S, Gao B, Meng X (2011) Perchlorate removal by quaternary amine modified reed. J Hazard Mater 189:54–61. https://doi.org/10.1016/j.jhazmat.2011.01.124
13. Barata PD, Costa AI, Granja P, Prata JV (2004) The synthesis of novel polymer-bound calix[4]arenes. React Funct Polym 61:147–151. https://doi.org/10.1016/j.reactfunctpolym.2004.04.005
14. Bashir A, Malik LA, Ahad S, Manzoor T, Bhat MA, Dar GN, Pandith AH (2019) Removal of heavy metal ions from aqueous system by ion-exchange and biosorption methods. Environ Chem Lett 17:729–754. https://doi.org/10.1007/s10311-018-00828-y
15. Bohart GS, Adams EQ (1920) Some aspects of the behavior of charcoal with respect to chlorine. J Am Chem Soc 42:523–544. https://doi.org/10.1021/ja01448a018
16. Burakov AE, Galunin EV, Burakova IV, Kucherova AE, Agarwal S, Tkachev AG, Gupta VK (2018) Adsorption of heavy metals on conventional and nanostructured materials for wastewater treatment purposes: a review. Ecotoxicol Environ Saf 148:702–712. https://doi.org/10.1016/j.ecoenv.2017.11.034
17. Calmon C (1986) Recent developments in water treatment by ion exchange. React Polym Ion Exch Sorbents 4:131–146. https://doi.org/10.1016/0167-6989(86)90008-5
18. Charles J, Bradu C, Crini NM, Sancey B, Winterton P, Torri G, Badot PM, Crini G (2016) Pollutant removal from industrial discharge water using individual and combined effects of adsorption and ion-exchange processes: chemical abatement. J Saud Chem Soc 20:185–194. https://doi.org/10.1016/j.jscs.2013.03.007
19. Chen M, Chen Y, Diao G (2010) Adsorption kinetics and thermodynamics of methylene blue onto *p*-tert-butyl-calix[4,6,8]arene-bonded silica gel. J Chem Eng Data 55:5109–5116. https://doi.org/10.1021/je1006696
20. Chen M, Shang T, Fang W, Diao G (2011) Study on adsorption and desorption properties of the starch grafted *p*-tert-butyl-calix[n]arene for butyl rhodamine B solution. J Hazard Mater 185:914–921. https://doi.org/10.1016/j.jhazmat.2010.09.107
21. Chowdhury MRI, Mulligan CN (2011) Biosorption of arsenic from contaminated water by anaerobic biomass. J Hazard Mater 190:486–492. https://doi.org/10.1016/j.hazmat.2011.03.070
22. Chu KH (2020) Breakthrough curve analysis by simplistic models of fixed bed adsorption: in defense of the century-old Bohart-Adams model. Chem Eng J 380:122513. https://doi.org/10.1016/j.cej.2019.122513
23. Crini G, Lichtfouse E (2019) Advantages and disadvantages of techniques used for wastewater treatment. Environ Chem Lett 17:145–155. https://doi.org/10.1007/s10311-018-0785-9
24. Crini G, Lichtfouse E, Wilson LD, Morin-Crini N (2019) Conventional and non-conventional adsorbents for wastewater treatment. Environ Chem Lett 17:195–213. https://doi.org/10.1007/s10311-018-0786-8

25. Cui Y, Chang X, Zhu X, Zou X (2008) Selective solid phase extraction of trace cadmium(II) and lead(II) from biological and natural water samples by floxacin-modified-silica gel. Int J Environ Anal Chem 88:857–868. https://doi.org/10.1080/03067310802208360
26. Cyganowski P, Dzimitrowicz A (2020) A mini-review on anion exchange and chelating polymers for applications in hydrometallurgy, environmental protection, and biomedicine. Polymers 12:784. https://doi.org/10.3390/polym12040784
27. Czikkely M, Neubauer E, Fekete I, Ymeri P, Fogarassy C (2018) Review of heavy metal adsorption processes by several organic matters from wastewaters. Water 10:1377. https://doi.org/10.3390/w10101377
28. Deblonde T, Leguille CC, Hartemann P (2011) Emerging pollutants in wastewater: a review of the literature. Int J Hyg Environ Health 214:442–448. https://doi.org/10.1016/j.ijheh.2011.08.002
29. Dondoni A, Marra A, Rossi M, Scoponi M (2004) Synthesis and characterization of calix[4]arene-based copolyethers and polyurethanes. Ionophoric properties and extraction abilities towards metal cations of polymeric calix[4]arene urethanes. Polymer 45:6195–6170. https://doi.org/10.1016/j.polymer.2004.06.012
30. Dubinin MM (1960) The potential theory of adsorption of gases and vapors for adsorbents with energetically non uniform surface. Chem Rev 60:235–266. https://doi.org/10.1021/cr60204a006
31. Ebrahimi A, Lakouraj MM, Hassantabar V (2020) Synthesis and characterization of amphiphilic star copolymer of polyaniline and polyacrylic acid based on calix[4]resorcinarene as an efficient adsorbent for removal of paraquat herbicide from water. Mater Today Commun 25:101523. https://doi.org/10.1016/j.mtcomm.2020.101523
32. Eddaif L, Shaban A, Telegdi J (2019) Sensitive detection of heavy metals ions based on the calixarene derivatives-modified piezoelectric resonators: a review. Int J Environ Anal Chem 99:824–853. https://doi.org/10.1080/03067319.2019.1616708
33. Engrand P, de Vains JBR (2002) A bifunctional calixarene designed for immobilization on a natural polymer and for metal complexation. Tetrahedron Lett 43:8863–8866. https://doi.org/10.1016/S0040-4039(02)02203-7
34. Espanol ES, Villamil MM (2019) Calixarenes: generalities and their role in improving the solubility, biocompatibility, stability, bioavailability, detection, and transport of biomolecules. Biomolecules 9:90. https://doi.org/10.3390/biom9030090
35. Fang Q, Chen B (2012) Adsorption of perchlorate onto raw and oxidized carbon nanotubes in aqueous solution. Carbon 50:2209–2219. https://doi.org/10.1016/j.carbon.2012.01.036
36. Freundlich HMF (1906) Uber die adsorption in Losungen. Z Phys Chem 57:385–470. https://doi.org/10.1515/zpch-1907-5723
37. Garg T, Hamilton SE, Hochard JP, Kresch EP, Talbot J (2018) (Not so) gently down the stream: river pollution and health in Indonesia. J Environ Econ Manage 92:35–53. https://doi.org/10.1016/j.jeem.2018.08.011
38. Gebreeyessus GD (2019) Status of hybrid membrane-ion-exchange systems for desalination: a comprehensive review. Appl Water Sci 9:135. https://doi.org/10.1007/s13201-019-1006-9
39. Gezici O, Tabakci M, Kara H, Yilmaz M (2006) Synthesis of *p*-tert-butylcalix[4]arene dinitrile bonded aminopropyl silica and investigating its usability as a stationary phase in HPLC. J Macromol Sci Pure Appl Chem 43:221–231. https://doi.org/10.1080/10601320500437060
40. Ghaedi M, Karami B, Ehsani S, Marahel F, Soylak M (2009) Preconcentration-separation of Co^{2+}, Ni^{2+}, Cu^{2+} and Cd^{2+} in real samples by solid phase extraction of a calix[4]resorcinarene modified Amberlite XAD-16 resin. J Hazard Mater 172:802–808. https://doi.org/10.1016/j.jhazmat.2009.07.065
41. Goswami S, Vig K, Singh DK (2009) Biodegradation of α and β endosulfan by Aspergillus sydoni. Chemosphere 75:883–888. https://doi.org/10.1016/j.chemosphere.2009.01.057
42. Gu Z, Deng B (2007) Use of iron-containing mesoporous carbon (IMC) for arsenic removal from drinking water. Environ Eng Sci 24:113–121. https://doi.org/10.1089/ees.2007.24.113
43. Gu S, Kang X, Wang L, Lichtfouse E, Wang C (2019) Clay mineral adsorbents for heavy metal removal from wastewater: a review. Environ Chem Lett 17:629–654. https://doi.org/10.1007/s10311-018-0813-9

44. Gupta VK, Jain CK, Ali I, Chandra S, Agarwal S (2002) Removal of lindane and malathion from wastewater using bagasse fly ash: a sugar industry waste. Water Res 36:2483–2490. https://doi.org/10.1016/S0043-1354(01)00474-2
45. Gustavo L, Reis TD, Robaina NF, Pacheco WF, Cassella RJ (2011) Separation of malachite green and methyl green cationic dyes from aqueous medium by adsorption on amberlite XAD-2 and XAD-4 resins using sodium dodecylsulfate as carrier. Chem Eng J 171:532–540. https://doi.org/10.1016/j.cej.2011.04.024
46. Ho YS, McKay G (1999) Pseudo second order model for sorption process. Process Biochem 34:451–465. https://doi.org/10.1016/S0032-9592(98)00112-5
47. Huang F, Anslyn EV (2015) Introduction: supramolecular chemistry. Chem Rev 115:6999–7000. https://doi.org/10.1021/acs.chemrev.5b00352
48. Huang JJS, Lin SC, Lowemark L, Liou SYH, Chang Q, Chang TK, Wei KY, Croudace IW (2019) Rapid assessment of heavy metal pollution using ion-exchange resin sachets and micro-XRF core-scanning. Sci Rep 9:6601. https://doi.org/10.1038/s41598-019-43015
49. Hube S, Eskafi M, Hrafnkelsdottir KF, Bjarnadottir B, Bjarnadottir MA, Axelsdottir S, Wu B (2020) Direct membrane filtration for wastewater treatment and resource recovery: a review. Sci Total Environ 710:136375. https://doi.org/10.1016/j.scitotenv.2019.136375
50. Jain VK, Handa A, Pandya R, Shrivastav P, Agrawal YK (2002) Polymer supported calix[4]arene-semicarbazone derivative for separation and preconcentration of La(III), Ce(III), Th(IV) and U(VI). React Funct Polym 5:101–110. https://doi.org/10.1016/S1381-5148(02)00030-5
51. Jumina J, Priastomo Y, Setiawan HR, Mutmainah M, Kurniawan YS, Ohto K (2020a) Simultaneous removal of lead(II), chromium(III) and copper(II) heavy metal ions through an adsorption process using C-phenylcalix[4]pyrogallolarene material. J Environ Chem Eng 8:103971. https://doi.org/10.1016/j.jece.2020.103971
52. Jumina J, Setiawan HR, Triono S, Kurniawan YS, Siswanta D, Zulkarnain AK, Kumar N (2020) The C-arylcalix[4]pyrogallolarene sulfonic acid: a novel and efficient organocatalyst material for biodiesel production. Bull Chem Soc Jpn 93:252–259. https://doi.org/10.1246/bcsj.20190275
53. Kamboh MA, Solangi IB, Sherazi STH, Memon S (2009) Synthesis and application of calix[4]arene based resin for the removal of azo dyes. J Hazard Mater 172:234–239. https://doi.org/10.1016/j.jhazmat.2009.06.165
54. Karadag D, Turan M, Akgul E, Tok S, Faki A (2007) Adsorption equilibrium: kinetics of reactive black 5 and reactive red 239 in aqueous solution onto surfactant-modified zeolite. J Chem Eng Data 52:1615–1620. https://doi.org/10.1021/je7000057
55. Kauspediene D, Kazlauskiene E, Gefeniene A, Binkkiene R (2010) Comparison of the efficiency of activated carbon and neutral polymeric adsorbent in removal of chromium complex dye from aqueous solutions. J Hazard Mater 179:933–939. https://doi.org/10.1016/j.jhazmat.2010.03.095
56. Kazakova EK, Morozova JE, Mironova DA, Konovalov AI (2012) Sorption of azo dyes from aqueous solutions by tetradodecyloxybenzylcalix[4]resorcinarene derivatives. J Incl Phenom Macro Chem 74:467–472. https://doi.org/10.1007/s10847-011-0075-7
57. Kitano H, Hirabayashi T, Ide M, Kyogoku M (2003) Complexation of bisphenol A with calix[6]arene-polymer conjugates. Macromol Chem Phys 204:1419–1427. https://doi.org/10.1002/macp.200350008
58. Kumar S, Jain S (2013) History, introduction, and kinetics of ion exchange materials. J Chem 957647. https://doi.org/10.1155/2013/957647
59. Kumar R, Sharma A, Singh H, Suating P, Kim HS, Sunwoo K, Shim I, Gibb BC, Kim JS (2019) Revisiting fluorescent calixarenes: from molecular sensors to smart materials. Chem Rev 119:9657–9721. https://doi.org/10.1021/acs.chemrev.8b00605
60. Kurniawan YS, Sathuluri RR, Iwasaki W, Morisada S, Kawakita H, Ohto K, Miyazaki M, Jumina J (2018) Microfluidic reactor for Pb(II) ion extraction and removal with amide derivative of calix[4]arene supported by spectroscopic studies. Microchem J 142:377–384. https://doi.org/10.1016/j.microc.2018.07.001

61. Kurniawan YS, Imawan AC, Sathuluri RR, Ohto K, Iwasaki W, Miyazaki M, Jumina J (2019) Microfluidics era in chemistry field: a review. J Idn Chem Soc 2:7–23. https://doi.org/10.34311/jics.2019.02.1.7
62. Kurniawan YS, Sathuluri RR, Ohto K, Iwasaki W, Kawakita H, Morisada S, Miyazaki M, Jumina J (2019) A rapid and efficient lithium-ion recovery from seawater with tripropyl-monoacetic acid calix[4]arene derivative employing droplet-based microfluidic reactor system. Sep Purif Technol 211:925–934. https://doi.org/10.1016/j.seppur.2018.10.049
63. Kurniawan YS, Ryu M, Sathuluri RR, Iwasaki W, Morisada S, Kawakita H, Ohto K, Maeki M, Miyazaki M, Jumina J (2019) Separation of Pb(II) ion with tetraacetic acid derivative of calix[4]arene by using droplet-based microreactor system. Indones J Chem 19:368–375. https://doi.org/10.22146/ijc.34387
64. Kurniawan YS, Anggraeni K, Indrawati R, Yuliati L (2020a) Functionalization of titanium dioxide through dye sensitizing method utilizing red amaranth extract for phenol photodegradation. IOP Conf Ser Mater Sci Eng 902:012029. https://doi.org/10.1088/1757-899X/902/1/012029
65. Kurniawan YS, Sathuluri RR, Ohto K (2020b) Droplet microfluidic device for rapid and efficient metal separation using host-guest chemistry. In: Ren Y (ed) Advances in microfluidic technologies for energy and environmental applications. IntechOpen, London, pp 1–19. https://doi.org/10.5772/intechopen.89846
66. Kurniawan YS, Priyangga KTA, Krisbiantoro PA, Imawan AC (2021) Green chemistry influences in organic synthesis: a review. J Mult App Nat Sci 1:1–12. https://doi.org/10.47352/jmans.v1i1.2
67. Kusumaningsih T, Jumina J, Siswanta D, Mustofa M, Ohto K, Kawakita H (2011) Synthesis, characterization and adsorption test of poly-tetra-*p*-propenyltetrahydroxycalix[4]arene for cadmium ion. Indones J Chem 11:186–190. https://doi.org/10.22146/ijc.21408
68. Lagergren S (1898) Zur theorie der sogenannten adsorption geloester stoffe. Kungliga Svenska Vetenskapsakad Handl 24:1–39
69. Langmuir I (1918) The adsorption of gases on plane surfaces of glass, mica and platinum. J Am Chem Soc 40:13611–21403. https://doi.org/10.1021/ja02242a004
70. Largitte L, Pasquier R (2016) A review of the kinetics adsorption models and their application to the adsorption of lead by an activated carbon. Chem Eng Res Des 109:495–504. https://doi.org/10.1016/j.cherd.2016.02.006
71. Lavendomme R, Zahim S, Leener GD, Inthasot A, Mattiuzzi A, Luhmer M, Reinaud O, Jabin I (2015) Rational strategies for the selective functionalization of calixarenes. Asian J Org Chem 4:710–722. https://doi.org/10.1002/ajoc.201500178
72. Levsen K, Behnert S, Pries B, Svoboda M, Winkeler HD, Zietlow J (1990) Organic compounds in precipitation. Chemosphere 21:1037–1038. https://doi.org/10.1016/0045-6535(90)90127-F
73. Luo P, Kang S, Apip, Zhou M, Lyu J, Aisyah S, Binaya M, Regmi RK, Nover D (2019) Water quality trend assessment in Jakarta: a rapidly growing Asian megacity. PLoS One 14:e0219009. https://doi.org/10.1371/journal.pone.0219009
74. Maheria KC, Chudasama UV (2007) Sorptive removal of dyes using titanium phosphate. Ind Eng Chem Res 46:6852–6857. https://doi.org/10.1021/ie061520r
75. Malik LA, Bashir A, Qureashi A, Pandith AH (2019) Detection and removal of heavy metal ions: a review. Environ Chem Lett 17:1495–1521. https://doi.org/10.1007/s10311-019-00891-z
76. Martinus Y, Astono W, Hendrawan D (2018) Water quality study of Sunter river in Jakarta, Indonesia. IOP Conf Ser Earth Environ Sci 106:012022. https://doi.org/10.1088/1755-1315/106/012022
77. Mclldowie MJ, Mocerino M, Ogden MI (2009) A brief review of Cn-symmetric calixarenes and resorcinarenes. Supramol Chem 22:13–39. https://doi.org/10.1080/10610270902980663
78. Memon S, Oguz O, Yilmaz A, Tabakci M, Yilmaz M, Ertul S (2002) Synthesis and extraction study of calix[4]arene dinitrile derivatives incorporated in a polymeric backbone with bisphenol A. J Polym Environ 9:97–101. https://doi.org/10.1023/A:1020256907414

79. Memon S, Akceylan E, Sap B, Tabakci M, Roundhill DM, Yilmaz M (2003) Polymer supported calix[4]arene derivatives for the extraction of metals and dichromate anions. J Polym Environ 11:67–74. https://doi.org/10.1023/A:1024223922541
80. Memon S, Yilmaz A, Roundhill DM, Yilmaz M (2004) Synthesis of polymeric calix[4]arene dinitrile and diamino-derivatives. Exploration of their extraction properties towards dichromate anion. J Macromol Sci Pure Appl Chem A41:433–447. https://doi.org/10.1081/MA-120028477
81. Memon S, Tabakci M, Yilmaz M, Roundhill DM (2005) A useful approach toward the synthesis and evaluation of extraction ability of thioalkyl calix[4]arenes appended with a polymer. Polymer 46:1553–1560. https://doi.org/10.1016/j.polymer.2004.12.019
82. Memon S, Memon N, Memon S, Latif Y (2011) An efficient calix[4]arene based silica sorbent for the removal of endosulfan from water. J Hazard Mater 186:1696–1702. https://doi.org/10.1016/j.jhazmat.2010.12.048
83. Memon S, Bhatti AA, Memon N (2013) New calix[4]arene appended amberlite XAD-4 resin with versatile perchlorate removal efficiency. Chem Eng Data 58:2819–2827. https://doi.org/10.1021/je400554q
84. Memon S, Bhatti AA, Bhatti AA (2014) Sorption of As(III) by calix[4]arene modified XAD-4 resin: kinetic and thermodynamic approach. J Iran Chem Soc 12:727–735. https://doi.org/10.1007/s13738-014-0531-6
85. Memon S, Memon S, Memon N (2014) A highly efficient *p*-tert-butylcalix[8]arene-based modified silica for the removal of hexachlorocyclohexane isomers from aqueous media. Desalin Water Treat 52:2572–2582. https://doi.org/10.1080/19443994.2013.794710
86. Memon FN, Memon S (2015) Sorption and desorption of basic dyes from industrial wastewater using calix[4]arene based impregnated material. Sep Sci Technol 50:1135–1146. https://doi.org/10.1080/01496395.2014.965831
87. Mendes AR, Gregorio CC, Barata PD, Costa AI, Prata JV (2005) Linear and crosslinked copolymers of *p*-tert-butylcalix[4]arene derivatives and styrene: new synthetic approaches to polymer-bound. React Funct Polym 65:9–21. https://doi.org/10.1016/j.reactfunctpolym.2005.01.006
88. Mir SH, Nagahara LA, Thundat T, Tabari PM, Furukawa H, Khosla A (2018) Review—organic-inorganic hybrid functional materials: an integrated platform for applied technologies. J Electrochem Soc 165:B3137-3156. https://doi.org/10.1149/2.0191808jes
89. Mirjavadi ES, Tehrani RMA, Khadir A (2019) Effective adsorption of zinc on magnetic nanocomposite of Fe_3O_4/zeolite/cellulose nanofibers: kinetic, equilibrium, and thermodynamic study. Environ Sci Pollut Res 26:33478–33493. https://doi.org/10.1007/s11356-019-06165-z
90. Mohan D, Pittmann CU (2007) Arsenic removal from water/wastewater using adsorbents: a critical review. J Hazard Mater 142:1–53. https://doi.org/10.1016/j.jhazmat.2007.01.006
91. Mohddin AT, Hameed BH (2010) Adsorption of methyl violet dye on acid modified activated carbon: isotherms and thermodynamics. J Appl Sci Environ Sani 5:151–160
92. Mollahosseini A, Khadir A, Saeidian J (2019) Core-shell polypyrrole/Fe_3O_4 nanocomposite as sorbent for magnetic dispersive solid-phase extraction of Al^{3+} ions from solutions: investigation of the operational parameters. J Water Proc Eng 29:100795. https://doi.org/10.1016/j.jwpe.2019.100795
93. Mouzdahir YE, Elmchaouri A, Mahboub R, Gil A, Korili SA (2007) Adsorption of methylene blue from aqueous solutions on a Moroccan clay. J Chem Eng Data 52:1621–1625. https://doi.org/10.1021/je700008g
94. Navarro RR, Sumi K, Matsumura M (1999) Improved metal affinity of chelating adsorbents through graft polymerization. Water Res 33:2037–2044. https://doi.org/10.1016/S0043-1354(98)00421-7
95. Nicolopoulou-Stamati P, Maipas S, Kotampasi C, Stamatis P, Hens L (2016) Chemical pesticides and human health: the urgent need for a new concept in agriculture. Front Public Health 4:148. https://doi.org/10.3389/fpubh.2016.00148

96. Nie R, Chang X, He Q, Hu Z, Li Z (2009) Preparation of *p*-tert[(dimethylamino)methyl]-calix[4]arene functionalized aminopropylpolysiloxane resin for selective solid-phase extraction and preconcentration of metal ions. J Hazard Mater 169:203–209. https://doi.org/10.1016/j.jhazmat.2009.03.084
97. Nithya R, Thirunavukkarasu A, Sathya AB, Sivashankar R (2021) Magnetic materials and magnetic separation of dyes from aqueous solutions: a review. Environ Chem Lett. https://doi.org/10.1007/s10311-020-01149-9
98. Nodeh HR, Kamboh MA, Ibrahim WAW, Jume BH, Sereshti H, Sanagi MM (2019) Equilibrium, kinetic and thermodynamic study of pesticides removal from water using novel glucamine-calix[4]arene functionalized magnetic graphene oxide. Environ Sci Process Impacts 21:714–726. https://doi.org/10.1039/C8EM00530C
99. Ofomaja AE, Ho YS (2008) Effect of temperatures and pH on methyl violet biosorption by Mansonia wood sawdust. Biores Tech 99:5411–5417. https://doi.org/10.1016/j.biortech.2007.11.018
100. Ohto K, Senba Y, Eguchi N, Shinohara T, Inoue K (1999) Solid phase extraction of metal ions on resins impregnated with carboxylates of phenolic oligomers. Solvent Extr Res Dev Jpn 6:101–112
101. Ohto K, Tanaka Y, Yano M, Shinohara T, Murakami M, Inoue K (2001) Selective adsorption of lead ion on calix[4]arene carboxylate resin supported by polyallylamine. Solvent Extr Ion Exch 19:725–741. https://doi.org/10.1081/SEI-100103817
102. Ohto K, Yamasaki T, Wakisaka S, Shinohara T, Inoue K (2003) Preparation of novel crosslinking type resins based on calix[4]arene tetracarboxylate with high selectivity and high loading capacity for lead ion. J Ion Exch 14:301–304
103. Ohto K (2010) Review of the extraction behavior of metal cations with calixarene derivatives. Solvent Extr Res Des, Jpn 17:1–18
104. Ovsyannikov A, Solovieva S, Antipin I, Ferlay S (2017) Coordination polymers based on calixarene derivatives: structures and properties. Coord Chem Rev 352:151–186. https://doi.org/10.1016/j.ccr.2017.09.004
105. Panda SK, Aggarwal I, Kumar H, Prasad L, Kumar A, Sharma A, Vo DVN, Thuan DV, Mishra V (2021) Magnetite nanoparticles as sorbents for dye removal: a review. Environ Chem Lett. https://doi.org/10.1007/s10311-020-01173-9
106. Panday KK, Prasad G, Singh VN (1985) Copper(II) removal from aqueous solutions by fly ash. Water Res 19:869–873. https://doi.org/10.1016/0043-1354(85)90145-9
107. Piri F, Mollahosseini A, Khadir A, Hosseini MM (2019) Enhanced adsorption of dyes on microwave-assisted synthesized magnetic zeolite-hydroxyapatite nanocomposite. J Environ Chem Eng 7:103338. https://doi.org/10.1016/j.jece.2019.103338
108. Piri F, Mollhosseini A, Khadir A, Hosseini MM (2020) Synthesis of a novel magnetic zeolite-hydroxyapatite adsorbent via microwave-assisted method for protein adsorption via magnetic solid-phase extraction. J Iran Chem Soc 17:1635–1648. https://doi.org/10.1007/s13738-020-01883-5
109. Pirillo S, Ferreira ML, Rueda EH (2007) Adsorption of alizarin, eriochrome blue black R, and fluorescein using different iron oxides as adsorbents. Ind Eng Chem Res 46:8255–8263. https://doi.org/10.1021/ie0702476
110. Prabawati SY, Jumina, Santosa SJ, Mustofa (2011) Synthesis of a series of calix[6]arene polymers from *p*-tert-butylphenol. ICBB2011 Proc 1:93–100. ISSN:2088–9771
111. Priyangga KTA, Kurniawan YS, Yuliati L (2020) Synthesis and characterizations of C-3-nitrophenylcalix[4]resorcinarene as a potential chemosensor for La(III) ions. IOP Conf Ser Mater Sci Eng 959:012014. https://doi.org/10.1088/1757-899X/959/1/012014
112. Qureshi I, Memon S, Yilmaz M (2009) Estimation of chromium(VI) sorption efficiency of novel regenerable *p*-tert-butylcalix[8]areneoctamide impregnated Amberlite resin. J Hazard Mater 164:675–682. https://doi.org/10.1016/j.jhazmat.2008.08.076
113. Qureshi I, Qazi MA, Bhatti AA, Memon S, Sirajuddin YM (2011) An efficient calix[4]arene appended resin for the removal of arsenic. Desalination 278:98–104. https://doi.org/10.1016/j.desal.2011.05.007

114. Rajmohan KS, Chandrasekaran R, Varjani S (2020) A review on occurrence of pesticides in environment and current technologies for their remediation and management. Indian J Microbiol 60:125–138. https://doi.org/10.1007/s12088-019-00841-x
115. Raman N, Sudharsan S, Pothiraj K (2012) Synthesis and structural reactivity of inorganic-organic hybrid nanocomposites: a review. J Saud Chem Soc 16:339–352. https://doi.org/10.1016/j.jscs.2011.01.012
116. Santhi T, Manonmani S, Smitha T (2010) Kinetics and isotherm studies on cationic dyes adsorption onto annona squmosa seed activated carbon. Inter J Eng Sci Tech 2:287–295
117. Sathuluri RR, Kurniawan YS, Kim JY, Maeki M, Iwasaki W, Morisada S, Kawakita H, Miyazaki M, Ohto K (2018) Droplet-based microreactor system for stepwise recovery of precious metal ions from real metal waste with calix[4]arene derivatives. Sep Sci Technol 53:1261–1272. https://doi.org/10.1080/01496395.2017.1366518
118. Savyasachi AJ, Kotova O, Shanmugaraju S, Bradberry SJ, OMaille GM, Gunnlaugsson T (2017) Supramolecular chemistry: a toolkit for soft functional materials and organic particles. Chem 3:764–811. https://doi.org/10.1016/j.chempr.2017.10.006
119. Shetty D, Boutros S, Skorjanc T, Garai B, Asfari Z, Raya J, Trabolsi A (2020) Fast and efficient removal of paraquat in water by porous polycalix[n]arenes (n = 4, 6, and 8). J Mater Chem A 8:13942–13945. https://doi.org/10.1039/D0TA01907K
120. Shih PK, Chiang LC, Lin SC, Chang TK, Hsu WC (2019) Application of time-lapse ion exchange resin sachets (TIERS) for detecting illegal effluent discharge in mixed industrial and agricultural areas Taiwan. Sustainability 11:3129. https://doi.org/10.3390/su11113129
121. Shinohara T, Wakisaka S, Ohto K, Inoue K (2000) Synthesis of novel type resin based on calix[4]arene carboxylate and selective separation of lead from zinc. Chem Lett 29:640–641. https://doi.org/10.1246/cl.2000.640
122. Sikosana ML, Sikhwivhilu K, Moutloali R, Madyira DM (2019) Municipal wastewater treatment technologies: a review. Procedia Manuf 35:1018–1024. https://doi.org/10.1016/j.promfg.2019.06.051
123. Sliwa W, Girek T (2010) Calixarene complexes with metal ions. J Incl Phenom Macrocycl Chem 66:15–41. https://doi.org/10.1007/s10847-009-9678-7
124. Sousa JCG, Riberio AR, Barbosa MO, Pereira FR, Silva AMT (2018) A review on environmental monitoring of water organic pollutants identified by EU guidelines. J Hazard Mater 344:146–162. https://doi.org/10.1016/j.jhazmat.2017.09.058
125. Sprynskyy M, Ligor T, Buszewski B (2008) Clinoptilolite in study of lindane and aldrin sorption processes from water solution. J Hazard Mater 151:570–577. https://doi.org/10.1016/j.jhazmat.2007.06.023
126. Sun S, Sidhu V, Rong Y, Zheng Y (2018) Pesticide pollution in agricultural soils and sustainable remediation methods: a review. Curr Pollut Rep 4:240–250. https://doi.org/10.1007/s40726-018-0092-x
127. Sun Y, Zhou S, Chiang PC, Shah KJ (2019) Evaluation and optimization of enhanced coagulation process: water and energy nexus. Water-Energy Nexus 2:25–36. https://doi.org/10.1016/j.wen.2020.01.001
128. Tabakci B, Beduk AD, Tabakci M, Yilmaz M (2006) Synthesis and binding properties of two polymeric thiacalix[4]arenes. React Funct Polym 66:379–386. https://doi.org/10.1016/j.reactfunctpolym.2005.08.013
129. Tabakci M, Erdemir S, Yilmaz M (2007) Preparation, characterization of cellulose-grafted with calix[4]arene polymers for the adsorption of heavy metals and dichromate anions. J Hazard Mater 148:428–435. https://doi.org/10.1016/j.jhazmat.2007.02.057
130. Tabakci M, Yilmaz M (2008) Sorption characteristics of Cu(II) ions into silica gel-immobilized calix[4]arene polymer in aqueous solutions: batch and column studies. J Hazard Mater 151:331–338. https://doi.org/10.1016/j.jhazmat.2007.05.077
131. Tabakci M, Yilmaz M (2008) Synthesis of a chitosan-linked calix[4]arene chelating polymer and its sorption ability toward heavy metals and dichromate anions. Bioresour Technol 99:6642–6645. https://doi.org/10.1016/j.biotech.2007.11.066

132. Tang Y, Liang S, Guo H, You H, Gao N, Yu S (2013) Adsorptive characteristic of perchlorate from aqueous solutions by MIEX resin. Colloids Surf A 417:26–31. https://doi.org/10.1016/j.colsurfa.2012.10.040
133. Temkin MJ, Pyzhev V (1940) Recent modifications to Langmuir isotherms. Acta Physiochim 12:217–222
134. Thirunavukkarasu OS, Viraraghavan T, Subramanian KS, Chaalal O, Islam MR (2005) Arsenic removal in drinking water-Impacts and novel removal technologies. Energy Sour 27:209–219. https://doi.org/10.1080/00908310490448271
135. Thomas HC (1944) Heterogeneous ion exchange in a flowing system. J Am Chem Soc 66:1664–1666. https://doi.org/10.1021/ja01238a017
136. Tuzen M, Saygi KO, Soylak M (2008) Solid phase extraction of heavy metal ions on environmental samples on multiwalled carbon nanotubes. J Hazard Mater 152:632–639. https://doi.org/10.1016/j.jhazmat.2007.07.026
137. Valenzuela EF, Menezes HC, Cardeal ZL (2020) Passive and grab sampling methods to assess pesticide residues in water: a review. Environ Chem Lett 18:1019–1048. https://doi.org/10.1007/s10311-020-00998-8
138. Wang L, Shi X, Jia P, Yang Y (2004) Preparation of calixarene-containing polymer with proton transport ability. J Polym Sci A Polym Chem 42:6259–6266. https://doi.org/10.1002/pola.20480
139. Wang S, Li L, Zhu ZH (2007) Solid-state conversion of fly ash to effective adsorbents for Cu removal from wastewater. J Hazard Mater 139:254–259. https://doi.org/10.1016/j.jhazmat.2006.06.018
140. Weber WJ, Morris JC (1963) Kinetics of adsorption on carbon from solution. J Sanit Eng Div Proc Am Soc Civil Eng 89:31–59
141. Woodberry P, Stevens G, Snape I, Stark S (2007) Removal of metal contaminants by ion-exchange resin columns, Thala valley tip, Casey station, Antartica. Solvent Extr Ion Exch 24:603–620. https://doi.org/10.1080/07366290600762108
142. Wu X, Wang Y, Xu L, Lv L (2010) Removal of perchlorate contaminants by calcined Zn/Al layered double hydroxides: equilibrium, kinetics, and column studies. Desalination 256:136–140. https://doi.org/10.1016/j.desal.2010.02.001
143. Xiang G, Huang Y, Luo Y (2009) Solid phase extraction of trace cadmium and lead in food samples using modified peanut shell prior to determination by flame atomic absorption spectrometry. Microchim Acta 165:237–242. https://doi.org/10.1007/s00604-008-0126-y
144. Xu D, Hein S, Loo SL, Wang K (2008) The fixed-bed study of dye removal on chitosan beads at high pH. Ind Eng Chem Res 47:8796–8800. https://doi.org/10.1021/ie800387z
145. Xu JH, Gao NY, Deng Y, Sui MH, Tang YI (2011) Perchlorate removal by granular activated carbon coated with cetyltrimethyl ammonium chloride. Desalination 275:87–92. https://doi.org/10.1016/j.desal.2011.02.036
146. Xu JH, Gao NY, Deng Y, Xia SQ (2013) Nanoscale iron hydroxide-doped granular activated carbon (Fe-GAC) as a sorbent for perchlorate in water. Chem Eng J 222:520–526. https://doi.org/10.1016/j.cej.2012.07.141
147. Yang Y, Swager TM (2007) Main-chain calix[4]arene elastomers by ring-opening metathesis polymerization. Macromolecules 40:7437–7440. https://doi.org/10.1021/ma071304
148. Yoon YH, Nelson JA (1984) Application of gas adsorption kinetics: a theoretical model for respirator cartridge service life. Am Ind Hyg Assoc J 45:509–516. https://doi.org/10.1080/15298668491400197
149. Yu B, Zhang Y, Shukla A, Shukla SS, Dorris KL (2000) The removal of heavy metal from aqueous solutions by sawdust adsorption-removal of copper. J Hazard Mater 80:33–42. https://doi.org/10.1016/S0304-3894(00)00278-8
150. Zhang X, Li A, Jiang Z, Zhang Q (2006) Adsorption of dyes and phenol from water on resin adsorbents: effect of adsorbate size and pore size distribution. J Hazard Mater 137:1115–1122. https://doi.org/10.1016/j.jhazmat.2006.03.061

151. Zhang GF, Zhan JY, Li HB (2011) Selective binding of carbamate pesticides by self-assembled monolayers of calix[4]arene lipoic acid: wettability and impedance dual-signal response. Org Lett 13:3392–3395. https://doi.org/10.1021/ol201143z
152. Zhu X, Cui Y, Chang X, Zou X, Li Z (2009) Selective solid-phase extraction of lead(II) from biological and natural water samples using surface-grafted lead(II)-imprinted polymer. Microchim Acta 164:125–132. https://doi.org/10.1007/s00604-008-0045-y

Luminescent Carbon Dots for Environmental Photocatalytic

Fernanda G. L. Medeiros Borsagli and Alessandro Borsagli

Abstract Innovative materials based on different natural sources with enhanced luminescence propertyare called carbon dots. Different synthesis methods of these incredible nanomaterials are changing their property. The photocatalytic potential of the carbon dots has a solution for wastewater treatment. Important characterizations to determine carbon dots' features Are low cost, biocompatible, facile synthesis, high-quantum yield (QY) nanomaterial, namely carbon dot. In regard to the scenario in which the proliferation of waste in waters caused by anthropogenic activities and natural disasters has become a primary global concern because of severe health and environmental harms, the search for new materials that provide eradication of waste present in water is a great challenge. In this sense, nanomaterials arising from natural sources is an incredible alternative, as the carbon dots. Many types of carbon dots were anaalyzed in the literature, as these nanomaterials have considerable potential for many purposes, including the photocatalytic process of organic and inorganic materials in water because they are low cost and biocompatible, which makes them eco-friendly materials to apply in water treatment. In this overview, we presented various researches that demonstrated different carbon dots syntheses, an extensive characterization by many techniques, and the photocatalytic process was reported based on different models, as different precursors, methodologies of syntheses, which changes all photodegradation results. Based on the literature, we have also demonstrated that these nanomaterials exhibit an incredible potential for photodegradation of diversity of organic materials, including benzene, pesticides, phenol and methylene blue, and others, in water, indicating their promising solution of low cost, biocompatible, photocatalytic nanomaterial for wastewater treatment. Finally, this overview shows that these luminescent materials presented the incredible potential of use in wastewater treatment.

F. G. L. Medeiros Borsagli (✉) · A. Borsagli
Institute of Engineering, Science and Technology, Universidade Federal Dos Vales Do Jequitinhonha E Mucuri/UFVJM, Av. 01, 4050 Cidade Universitária Zip code 39440-039, Janaúba-MG, Brazil
e-mail: fernanda.borsagli@ufvjm.edu.br

E. Lichtfouse et al. (eds.), *Inorganic-Organic Composites for Water and Wastewater Treatment*, Environmental Footprints and Eco-design of Products and Processes,
https://doi.org/10.1007/978-981-16-5928-7_6

Keywords Carbon dots · Photocatalytic process · Wastewater treatment · Characterization · Organic and inorganic pollutants · Eco-friendly · Luminescence

Abbreviation

AFM	Atomic Microscopy Force
ATP	Adenosine 5′-triphosphate (ATP)
CDs or Cdots	Carbon Dots
DLS	Dynamic Light Scattering
FTIR	Infrared Spectroscopy
FL	Fluorescence Spectroscopy
HRTEM	Transmission Electronic Microscopy with Higher Resolution
PL	Photoluminescence or Photoluminescence Spectroscopy
UV-Visible	Ultraviolet-Visible Spectroscopy
XRD	X-Ray Diffraction
XPS	X-Ray Photoelectron Spectroscopy

1 Introduction

Water is one of the essential substances present in nature that is fundamental for the equilibrium of animal and vegetable life; even so, the ecosystem. In addition, this element is connected to the development of each society and culture around the world. Moreover, it directly impacts many human activities, like many industries, agriculture, livestock, energy, and governments. Hence, the connection between water and sustainable progress is quite complex, numerous, and deep profound. Additionally, the water quality has increased a lot of its deterioration due to anthropogenic actives [2, 72]. In this way, the gigantic toxicity and significant pollutants resistance of many materials, mainly in the aquatic's environments, as pigments, phenolics, drugs, fungicide, pesticide, and insecticide, caused a major concern of reduction of potable water on the planet [72].

On the other wise, many chemical, physical, and physicochemical water treatments are available. The choice depends on the type of compound, the treatment methodology and its cost. Also, the material used for water treatment changes all analyses, results, and cost of the process [5, 6, 51]. In this sense, the use of different natural sources, as natural polymers, biological materials, biomass, others, is fascinating based on various applications. Moreover, these natural sources show an excellent thermal conversion and do not release toxic compounds in the environment, are available in large quantities, or particular wastes from manufacturing operations may have great potential to be used as low-cost biomass, as they represent practically unexploited resources, are broadly available and are environmentally friendly [6, 51, 61].

In this way, the use of these natural sources in many applications has increased in the past few years. In such a way, these materials have been widely used in nanomaterials production, as stabilizers for nanoparticles or even in their production. Thusly, distinct nanomaterials, semiconductor nanocrystals, namely, quantum dots (QDs), are inorganic particles in the range of 1–10 nm. Their electronic and optical characteristics vary a lot according to crystal size as for the physical dimension. These unique ultra-small semiconductor nanocrystals have a short time ago come up as a progressive materials class, which possess unique characteristics, [67, 74, 103]. Therefore, many of these quantum dots are toxic to human health. In this sense, a recently produced nanoparticle class based on these natural sources has drawn the attention of many research pieces, the carbon quantum dots or knowledge carbon dots (Cdots or CD). This class of nanoparticles is a new class of ultra-small semiconductors with 10 nm of diameter approximately, and it can overcome the traditional quantum dots restriction [8].

The advantages of the carbon dots are the simple synthesis, low cost, and outstanding biocompatibility. Their applications range from sensing, catalysis, nanomedicine, energy conversion, and bioimaging because of the photoluminescence (PL) emission, which implies good luminescence (Fig. 1) [90] (Zhang and Hu 106). In this context, the ultra-small semiconductor nanocrystal carbon dot has many applications; one of these is related to cell labeling, cancer cell diagnosis, and photocatalytic activity for wastewater treatment.

Many methods are used to produce carbon dots. One of them includes top-down approaches, where materials based on graphite or carbon nanotubes are exposed to laser ablation or electrochemical process. Another option is via bottom-up methodology, where polysaccharides or similar are used to produce the carbon dots using external energy, as ultrasonication, microwave, or heating [49, 59, 81]. The advantage

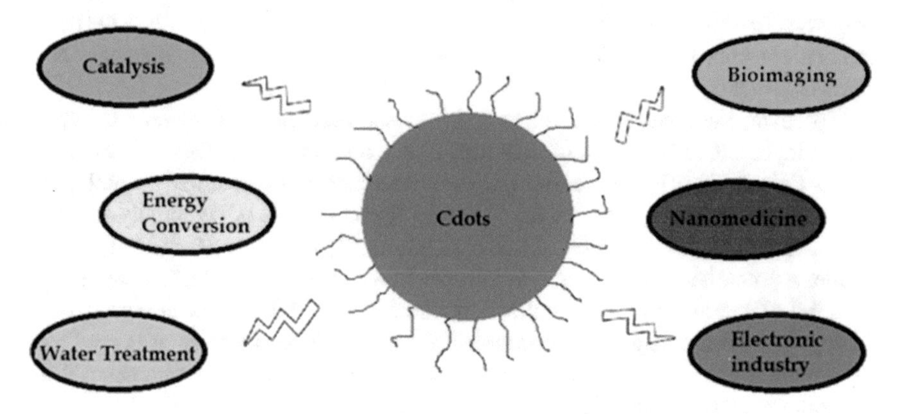

Fig. 1 Schematic presentation of the application of carbon dots in Science, Engineering and Industry (Produced by the author)

goes to an excellent solubility in a water medium, incredible luminescence in the presence of visible light, little toxicity, excellent biological compatibility, considerable stability, to facility functionalization [18, 33, 86].

Moreover, the different synthesis of these carbon dots may modify their size and properties. Besides, the introduction of chemical groups at the graphitic sheet edges also adjusted their characterization and properties, including the photoluminescence [10, 41, 75, 91]. Furthermore, their luminescence property may explain based on defects in its surface, boundary structure, doping, and triple fundamental carbene state as free zig-zag sites, recombination of hole, and electrons [18, 57]. Another fascinating property perceived in some carbon dots is the anti-Stoke transition based on the excitation or conversion of multiple photons [11, 95].

In this sense, this chapter overviews the researches about carbon dots in the last years. This new class of nanomaterials brings lots of advantages over many types of nanoparticles. The possibility of using these natural nanomaterials in water treatment has increased the attention in the recent decade. This overview intends to demonstrate different carbon dots synthesis, as its intriguing properties provide diverse applications, mainly with regard to using their luminescent potential for water treatment.

2 Carbon Dots Synthesis

The literature exhibited various researches that produced different carbon dots, hydrothermal method, carbonization/pyrolysis, microwave, and others. The carbon dots were mostly performed based on hydrothermal methods, probably, based on the simple equipment, few parameters, facile synthesis. The microwave synthesis demonstrated a better performance in terms of time, as most of this type is made in a few minutes, and a facile synthesis [63]. Therefore, regardless of the method used, all synthesis revealed carbon dots with a remarkable morphology (1–10 nm), properties (mainly, the photoluminescence excitation and emission at 200–400 nm), and achievement [47], which are the requirements for the use of nanomaterials in water treatment.

Moreover, the source used for each synthesis, regardless of method, chemical groups involved, the concentration of precursors, and time of synthesis changing all results (Wu et al. 2011). Also, most of the researches was performed on biological applications. However, a few works showed a great result of selectivity of some ions, like mercury (Hg^{2+}) [89], chromates ($Cr_2O_7^{2-}$) [55], iron (Fe^{3+}) [66], which implies a potential for photocatalytic activity in water treatment. Also, an incredible potential of conjugated, functionalization, energy transfer, luminescence properties, doping, low cost, outstanding biocompatibility, and facile synthesis, it is predicted the potential use of these materials as an alternative luminescent nanomaterial for environmental photocatalytic.

In the past years, a crescent number of different routes of carbon dots were performed in the literature. Although the proposal of this overview is bringing most of the data compiled in the literature, it is impossible to demonstrate all synthesis. In this way, it was compiled the most intriguing researches about carbon dots synthesis in the last years.

2.1 Hydrothermal Method

Most of the carbon dots syntheses are performed on the top-down method or bottom-up methods. The differences of both are based on the cost and method of producing the carbon dots. Newly, chemical routes based on the hydrothermal synthesis, considering one of the most straightforward and low-cost methods because using a cheap apparatus, it is performed with basic manipulation, it has small energy expenditure, excellent selectivity, and its performance is one single step (Liu et al. 107) [40]. This approach has been used in the past years by many scientists worldwide [24, 40, 64, 92] (Liu et al. 107).

In this way, Liang et al. [40] performed a simple, greener, and facile hydrothermal carbon dots synthesis using only water and gelatine (Fig. 2). Its synthesis produced a carbon dot with a small size (Fig. 3) and photoluminescence (PL) emission at 430 nm. The synthesis consists of dissolving the gelatine, whose source is the skin and bones of animals, in water, then, the solution goes to the autoclave at the temperature of 200 °C for 3 h. In sequence, the yellow solution is centrifugated for 30 min to remove any precipitate, and carbon dots by hydrothermal synthesis was produced.

In addition, Zhang et al. [93] used hydrothermal synthesis based on the fish scale of grass carp precursors. Similar to Liang et al. [40], the fish scale precursor was dissolved in deionized water. In sequence, a Teflon stainless steel was used at a temperature of 200 °C in an autoclave for 20 h. Then, it was centrifugated for 15 min to remove the precipitate, and an aqueous Carbon dot was produced with distinct chemical groups, like OH, NH_2, C = O, C≡N, C-N, C-O (Fig. 4). The Lidocaine hydrochloride (2-(diethylamino)-N-(2,6-dimethy-lphenyl)-acetamide hydrochloride (LH) was the main target to detect based on carbon dots fluorescence. Similarly, Saud et al. [65] performed carbon dots synthesis based on hydrothermal methodology, in their research was used acid citric and urea were dissolved in water at 180 °C for

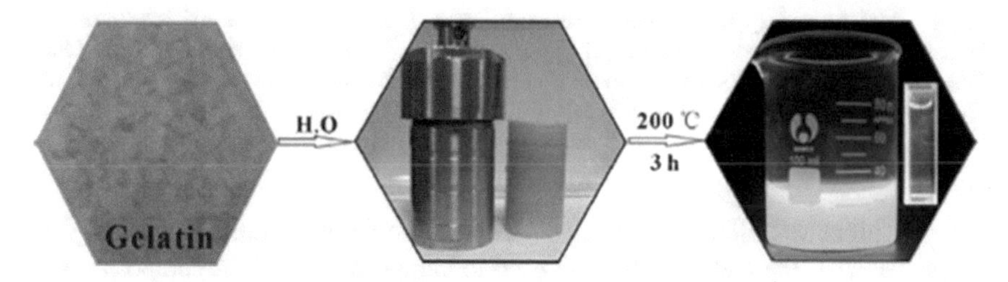

Fig. 2 Graphical abstract of the photoluminescent carbon dots synthesis based on commercial gelatine using a hydrothermal treatment. (By Liang et al. [40], License number 4960230818190 provided by Elsevier and Copyright Clearance Center)

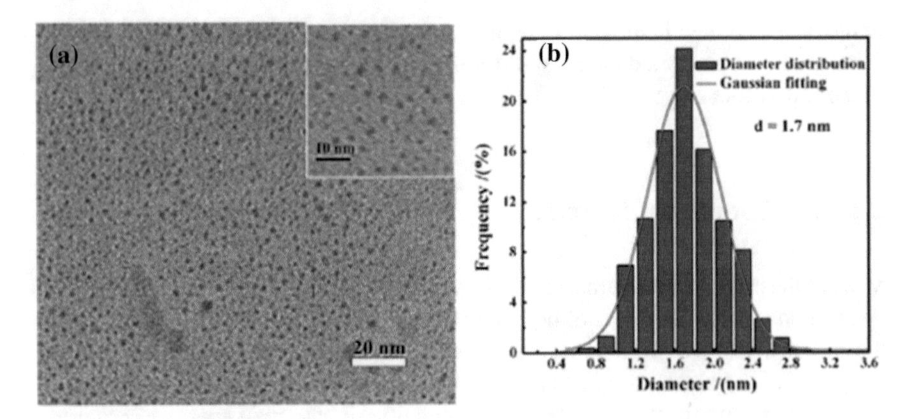

Fig. 3 **a** Images by High-Resolution Transmission Electronic Microscopy (HRTEM) (inset) and **b** Carbon dots Size distribution. (By Liang et al. [40], License number 4960230818190 provided by Elsevier and Copyright Clearance Center)

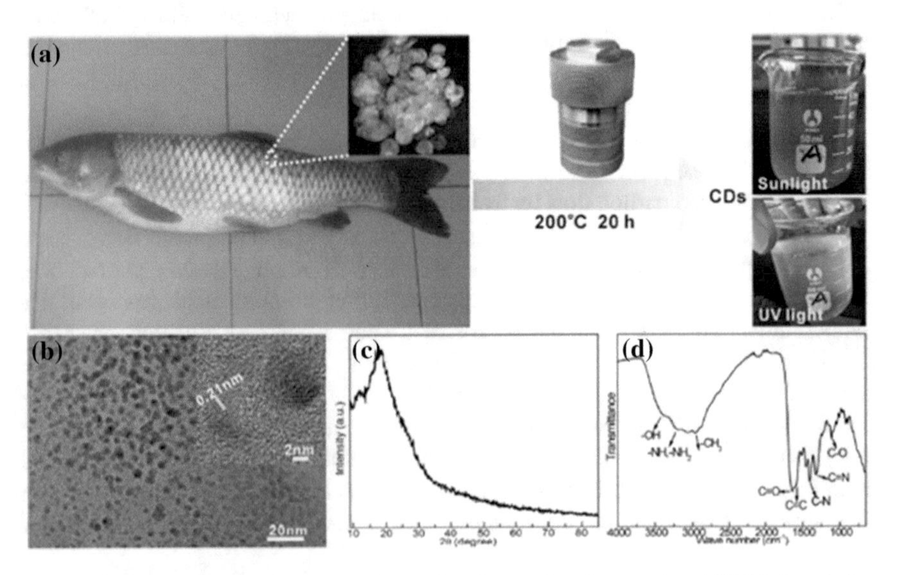

Fig. 4 **a** Fish scales preparation schematic of carbon dots, carbon dots results at (**b**) High-Resolution Transmission Electronic Microscopy (HRTEM) and Transmission Electronic Microscopy (TEM) (inset), **c** X-Ray Diffraction (XRD) and **d** Fourier Transform Infrared Spectroscopy (FTIR). (By Zhang et al. [93], License number 4960231003729 provided by Elsevier and Copyright Clearance Center)

5 h into autoclave Teflon stainless steel. After the complete filtration, separating the larges particles, the carbon dots were used to compose a nanofiber composite with TiO_2 to apply into the photocatalytic process.

Shangguan et al. [66] also made new carbon dots using adenosine 5′-triphosphate (ATP) by hydrothermal synthesis at 220 °C for 6 h. The adenosine 5′-triphosphate was

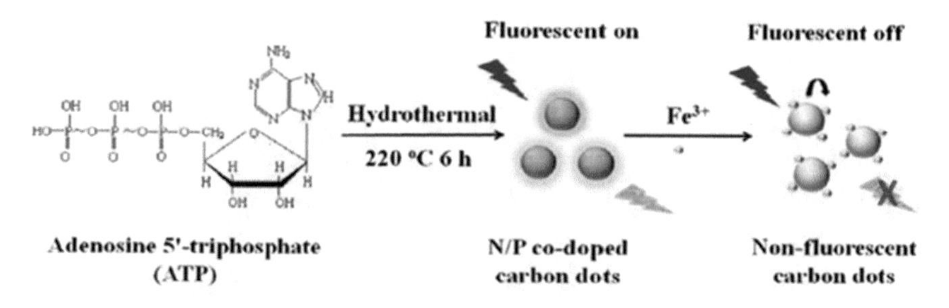

Fig. 5 Highly Fluorescent N/P Co-doped Carbon Dots Synthesis for Fe^{3+} Detection (By Shangguan et al. [66], Reprinted (adapted) with permission from (By Shangguan et al. [66]. Copyright (2020) American Chemical Society)

used to provide a source of nitrogen, phosphorous, and carbon simultaneous. These carbon dots showed excellent fluorescence properties, low toxicity, chemical stability, incredible quantum yield. Added these nanomaterials presented a remarkably Fe^{3+} selective in the presence of Ethylenediamine tetra acetic acid (EDTA) doping the carbon dots (Fig. 5).

Further, Yuan et al. [104] prepared carbon dots by a hydrothermal method based on wheat straw on autoclave at the temperature of 250 °C for 10 h. In this synthesis, the supernatant was discharged, and the solution was dialyzed for 10 min. The pellets obtained were dissolved in deionized water for later use. The carbon dots produced showed a photoluminescence excitation at 360 nm and (1.7 ± 0.2) nm of average size based on Dynamic Light Scattering (DLS) (Fig. 6).

The advantage of this methodology is the facility of synthesis and the need for only a few pieces of equipment. Nevertheless, the production time is too long, which implies an increase in energy cost and difficulty in their reproduction on a large scale. However, it is still the most used methodology to produce carbon dots in the literature.

2.2 *Carbonization or Pyrolysis Synthesis*

Moreover, the hydrothermal synthesis, other methods have been performed based on carbonization using natural sources, as watermelon peel [96], carbohydrates [56], low-molecular-weight alcohols [15], water hyacinth leaves [60], citric acid, and L-tyrosine methyl Ester produced carbon dots based on low-temperature carbonization using watermelon peel [96, 98]. In their research, at the temperature of 220 °C, the watermelon peel was carbonized for 2 h in an air atmosphere (Fig. 7). The carbon dots showed a size of approximately 2 nm and incredible luminescence.

Prathumsuwan et al. [60] produced carbon dots using water hyacinth leaves by carbonization based on acid-treated pyrolysis to detected borax (sodium tetraborate

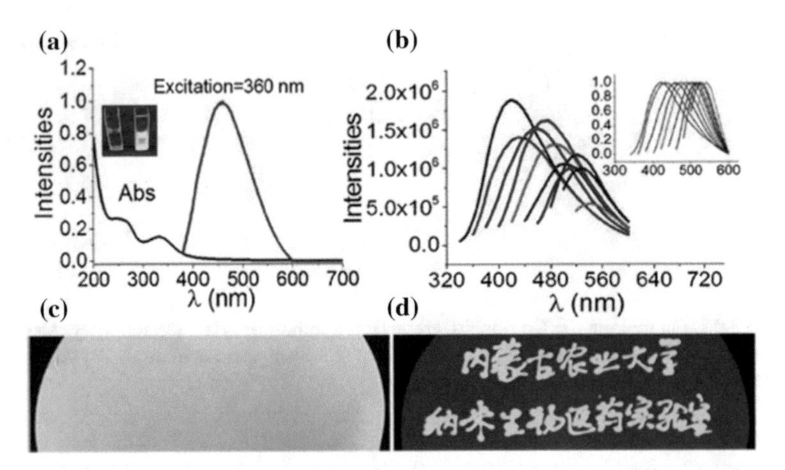

Fig. 6 **a** Bands at 247 and 334 nm of the carbon dots at Ultraviolet-Visible Spectroscopy (UV-Vvis) absorption (black line), and the carbon dots narrow and symmetrical photoluminescence (PL) (red line) band. Inset digital images of carbon dots dispersed in pure water at daylight (left) and 365-nm Ultraviolet–Visible (UV) light (right), respectively. **b** Carbon dots photoluminescence spectra at different excitation wavelengths in the range of 320–500 nm (slit = 5 nm) (PL-normalized emission spectra inset). Characters' optical images using the C-dots as ink on the UV-non-resistant paper at (**c**) white light and **d** UV light. (By Yuan et al. [104], License number 4960311333 9822 provided by Elsevier and Copyright Clearance Center)

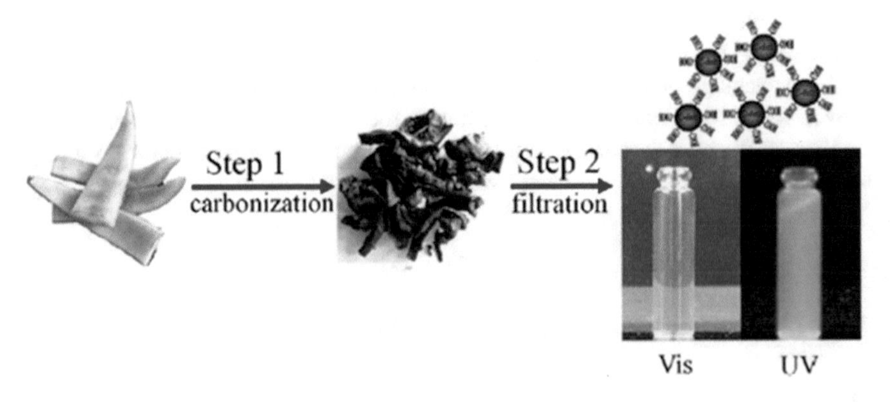

Fig. 7 Graphical abstract of water-soluble fluorescent carbon dots synthesis based on watermelon peel (By Zhou et al. [96], License number 4960330931030 provided by Elsevier and Copyright Clearance Center)

decahydrate ($Na_2B_4O_7.10H_2O$)), a chemical reagent used to several industrial products, including sanitary products for houses (Fig. 8). Their carbon dots showed a 3–4 nm size and presented an excellent sensibility with borax and borax conjugated with other reagents and an incredible borax recovery (Table 1).

Also, Zhou et al. (2018) produced a new bio-thiol based on carbon dots. In their research, the carbon dots stabilizer with L-tyrosine methyl Ester (Try-CD)

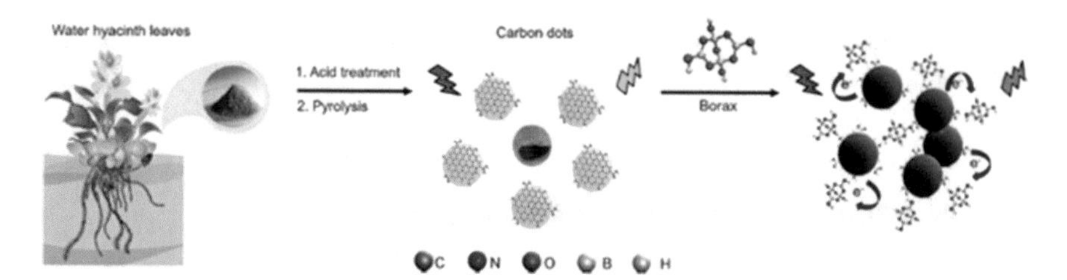

Fig. 8 Carbon dots synthesis using acid treatment and pyrolysis of water hyacinth leaves and borax sensing application. (By Prathumsuwan et al. [60], License number 4960340738285 provided by Elsevier and Copyright Clearance Center)

Table 1 Borax Relative Standard Deviation (RSD) and Recovering at fish ball samples (adapted from Prathumsuwan et al. [60], License number 4960340738285 provided by Elsevier and Copyright Clearance Center)

Added (μM)	Found (μM)	Recovery ± SD[a] (%)	RSD[a] (%)
10	10.2	101.8 ± 0.6	0.59
30	30.3	101.0 ± 0.8	0.80
50	49.4	98.8 ± 1.1	1.12

[a] Results after nine independent measurements (Adapted By Prathumsuwan et al. [60], License number 4960340738285 provided by Elsevier and Copyright Clearance Center)

was performed using citric acid by incomplete pyrolysis conjugated a hydrothermal method (Fig. 9). The Try-CD showed a size of 2.2 nm and excellent luminescence. Similarly, [16] performed a nitrogen dopped-carbon dots (N-CD) based on electrochemical carbonization of ethanolamine for cysteine detection. Their work presented

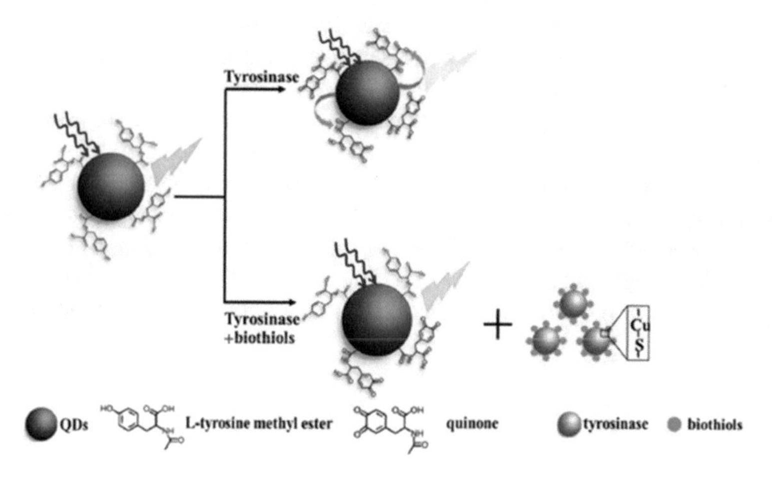

Fig. 9 The interaction among Tyr-CDs, tyrosinase, and biothiols graphical abstract. (By c), License number 4960771261694 provided by Elsevier and Copyright Clearance Center)

an N-CD with a remarkable cysteine selectivity, nanoparticles size of 3 nm, and photoluminescence at 320–480 nm.

2.3 Microwave and Other Methodologies

Although the hydrothermal and carbonization synthesis, other types were used in the literature. One of them is microwave synthesis; this method consists of producing carbon dots under microwave irradiation at a specific time that varies from a few minutes to an hour. Hence, [89] produced new carbon dots using microwave irradiation for 5 min using ethylenediamine and citric acid as a forerunner to detected Hg^{2+}. In their work, the carbon dots showed a narrow distribution of (2.5 ± 0.5) nm, a great quantum yield (QY), and the presence of chemical elements like C, N, O, and limited H (4%, calculated).

Similarly, [14] made a carbon dot based on microwave irradiation under chitosan hydrogel. In their research, the chitosan was dissolved into 0.1 M acetic acid under microwave irradiation for 5 min. These researches focus into determine the characteristics of new carbon dots and properties on different pH. The results were fascinating; the particles sizes were among 0.6–8.7 nm, the photoluminescence (PL) emission was among 300–400 nm, decreasing the intensity for higher wavelengths at different pH produced, although the carbon dots solutions show distinct color (from dark yellow to white as the pH increase from 1 to 5).

Furthermore, [19] composed thiol-functionalized carbon dots based on microwave synthesis to detect mercury ion (Hg^{2+}). It was used polyethylene glycol and chitosan gel functionalized with 1,4-Dithiothreitol (DTT as a precursor of the thiol group. The photoluminescence showed luminescence in the range of 280–412 nm; the average size was around 8 nm and an incredible Hg^{2+} selectivity.

Unlike microwave synthesis, [88] used a solvothermal method to produce carbon dots-embedded zincone microspheres to detect chromates ($Cr_2O_7^{2-}$) in water. Their photoluminescence properties were dependent on CD concentration at 345–525 nm (Fig. 10). Besides, it was showed excellent selectivity for $Cr_2O_7^{2-}$ compared to other ions (Fig. 11).

In addition to microwave synthesis and solvothermal methods, many other syntheses are shown in the literature, as electrochemical synthesis, physicochemical synthesis, and others. The main differences among them are the time of production, equipment, sources, and energy used, contributing to different results in the class of cost and facility of production. In this overview, the most used in the literature were presented.

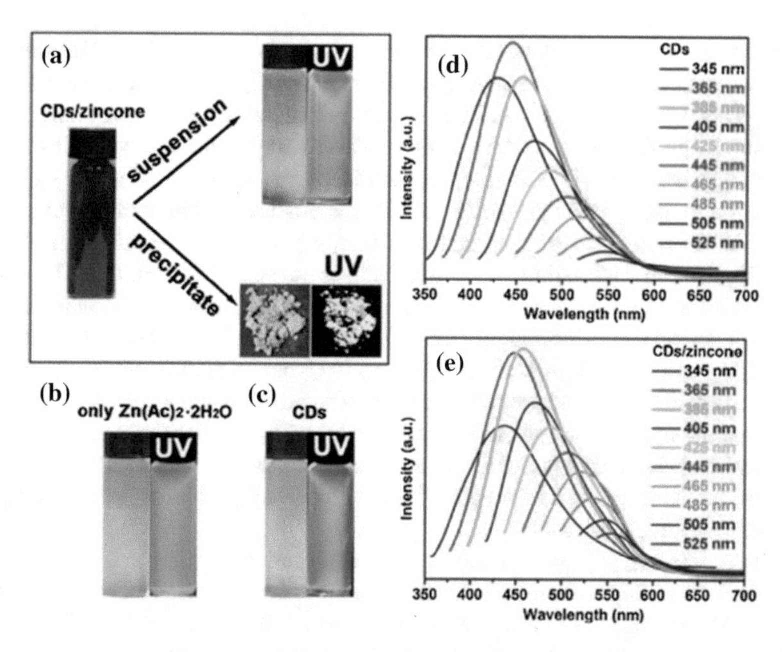

Fig. 10 **a** Mixture of carbon dots/zincone synthesis at ambient light (left); Images of the residual suspension (right-top) and the carbon dots/zincone precipitates separated (right) at ambient and UV light. Images of $Zn(Ac)_2{\cdot}2H2O$ (**b**) and only HMTA (**c**). Photoluminescence of carbon dots spectra (**d**) and the carbon dots/zincone solution at distinct excitation wavelengths (**e**). (By Xue et al. [88], License number 4960360762015 provided by Elsevier and Copyright Clearance Center)

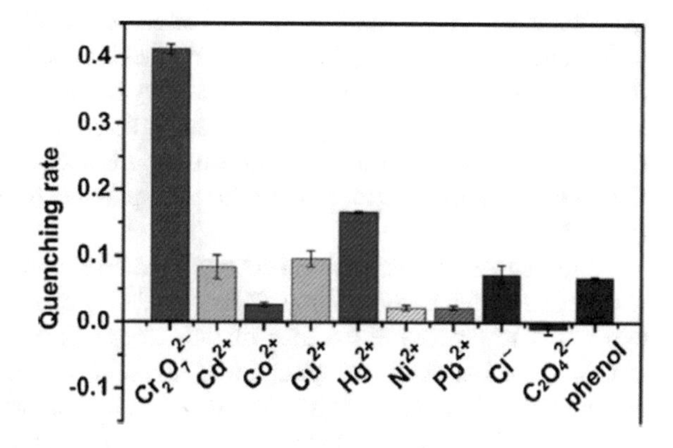

Fig. 11 Analysis of carbon dots/zincone selectivity for different ions and phenol (Concentration = 100 μM). (By Xue et al. [88], License number 4960360762015 provided by Elsevier and Copyright Clearance Center)

3 Characterization and Properties of Carbon Dots

A nanomaterial characterization implies a diversity of technique, equipment, and preparation to determine the potential characteristics and properties beyond the human senses, requiring immense sensitivity, short of many applied physics and chemistry laws, to most materials. On this wise, techniques such as Transmission Electronic Microscopy with higher resolution (HRTEM), Photoluminescence Spectroscopy (PL), X-Ray Diffraction (XRD), X-Ray Photoelectron Spectroscopy (XPS), Dynamic Light Scattering (DLS), Atomic Microscopy Force (AFM), Ultraviolet–Visible Spectroscopy (UV-Visible), Infrared Spectroscopy (FTIR), were used to indicate the unimaginable properties and characteristics of these carbon dots.

One of the most critical characteristics that imply these materials' potential for many applications, including water treatment, is the size of carbon dots. Their size provides intrinsic properties related to being applied in many areas, as catalysis [35], electrical conductivity, bioimaging [26, 30, 45], selectivity sensor [58], theragnostic cells [105], agricultural [34], pharmaceutical [43], batteries [22], etc. In the case of water treatment, the photocatalytic process associated with catalysis is one of the most exciting treatments. Therefore, this amusing mechanism implies some characteristics that may be modulated by the synthesis and precursors used in the process. Nowadays, this practice has achieved much attention, essentially due to process potential, not producing residues [4, 55].

Hence, the morphology of the carbon dots and their size are evaluated by different microscopy, as Transmission Microscopy (TEM) or High-Resolution Transmission Microscopy (HRTEM), or Atomic Force Microscopy (AFM). Saud et al. [65] performed carbon dots synthesis based on hydrothermal methodology. In their research, the carbon dots were used to form a nanofiber composite with TiO_2 to apply to degradation based on the sunlight of methylene blue. They compared the nanofiber composite with carbon dots without them aiming to study the nanoparticles' distribution (Fig. 12). Comparing the TiO_2 nanofibers and the carbon dots/TiO_2 nanofiber, the distribution of carbon dots into nanofiber was very homogeneous under the TiO_2 nanofibers surface. In addition, the size of carbon dots was in the range of 3–4 nm.

Thusly, Iqbal et al. [29] developed a synthesis of carbon dots doping the nanoparticle with nitrogen (N-CD) using melamine and anhydrous citric acid as precursors. Their N-CD showed a yellow solution that confirmed the presence of carbon dots based on High-Resolution Transmission Electronic Microscopy characterization, showing a homogenous dispersion and spherical formation with a 3 nm average (Fig. 13).

Although the Transmission Electronic Microscopy or High-Resolution Transmission Electronic Microscopy (TEM/HRTEM) analysis, other techniques were performed to determine the size of the nanoparticle, however important characteristics, such as chemical groups and luminescence. The luminescence of these nanomaterials is one of the characteristics that the researchers sought because of the potential for many applications, mainly water treatment. This property is associated with electron radiative combinations and defects in these materials' surfaces [7, 54, 57, 78, 79]. Most carbon dots showed photoluminescence (PL) emission in the range of blue and green spectra [1, 100]. Along these lines [35] performed a facile synthesis to produce the carbon dots based on alkali-assisted electrochemical oxidant fabrication, using graphite rods as precursors. The carbon dots presented a color emission of blue, green, brown, and yellow using a fluorescence microscope (FL), and it was used for the photocatalytic activity.

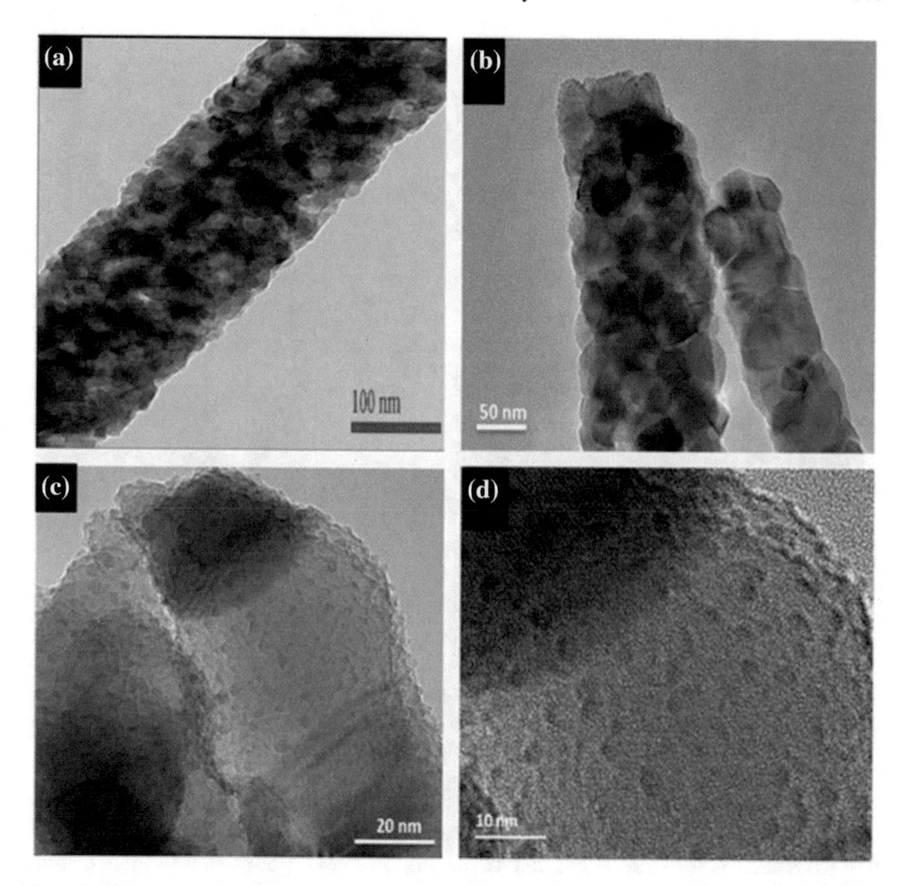

Fig. 12 TiO_2 nanofiber Images in Transmission Electronic Microscopy (TEM) analysis of (**a**), TEM images of carbon dots/TiO_2 composite nanofiber (**b**), and carbon dots/TiO_2 composite nanofiber High-Resolution Transmission Electronic Microscopy (HRTEM) images (**c**, **d**). (By Saud et al. [65], License number 4960230347225 provided by Elsevier and Copyright Clearance Center)

In this way, Wang et al. [76] developed a carbon dot doping with nitrogen using m-aminobenzoic derivative for a biosensor of Fe^{3+}. Their work reported an excellent sensitivity of Fe^{3+} depended on pH. Besides, Sun et al. [68] developed new carbon dots using oligomeric ethylene glycol diamine that presented a quenching of PL when together with Ag^+, 4-nitrotoluene, 2,4-dinitrotoluene and *N,N*-diethyl aniline. Hence, Liu et al. [45] produced new carbon dots based on the carbonization of chitosan to obtain an amino-functionalized Carbon Dot. In their work, the luminescence of these materials presented a quantum yield of 4.34% excitation. Therefore, this quantum yield (QY) was dependent on pH, showing more stability in the pH range of 5–9 (Fig. 14). According to the authors, the photoluminescence emission emerged to a resonant wavelength in acid conditions, as the basic medium, the reduced wavelength implicated that the carbon dots functionalized with the amino group were incredibly sensitive to the pH of the solution.

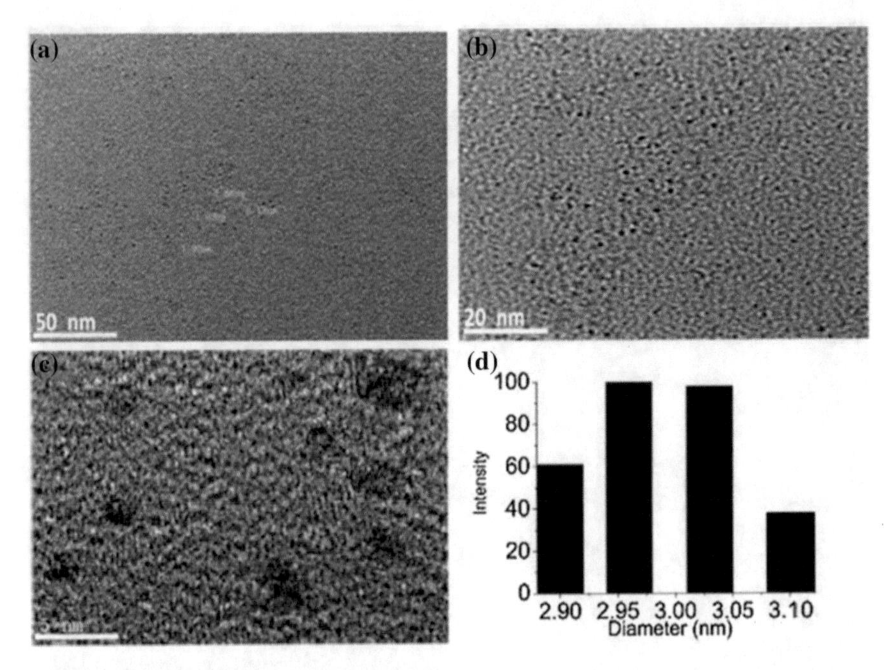

Fig. 13 Transmission Electronic Microscopy images (**a**, **b**) and High-Resolution Transmission Electronic Microscopy (C) of N-CDs. Diameter distribution of N-CDs. (By Iqbal et al. [29], License number 4961460644964 provided by Elsevier and Copyright Clearance Center)

Beyond the photoluminescence (PL) characterization, other important techniques that determine some crucial characteristics, as chemical groups that influence some properties in these potential nanomaterials, are the infrared spectroscopy, mainly the Fourier Transform Infrared Spectroscopy (FTIR). Infrared Spectroscopy is a critical analysis to evaluate the chemical groups present in the materials. In the case of carbon dots, the type of synthesis and reagents involved in the carbon dots production change the chemical groups in the surface of carbon dots, which may impact the photoluminescence mechanism, consequently, the photocatalytic activity in water treatment, as may change the bandgap transition or implied the existence of various surface defects [50].

Liu et al. (107) synthesized a hydrothermal carbon dot using ammonia to detect Cu^{2+} in water. The Infrared Spectroscopy analysis exhibited bands at 1309 to 1650 cm^{-1} relative to the presence of aromatic CN heterocycles, also at 2930 and 3402 cm^{-1} band associated with O–H and N–H groups. This technique conjugated to others proves the existence of nitrogen-doped carbon dots with great photoluminescence that was selective to Cu^{2+} in water. Xiao et al. (110) synthesized a new Carbon dot using microwave methodology based on chitosan for potential water treatment. In their research, amino-functionalized carbon dots were produced to bands related to O–H (at 3447 cm^{-1}), N–H bending vibrations (at 1644 and 1593 cm^{-1}), C-H (at

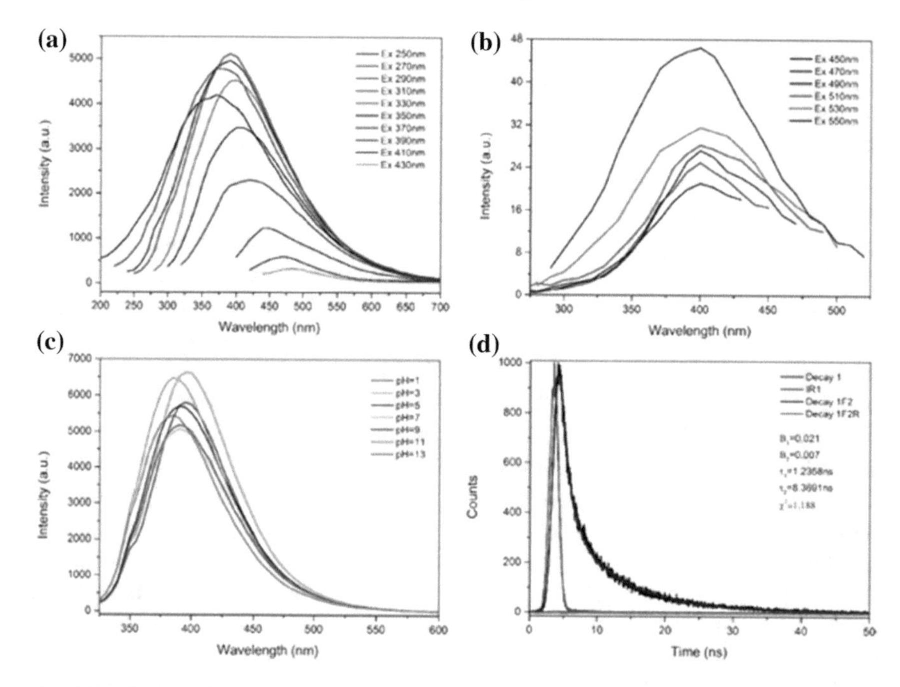

Fig. 14 **a** Photoluminescence (PL) emission spectra and **b** Photoluminescence (PL) up-conversion of CDs. **c** Photoluminescence (PL) spectra of CDs at distinct pH solutions and **d** Time-resolved PL decay and fitting curves. (By Liu et al. [45], Creative Commons, this is an open-access article distributed under the terms of the Creative Commons CC BY license, which permits unrestricted use, distribution, and reproduction in any medium, provided the original work is properly cited. You are not required to obtain permission to reuse this article. To request permission for a type of use not listed, please contact Springer Nature)

1130 cm^{-1}) groups around the carbon dot surface, indicating the chitosan degradation and dehydration based on pyranose ring decomposition, similarly to report to [101]. Along these lines, [69] made colloidal carbon dots spheres with noble metals. In their work, bands at 1585 cm^{-1} and 1390 cm^{-1} related to silver-loaded spheres were visualized, indicating the in-plane vibrations of crystalline graphite, also the disorder of amorphous Carbon.

X-Ray Photoelectron Spectroscopy (XPS) is another interesting technique that indicates the presence of chemical groups and their state, as oxidation, which can induce luminescence effects that improve photocatalytic activity. Wu et al. [82] made carbon dots based on carboxymethylcellulose by hydrothermal synthesis. Their work used some techniques, one of them was X-ray Photoelectron Spectroscopy. This characterization was used to indicate some characteristics that improve electrophilicity, which is an essential property of conductive materials. The X-Ray Photoelectron Spectroscopy confirmed the presence of N protonated in the characteristic centered band of quaternary at 400.7–402.3 eV, indicating that hydrochars improve their conductivity. In addition, the technique showed the N existence in the format of N-groups, which is a stabler group than O-groups.

Likewise, X-Ray Diffraction (XRD) may be used to determine the crystalline structure of carbon components or the characteristic amorphous structure of carbon dots after synthesis. [70] done a porous carbon by chemical activation, which produced some carbon dots in the synthesis. The X-Ray Diffraction analysis indicates the presence of amorphous Carbon, probably indicating the carbon dots' presence.

Added to these different techniques along these lines, another fascinating characterization used to demonstrate these carbon nanomaterials' critical characteristics is the ultraviolet–visible spectroscopy, UVVisible. Most of the carbon dots presented in the literature exhibit broad wavelength at 250–350 nm in the ultraviolet–visible analysis [3]. Hence, Wu et al. [83] made carbon dots based on hydrothermal methodology using bleach hard Kraft pulp as precursors for potential applications as a photosensitizer. The ultraviolet–visible analysis allowed the determination of the weak and large band at 250–300 nm associated with intrinsic complex electron transition over the surface. Li et al. (109) produced a new composite based on carbon dots and ZnO based on sol–gel methodology using spin-coating processing. In their research, the nanocomposite showed an absorption centered band at 304 nm with a diminishing curve, similar to graphene quantum dots produced in the literature [37], which may associate with small particle size and defects in the surface (Fig. 15).

The precursor of carbon dots may change the surface of these nanomaterials, and these modifications appeared in the spectroscopy analysis, as ultraviolet–visible analysis based on the emission spectra [31]. Moreover, carbonyl and amino functionalization in some carbon dots showed spectra at ultraviolet–visible analysis, promoting red shift depends on the alterations in the HOMO–LUMO energy levels in the carbon dots [38].

4 Photocatalytic Activity of Carbon Dots

Despite the fundamental advances in the recent decades in the range of water treatment and depuration processes, the growth of innovative materials based on different nanoparticles combining high photocatalytic capacity, low cost, biocompatibility, and environmentally friendly features represents a decisive challenge to be overcome by researchers and professionals. Nonetheless, one of the significant challenges in photocatalytic treatment is discovering a potent photocatalytic material [28, 73]. Thus, the carbon dots are one of the most innovative materials with significant fluorescence applied in the water treatment, as the photocatalytic process (Li et al. 109) [27].

Although many nanomaterials have a large bandgap, like TiO_2, ZnO, CdS, etc., the Carbon dots have the advantage of biocompatibility, environment-friendly, facile synthesis, residues as sources, incredible luminescence, and absurd photolytic potential [12, 32, 62]. On this line, [9] produced a new Carbon Dot with functionalization over their surface using polyethylene glycol (PEG) (Fig. 16). Their innovative nanomaterial was capable of catalyzing and photo regenerating H_2 from water.

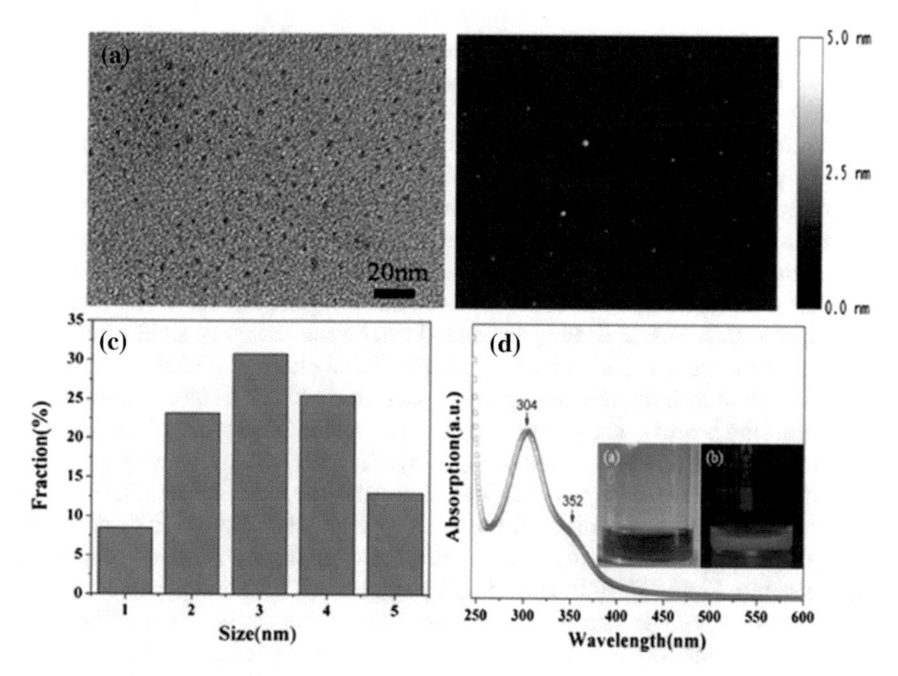

Fig. 15 Transmission Electronic Microscopy and Atomic Force Microscopy images of carbon dots (**a**, **b**), respectively, **c** Diameter distributions, **d** Ultraviolet–visible absorption spectra at visible and Ultraviolet (UV) irradiation (365 nm), respectively. (By Liu et al. (108), License number 4964220569467 provided by Elsevier and Copyright Clearance Center)

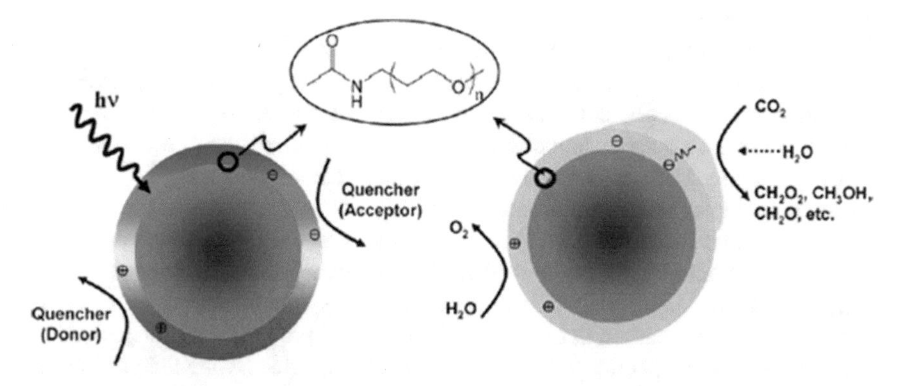

Fig. 16 Carbon dots based on PEG-functionalized before (left) and after (right) metal coating. (By Cao et al. [9], Copyright (2020) American Chemical Society)

The functionalization with poly(ethylene glycol) diamine (PEG) provides carboxylic groups that change the size and, consequently, the nanomaterial's photoluminescence property.

Following this line, Liu et al. (46) made a metal-free nanoparticle based on carbon dots for photocatalytic solar water splitting. In their research, the quantum yield efficiency was 16% in the band at 420 nm with 2% of solar energy conversion, and the photocatalytic showed incredible stability. Moreover, when the concentration of carbon dots was changed in the ultrathin layer, the bandgap changed too, mainly because of the agglomeration of carbon dots. Furthermore, the Carbon Dot showed a synergic with $BiVO_4$ indicating a potential solar water energy capability based on photocatalytic activity. Similarly, Ye et al. [102] designed a new Carbon Dot to be used as an ultrathin intralayer among $BiVO_4$ and NiOOH/FeOOH layers based on the water-splitting process. The photoluminescence emission was among 340–500 nm, similar to other studies in the literature [48, 53, 71, 85, 97, 99].

Wang et al. [77] performed a Carbon Dot using overcooked barbecue meat as a precursor. Their carbon material presented optical similar to other carbon dots performed in the literature using other sources. Furthermore, the carbon dots showed a high Quantum Yield (QY) (40% into the green region) and an amusing photocatalytic activity reducing silver and gold ions over carbon dots surface. Also, Wu et al. [83] made carbon dots based on pentosan using a hydrothermal methodology. In their work, the photocatalytic activity was tested to methylene blue (MB), an azo dye, generally used as a model of cationic dye in water [6, 51], using visible light irradiation. This nanomaterial, combined with TiO_2, showed a 100% Methylene Blue (MB) degradation, mostly because of free electron and energy transfer to the TiO_2 conduction band (Fig. 17).

In this way, [80] designed carbon dots functionalized with Polyethylene Glycol (PEG) using gold as covering. This hybrid nanomaterial presented an outstanding absorption at visible light and a high photoluminescence property. Along these lines, Li et al. (109) developed carbon dots to use as selective oxidation of benzyl alcohol to benzaldehyde. Their research showed that these carbon dots settled 92% of the conversion rate with an incredible selectivity.

Emanuele and collaborators [17] studied various graphitic and amorphous carbon dots based on fructose, glucose, and citric acid, comparing their structure and photocatalytic activity. Their research demonstrated that the methodology to produce the carbon dots affected carbon dots' structural and optical properties significantly. This changing affects the photon transferability, which adjustment the methyl viologen photocatalytic activity of these nanomaterials. Thusly, Yu et al. (111) developed a nanocomposite using carbon dots and TiO_2 as precursors into the nanosheet. Their investigation validated that the nanosheet's photocatalytic property was boosted with carbon dots incorporation using rhodamine B as a model under visible light irradiation (Fig. 18).

In this line of produce new nanocomposites, [23] performed a nanocomposite using carbon dots and TiO_2. Their research developed a new nanocomposite based on electrochemical synthesis to accelerate the photodegradation of methylene blue. The research demonstrated that the Methylene Blue degradation was directly affected

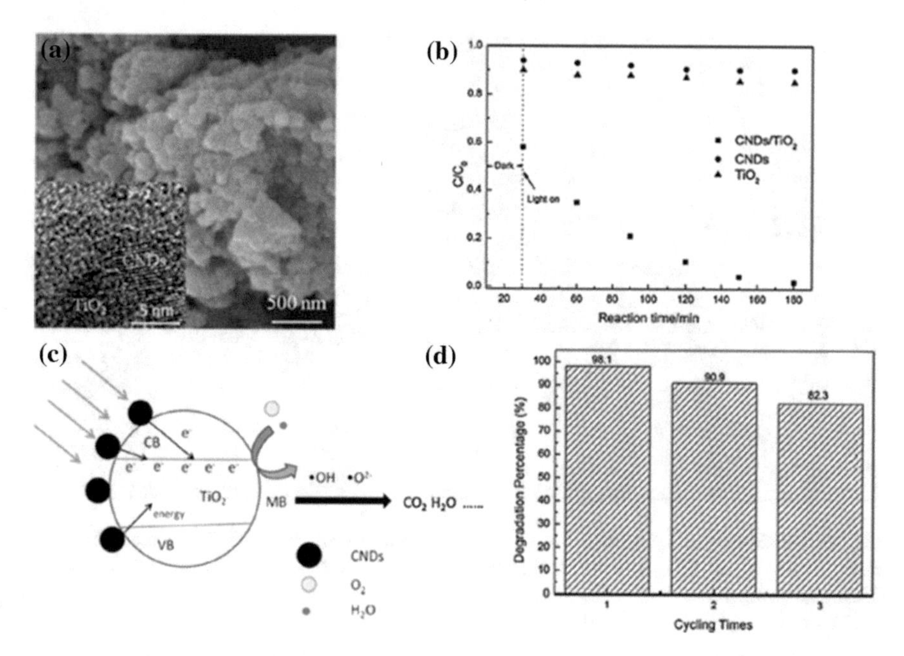

Fig. 17 **a** CNDs/TiO_2 composite Scanning Electronic Microscopy (SEM) and High-Resolution Transmission Electronic Microscopy (HRTEM) images; **b** % Methylene Blue degradation at time; **c** CNDs/TiO_2 photocatalytic process at visible light and **d** Reuse results. (By Wu et al. [83], License number 4964920164029 provided by Elsevier and Copyright Clearance Center)

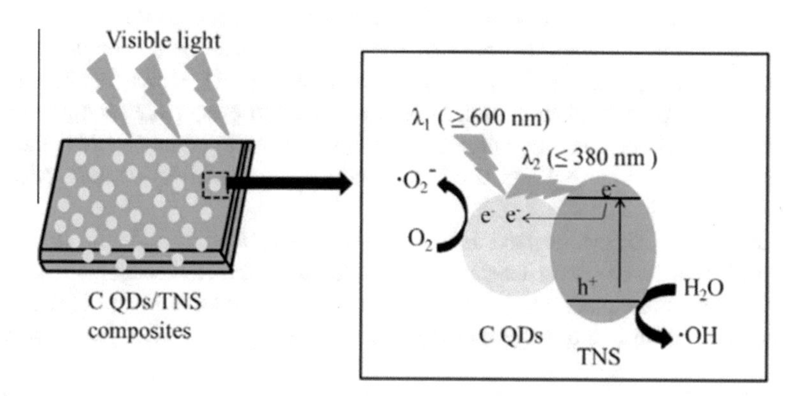

Fig. 18 Photocatalytic Mechanism of carbon dots/TNS composites at visible light. (By Yu et al. (111), License number 4965420327740 provided by Elsevier and Copyright Clearance Center)

by the crystallization of Carbon Dot/TiO_2. Even so, they reached 90% of Methylene Blue degradation after 120 min. Complementing these studies based on CD/TiO_2 nanocomposites, [13] processed this new nanocomposite using a sol–gel methodology to act in the photocatalytic process of Rhodamine B and cefradine. This new nanomaterial announced an increasing degradation of these two wastes water

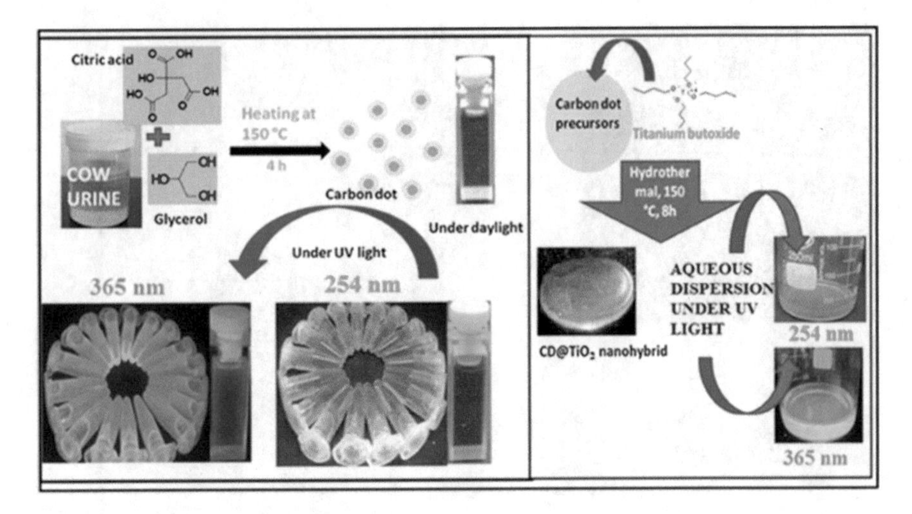

Fig. 19 Illustration of synthesis of **a** Carbon Dot and **b** Carbon Dot@TiO_2 nanohybrid. (By Hazarika and Karak [20], License number 4965430966051 provided by Elsevier and Copyright Clearance Center)

when carbon dots was incorporated into TiO_2; mainly, this enhancement occurred to electron–hole pairs.

Integrating this nanocomposite line using TiO_2, [20] performed a new carbon dots/TiO_2 based on hydrothermal methodology using citric acid and glycerol as precursors. This new material was used for organic pollutants degradation, as phenol and benzene in water (Figs. 19 and 20). They indicated that the carbon dots incorporation into TiO_2 reduced the electron recombination rate, which improved the efficiency of photodegradation, mainly because of anatase structure predomination into TiO_2 crystalline structure, when Carbon Dot was incorporated, as this phase in TiO_2 crystalline structure presents a better photocatalytic activity demonstrated in various works [52, 87, 94].

Additionally, [25] synthesized new carbon dots using a facile and efficient route. Their study was developed to determine the photoluminescent property, mainly in the organic photocatalytic activity in water. They demonstrated that O and N radicals' presence on the surface of carbon dots affected the increase of photoluminescence (PL).

Although these many important characteristics of the photoluminescence property of carbon dots and the mechanism involved in the photocatalytic process of these nanomaterials, the kinetics of the photocatalytic process is a critical parameter in the analysis. The kinetic shows the mechanism involved in the photocatalytic activity [42]. Most of the researches developed in the literature based on carbon dots did not study this parameter. They focus on the carbon dots properties, synthesis, and the photocatalytic process is just an application-focused hook. However, the kinetic is most studied following the kinetic equations, as pseudo-first-order and pseudo-second-order [21], even in the carbon dots photocatalytic activity. In this

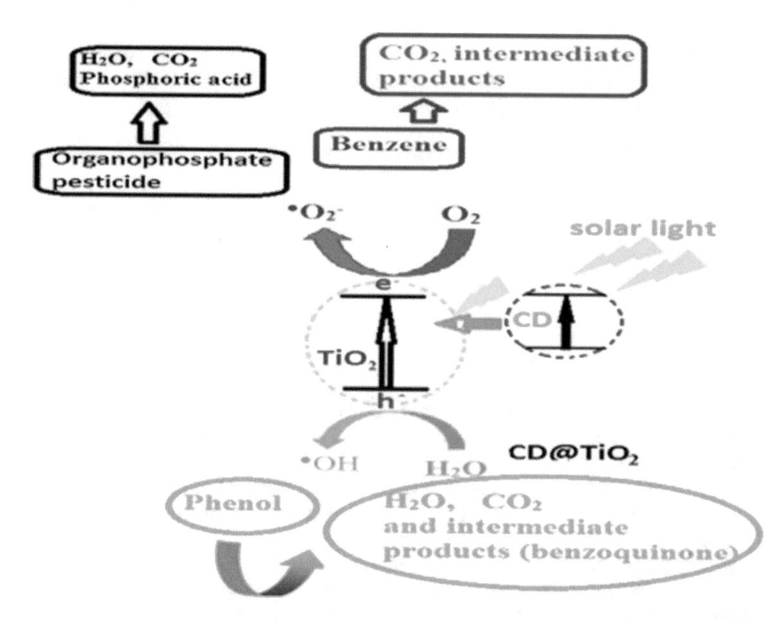

Fig. 20 Organics Photocatalytic process by Carbon Dot@TiO_2 nanohybrid. (By Hazarika and Karak [20], License number 4965440767721 provided by Elsevier and Copyright Clearance Center)

sense, Hazarika and Karak [20] studied the photocatalytic degradation kinetics of some organic chemicals, like benzene, pesticide, and phenol using carbon dots/TiO_2. In their work, the pseudo-first-order was the best fitting of benzene, and phenol degradation demonstrated a typical photodegradation behavior of this nanomaterial.

5 Conclusion

This overview exposed a range of researches based on carbon dots. These unequaled nanomaterials present various properties and characteristics that allow them many applications, including water treatment. The different synthesis affects their characteristics, which implies different properties, as luminescence, altering the photocatalytic activity, proposal of this overview. Many types of techniques are used to determine their characteristics, such as Infrared Spectroscopy, Ultraviolet–Visible Spectroscopy, X-ray Spectroscopy, Transmission Electronic Microscopy with Higher Resolution, and others. Moreover, these nanomaterials exhibit an incredible potential for photodegradation of diversity of organic and inorganic materials in water, indicating their promising solution of low cost, biocompatible, photocatalytic nanomaterial for wastewater treatment.

Acknowledgements The authors acknowledge the financial support, and they express their gratitude to the *Universidade Federal dos Vales do Jequitinhonha e Mucuri* (UFVJM)., and the BIOSEM from the Universidade Federal dos Vales do Jequitinhonha e Mucuri (UFVJM).

Author Disclosure Statement The authors indicate that they have no competing interests.

Author Contributions All authors wrote the manuscript and approved its final version.

References

1. Adedokun O, Roy A, Awodugba AO, Devi PS (2017) Fluorescent carbon nanoparticles from citrus sinensis as efficient sorbents for pollutant dyes. Luminescence 32:62–70. https://doi.org/10.1002/bio.3149
2. Ali Q, Asharaf M, Athar HUR (2007) Exogenously applied proline at different growth stages enhances growth of two maize cultivars grown under water deficit conditions. Pak J Bot 39:1133–1144
3. Baker SN, Baker GA (2010) Luminescent carbon nanodots: emergent nanolights. Angew Chem Int Ed 49:6726–6744. https://doi.org/10.1002/anie.200906623
4. Baliarsingh N, Parida KM, Pradhan GC (2014) Effects of Co, Ni, Cu, and Zn on photophysical and photocatalytic properties of carbonate intercalated MII/Cr LDHs for enhanced photodegradation of methyl orange. Ind Eng Chem Res 53:3834–3841. https://doi.org/10.1021/ie403769b
5. Borsagli FGLM, Borsagli A (2019) Chemically modified chitosan bio-sorbents for the competitive complexation of heavy metals ions: a potential model for the treatment of wastewaters and industrial spills. J Polym Environ 27:1542–1556. https://doi.org/10.1007/s10924-019-01449-4
6. Borsagli FGLM (2019) A green 3D scaffolds based on chitosan with thiol group as a model for adsorption of hazardous organic dye pollutants. Des Water Treat 169:395–411. https://doi.org/10.5004/dwt.2019.24709
7. Bourlinos AB, Stassinopoulos A, Anglos D, Zboril R, Karakassides M, Giannelis EP (2008) Surface functionalized carbogenic quantum dots. Small 4:455. https://doi.org/10.1002/smll.200700578
8. Campos BB, Contreras-Caceres R, Bandosz TJ, Jimenez-Jimenez J, Rodríguez-Castellon E, Da Silva JCGE, Algarra M (2016) Carbon dots as fluorescent sensor for detection of explosive nitrocompounds. Carbon 106:171–178. https://doi.org/10.1016/j.carbon.2016.05.030
9. Cao L, Sahu S, Anilkumar P (2011) Carbon nanoparticles as visible-light photocatalysts for efficient CO_2 conversion and beyond. J Am Chem Soc 133:4754–4757. https://doi.org/10.1021/ja200804h
10. Chandra S, Patra P, Pathan SH, Roy S, Mitra S, Layek A, Bhar R, Pramanik P, Goswami DA (2013) Luminescent S-doped carbon dots: an emergent architecture for multimodal applications. J Mater Chem B 18:2375–2382. https://doi.org/10.1039/C3TB00583F
11. Chen G, Qiu H, Prasad PN, Chen X (2014) Upconversion nanoparticles: design, nanochemistry, and applications in theranostics. Chem Rev 114:5161–5214. https://doi.org/10.1021/cr400425h
12. Chen X, Shen S, Guo L, Mao SS (2010) Semiconductor-based photocatalytic hydrogen generation. Chem Rev 110:6503. https://doi.org/10.1021/cr1001645
13. Chen J, Shu J, Anqi Z, Juyuan H, Yan Z, Chen J (2016) Synthesis of carbon quantum dots/TiO_2 nanocomposite for photodegradation of Rhodamine B and cefradine. Diam Relat Mat 70:137–144. https://doi.org/10.1016/j.diamond.2016.10.023
14. Chowdhury D, Gogoi N, Majumdar G (2012) Fluorescent carbon dots obtained from chitosan gel. RSC Adv 2:12156–12159. https://doi.org/10.1039/C2RA21705H

15. Deng J, Lu Q, Mi N, Li H, Liu M, Xu M, Tan L, Xie Q, Zhang Y, Yao S (2014) Electrochemical synthesis of carbon nanodots directly from alcohols. Chem-Eur J 20:4993. https://doi.org/10.1002/chem.201304869
16. Deng J, Lu Q, Hou Y, Liu M, Li H, Zhang Y, Yao S (2015) Nanosensor composed of nitrogen-doped carbon dots and gold nanoparticles for highly selective detection of cysteine with multiple signals. Anal Chem 87:2195–2203. https://doi.org/10.1021/ac503595y
17. Emanuele A, Cailotto S, Campalani C, Branzi L, Raviola C, Ravelli D, Cattaruzza E, Trave E, Benedetti A, Selva M, Perosa A (2020) Precursor-dependent photocatalytic activity of carbon dots. Molecules 25:101. https://doi.org/10.3390/molecules25010101
18. Esteves da Silva JCG, Goncalves HMR (2011) Analytical and bioanalytical applications of carbon dots. TRAC-Trends Anal Chem 30:1327–1336. https://doi.org/10.1016/j.trac.2011.04.009
19. Gupta A, Chaudhary A, Mehta P, Dwivedi C, Khan S, Verma NC, Nandi CK (2015) Nitrogen-doped, thiol-functionalized carbon dots for ultrasensitive Hg(II) detection. Chem Commun 51:10750. https://doi.org/10.1039/C5CC03019F
20. Hazarika D, Karak N (2016) Photocatalytic degradation of organic contaminants under solar light using carbon dot/titanium dioxide nanohybrid, obtained through a facile approach. Appl Surf Sci 376:276–285. https://doi.org/10.1016/j.apsusc.2016.03.165
21. Ho Y, McKay G (1999) Pseudo-second order model for sorption processes. Process Biochem 34:451–465. https://doi.org/10.1016/S0032-9592(98)00112-5
22. Hou H, Banks CE, Jing M, Zhang Y, Ji X (2015) Carbon quantum dots and their derivative 3D porous carbon frameworks for sodium-ion batteries with ultralong cycle life. Adv Mater 27:7861–7866. https://doi.org/10.1002/adma.201503816
23. Hou Y, Lu Q, Wang H, Li H, Zhang Y, Zhang S (2016) One-pot electrochemical synthesis of carbon dots/TiO_2 nanocomposites with excellent visible light photocatalytic activity. Mater Let 173:13–17. https://doi.org/10.1016/j.matlet.2016.03.003
24. Hsu P-C, Chang H-T (2012) Synthesis of high-quality carbon nanodots from hydrophilic compounds: role of functional groups. Chem Commun 48:3984–3986. https://doi.org/10.1039/C2CC30188A
25. Hu S, Tian R, Dong Y, Yang J, Liu J, Chang Q (2013) Modulation and effects of surface groups on photoluminescence and photocatalytic activity of carbon dots. Nanoscale 5:11665. https://doi.org/10.1039/C3NR03893A
26. Huang X, Zhang F, Zhu L, Choi KY, Guo N, Guo J, Tackett K, Anilkumar P, Liu G, Quan Q (2013) Effect of injection routes on the biodistribution, clearance, and tumor uptake of carbon dots. ACS Nano 7:5684–5693. https://doi.org/10.1021/nn401911k
27. Huggins T, Wang HM, Kearns J, Jenkins P, Ren ZYJ (2014) Biochar as a sustainable electrode material for electricity production in microbial fuel cells. Bioresour Technol 157:114–119. https://doi.org/10.1016/j.biortech.2014.01.058
28. Inoue T, Fujishima A, Konishi S, Honda K (1979) Photoelectrocatalytic reduction of carbon dioxide in aqueous suspensions of semiconductor powders. Nature 277:637. https://doi.org/10.1038/277637a0
29. Iqbal A, Iqbal K, Xu L, Li B, Gong D, Li X, Guo Y, Liu W, Qin W, Guo H (2018) Heterogeneous synthesis of nitrogen-doped carbon dots prepared via anhydrous citric acid and melamine for selective and sensitive turn on-off-on detection of Hg (II), glutathione and its cellular imaging. Sensor Actuat B-Chem 255:1130–1138. https://doi.org/10.1016/j.snb.2017.08.130
30. Jiang K, Sun S, Zhang L, Lu Y, Wu A, Cai C, Lin H (2015) Red, green, and blue luminescence by carbon dots: full-color emission tuning and multicolor cellular imaging. Angew Chem Int Ed 54:5360–5363. https://doi.org/10.1002/anie.201501193
31. Kim S, Hwang SW, Kim M-K, Shin DY, Shin DH, Kim CO, Yang SB, Park JH, Hwang E, Choi S-H, Ko G, Sim S, Sone C, Choi HJ, Bae S, Hong BH (2012) Anomalous behaviors of visible luminescence from graphene quantum dots: interplay between size and shape. ACS Nano 6:8203–8208. https://doi.org/10.1021/nn302878r
32. Kudo A, Miseki Y (2009) Heterogeneous photocatalyst materials for water splitting. Chem Soc Rev 38:253. https://doi.org/10.1039/B800489G

33. Li H, Zang Z, Liu Y, Lee ST (2012) Carbon nanodots: synthesis, properties and applications. J Mater Chem 22:24230–24253. https://doi.org/10.1039/C2JM34690G
34. Li H, Yan X, Lu G, Su X (2018) Carbon dot-based bioplatform for dual colorimetric and fluorometric sensing of organophosphate pesticides. Sensor Actuat B-Chem 260:563–570. https://doi.org/10.1016/j.snb.2017.12.170
35. Li H, He X, Kang Z, Huang H, Liu Y, Liu J, Lian S, Him C, Tsang A, Yang X, Lee S-T (2010) Water-soluble fluorescent carbon quantum dots and photocatalyst design. Angew Chem Int Ed 49:4430–4434. https://doi.org/10.1002/anie.200906154
36. Li Y, Hu Y, Zhao Y, Shi GQ, Deng LE, Hou YB, Qu LT (2011) An electrochemical avenue to green-luminescent graphene quantum dots as potential electron-acceptors for photovoltaics. Adv Mater 23:776. https://doi.org/10.1002/adma.201003819
37. Li Y, Zhang BP, Zhao JX, Ge ZH, Zhao XK, Zou L (2013) ZnO/carbon quantum dots heterostructure with enhanced photocatalytic properties. Appl Surf Sci 279:367–373. https://doi.org/10.1016/j.apsusc.2013.04.114
38. Li H, Liu R, Lian S, Liu Y, Huang H, Kang Z (2013) Near infrared light controlled photocatalytic activity of carbon quantum dots for highly selective oxidation reaction. Nanoscale 5:3289–3297. https://doi.org/10.1039/C3NR00092C
39. Li Y, Shu H, Niu X, Wang J (2015) Electronic and optical properties of edge-functionalized graphene quantum dots and the underlying mechanism. J Phys Chem C 119:24950–24957. https://doi.org/10.1021/acs.jpcc.5b05935
40. Liang Q, Ma W, Shi Y, Li Z, Yang X (2013) Easy synthesis of highly fluorescent carbon quantum dots from gelatin and their luminescent properties and applications. Carbon 60:421–428. https://doi.org/10.1016/j.carbon.2013.04.055
41. Lim SY, Shen W, Gao Z (2015) Carbon quantum dots and their applications. Chem Soc Rev 41:362–381. https://doi.org/10.1039/C4CS00269E
42. Limburg B, Wermink J, Van Nielen SS, Kortlever R, Koper MTM, Bouwman E, Bonnet S (2016) Kinetics of photocatalytic water oxidation at liposomes: membrane anchoring stabilizes the photosensitizer Kinetics of photocatalytic water oxidation at liposomes: membrane anchoring stabilizes the photosensitizer. ACS Catal 6:5968–5977. https://doi.org/10.1021/acscatal.6b00151
43. Liu R, Wu D, Liu S, Koynov K, Knoll W, Li Q (2009) An aqueous route to multicolor photoluminescent carbon dots using silica spheres as carriers. Angew Chem Int Ed 48:4598–4601. https://doi.org/10.1002/anie.200900652
44. Liu S, Tian J, Wang L, Zhang Y, Qin X, Luo Y, Asiri AM, Al-Youbi AO, Sun X (2015a) Hydrothermal treatment of grass: a low-cost, green route to nitrogen-doped, carbon-rich, photoluminescent polymer nanodots as an effective fluorescent sensing platform for label-free detection of Cu(II) ions. Adv Mater 24:2037–2041. https://doi.org/10.1002/adma.201200164
45. Liu J, Liu Y, Liu N, Han Y, Zhang X, Huang H, Lifshitz Y, Lee S-T, Zhong J, Kang Z (2015b) Metal-free efficient photocatalyst for stable visible water splitting via a two-electron pathway. Science 347:970–974. https://doi.org/10.1126/science.aaa3145
46. Liu X, Pang J, Xu F, Zhang X (2016) Simple approach to synthesize amino-functionalized carbon dots by carbonization of Chitosan. Sci Rep 6:31100. https://doi.org/10.1038/srep31100
47. Liu ML, Chen BB, Li CM, Huang CZ (2019) Carbon dots: synthesis, formation mechanism, fluorescence origin and sensing applications. Green Chem 3:449–471. https://doi.org/10.1039/C8GC02736F
48. Long B, Huang Y, Li H, Zhao F, Rui Z, Liu Z, Tong Y, Ji H (2015) Carbon dots sensitized BiOI with dominant 001 facets for superior photocatalytic performance. Ind Eng Chem Res 54:12788–12794. https://doi.org/10.1021/acs.iecr.5b02780
49. Lopez C, Zougagh M, Algarra M, Rodríguez-Castellon E, Campos BB, Esteves da Silva JCG, Jimenez-Jimenez JJ, Ríos A (2015) Microwave-assisted synthesis of carbon dots and its potential as analysis of four heterocyclic aromatic amines. Talanta 132:845–850. https://doi.org/10.1016/j.talanta.2014.10.008

50. Luo PG, Yang F, Yang S-T, Sonkar SK, Yang L, Broglie JJ, Liu Y, Sun Y-P (2014) Carbon-based quantum dots for fluorescence imaging of cells and tissues. RSC Adv 4:10791–10807. https://doi.org/10.1039/C3RA47683A
51. Medeiros Borsagli FGL, Ciminelli VST, Ladeira CL, Haas DJ, Lage AP, Mansur HS (2019) Multi-functional eco-friendly 3D scaffolds based on N-acyl thiolated chitosan for potential adsorption of methyl orange and antibacterial activity against *Pseudomonas* aeruginosa. J Environ Chem Eng 7:103286. https://doi.org/10.1016/j.jece.2019.103286
52. Ming H, Ma Z, Liu Y, Pan K, Yu H, Wang F, Kang Z (2012) Large scale electrochemical synthesis of high quality carbon nanodots and their photocatalytic property. Dalton Trans 41:9526–9531. https://doi.org/10.1039/C2DT30985H
53. Nan F, Kang Z, Wang J, Shen M, Fang L (2015) Carbon quantum dots coated BiVO4 inverse opals for enhanced photoelectrochemical hydrogen generation. Appl Phys Lett 106:37. https://doi.org/10.1063/1.4918290
54. Pathak P, Harruff BA, Wang X, Wang H, Luo PG, Yang H, Chen B, Veca LM, Xie S-Y (2006) Quantum-sized carbon dots for bright and colorful photoluminescence. J Am Chem Soc 128:7756. https://doi.org/10.1021/ja062677d
55. Paušová Š, Krýsa J, Jirkovský J, Forano C, Mailhot G, Prevot V (2015) Insight into the photocatalytic activity of ZnCr-CO_3LDH and derived mixed oxides. Appl Catal B Environ 170–171:25–33. https://doi.org/10.1016/j.apcatb.2015.01.029
56. Peng H, Travas-Sejdic J (2009) Simple aqueous solution route to luminescent carbogenic dots from carbohydrates. J Chem Mater 21:5563. https://doi.org/10.1021/cm901593y
57. Peng J, Gao W, Gupta BK, Liu Z, Romero-Aburto R, Ge LH, Song L, Alemany LB, Zhan XB, Gao GH, Vithayathil SA, Kaipparettu BA, Marti AA, Hayashi T, Zhu JJ, Ajayan PM (2012) Graphene quantum dots derived from carbon fibers. Nano Lett 12:844–849. https://doi.org/10.1021/nl2038979
58. Pooja D, Saini S, Thakur A, Kumar B, Tyagi S, Nayak MK (2017) A "Turn-On" thiol functionalized fluorescent carbon quantum dot based chemosensory system for arsenite detection. J Hazard Mater 328:117–126. https://doi.org/10.1016/j.jhazmat.2017.01.015
59. Prathik R, Chen PC, Periasamy AP, Chen YN, Chang HT (2015) Photoluminescent carbon nanodots: synthesis, physicochemical properties and analytical applications. Mater Today 18:447–458. https://doi.org/10.1016/j.mattod.2015.04.005
60. Prathumsuwan T, Jaiyong P, In I, Paoprasert P (2019) Label-free carbon dots from water hyacinth leaves as a highly fluorescent probe for selective and sensitive detection of borax. Sensor Actuat B-Chem 299:126936. https://doi.org/10.1016/j.snb.2019.126936
61. Raju KM, Raju MP, Mohan YM (2008) Synthesis of superabsorbent copolymers as water manageable materials. Polym Int 110:2453–2460. https://doi.org/10.1002/pi.1145
62. Roy SC, Varghese OK, Paulose M, Grimes CA (2010) Toward solar fuels: photocatalytic conversion of carbon dioxide to hydrocarbons. ACS Nano 4:1259. https://doi.org/10.1021/nn9015423
63. Sagbas S, Sahiner N (2019) 22-Carbon dots: preparation, properties, and application. Nanocarbon and its Composites. Ed. Elsevier, 1st, pp 651–676. ISBN: 9780081025109
64. Sahu S, Behera B, Maiti TK, Mohapatra S (2012) Simple one-step synthesis of highly luminescent carbon dots from orange juice. Application as excellent bio-imaging agents. Chem Commun 48:8835–8837. https://doi.org/10.1039/C2CC33796G
65. Saud PS, Pant B, Al-M A, Ghouri ZK, Park M, Kim H-Y (2015) Carbon quantum dots anchored TiO_2 nanofibers: effective photocatalyst for waste water treatment. Ceram Int 41:11953–11959. https://doi.org/10.1016/j.ceramint.2015.06.007
66. Shangguan J, Huang J, He D, He X, Wang K, Ye R, Yang X, Qing T, Tang J (2017) Highly Fe^{3+}-selective fluorescent nanoprobe based on ultrabright N/P codoped carbon dots and its application in biological samples. Anal Chem 89:7477–7484. https://doi.org/10.1021/acs.analchem.7b01053
67. Shen S, Wang Q (2013) Rational tuning the optical properties of metal sulfide nanocrystals and their applications. Chem Mater 25:1166. https://doi.org/10.1021/cm302482d

68. Sun C, Zhang Y, Kalytchuk S, Wang Y, Zhang X, Gao W, Zhao J, Cepe K, Zboril R, Yu WW, Rogach AL (2015) Down-conversion monochromatic light-emitting diodes with the color determined by the active layer thickness and concentration of carbon dots. J Mater Chem C 3:6613–6615. https://doi.org/10.1039/C5TC01379H
69. Sun XM, Li YD (2004) Colloidal carbon spheres and their core/shell structures with noble-metal nanoparticles. Angew Chem Int Ed 43:597–601. https://doi.org/10.1002/anie.200352386
70. Sun RQ, Sun LB, Chun Y, Xu QH (2008) Catalytic performance of porous carbons obtained by chemical activation. Carbon 46:1757–1764. https://doi.org/10.1016/j.carbon.2008.07.029
71. Tang D, Zhang H, Huang H, Liu R, Han Y, Liu Y, Tong C, Kang Z (2013) Carbon quantum dots enhance the photocatalytic performance of BiVO4 with different exposed facets. Dalton Trans 42:6285–6289. https://doi.org/10.1039/C3DT50567G
72. Unesco (2016) Water, people and cooperation: 50 years of water programmes for sustainable development at UNESCO (ISBN: 978–92–3–100128–4)
73. Usubharatana P, McMartin D, Veawab A, Tontiwachwuthikul P (2006) Photocatalytic process for CO_2 emission reduction from industrial flue gas streams. Ind Eng Chem Res 45:2558. https://doi.org/10.1021/ie0505763
74. Whitesides GM (2015) Nanoscience, Nanotechnology, and Chemistry. Small 1:172. https://doi.org/10.1002/smll.200400130
75. Wang Y, Hu A (2014) Carbon quantum dots: synthesis, properties and applications. J Mater Chem C 2:6921–6939. https://doi.org/10.1039/C4TC00988F
76. Wang R, Wang X, Sun Y (2017) One-step synthesis of self-doped carbon dots with highly photoluminescence as multifunctional biosensors for detection of iron ions and pH. Sensor Actuat B-Chem 241:73–79. https://doi.org/10.1016/j.snb.2016.10.043
77. Wang J, Sahu S, Sonkar SK, Tackett KN II, Sun KW, Liu Y, Maimaiti H, Anilkumar P, Sun Y-P (2013) Versatility with carbon dots from overcooked BBQ to brightly fluorescent agents and photocatalysts. RSC Adv 3:15604–15607. https://doi.org/10.1039/C3RA42302F
78. Wang X, Cao L, Lu F, Meziani MJ, Li H, Qi G, Zhou B, Harruff BA, Sun Y-P (2009) Photoinduced electron transfers with carbon dots. Chem Commun 2009:3774. https://doi.org/10.1039/B906252A
79. Wang X, Cao L, Yang S-T, Lu F, Meziani MJ, Tian L, Sun KW, Bloodgood MA, Sun Y-P (2010) Bandgap-like strong fluorescence in functionalized carbon nanoparticles. Angew Chem Int Ed 49:5310. https://doi.org/10.1002/anie.201000982
80. Wang H, Wei Z, Matsui H, Zhou S (2014) Fe_3O_4/Carbon quantum dots hybrid nanoflowers for highly active and recyclable visible-light driven photocatalyst. J Mater Chem A 2:15740–15745. https://doi.org/10.1039/C4TA03130J
81. Wei XuY, Hao Y, Yin XB, He XW (2014) Ultrafast synthesis of nitrogen doped carbon dots via neutralization heat for bioimaging and sensing applications. RSC Adv 4:44504–44508. https://doi.org/10.1039/C4RA08523J
82. Wu Q, Li W, Wu YJ, Huang ZH, Liu SX (2014) Pentosan-derived water-soluble carbon nano dots with substantial fluorescence: properties and application as a photosensitizer. Appl Surf Sci 315:66–72. https://doi.org/10.1016/j.apsusc.2014.06.127
83. Wu Q, Li W, Tan J, Wu Y, Liu S (2015) Hydrothermal carbonization of carboxymethylcellulose: one-pot preparation of conductive carbon microspheres and water-soluble fluorescent carbon nanodots. Chem Eng J 266:112–120. https://doi.org/10.1016/j.cej.2014.12.089
84. Wu ZL, Liu ZX, Yuan YH (2017) Carbon dots: materials, synthesis, properties and approaches to long-wavelength and multicolor emission. J Mater Chem B 21:3794–3809. https://doi.org/10.1039/C7TB00363C
85. Xie S, Su H, Wei W, Li M, Tong Y, Mao Z (2014) Remarkable photoelectrochemical performance of carbon dots sensitized TiO_2 under visible light irradiation. J Mater Chem A 2:16365–16368. https://doi.org/10.1039/C4TA03203A
86. Xu X, Ray R, Gu YL, Ploehn HJ, Gearheart L, Raker K, Scrivens W (2004) Electrophoretic analysis and purification of fluorescent single-walled carbon nanotube fragments. J Am Chem Soc 126:12736–12737. https://doi.org/10.1021/ja040082h

87. Xu YJ, Zhuang Y, Fu X (2010) New insight for enhanced photocatalytic activity of TiO2 by … of benzene and methyl orange. J Phys Chem C 114:2669–2676. https://doi.org/10.1021/jp909855p
88. Xue D, Yu F, Zhang Z, Yang Y (2019) One-step synthesis of carbon dots embedded zincone microspheres for luminescent detection and removal of dichromate anions in water. Sensor Actuat B-Chem. 279:130–137. https://doi.org/10.1016/j.snb.2018.09.101
89. Yang YH, Cui JH, Zheng MT, Hu CF, Tan SZ, Xiao Y, Yang Q, Liu Y (2012) One-step synthesis of amino-functionalized fluorescent carbon nanoparticles by hydrothermal carbonization of chitosan. Chem Commun 48:380–382. https://doi.org/10.1039/C1CC15678K
90. Ye K-H, Wang Z, Gu J, Xiao S, Yuan Y, Zhu Y, Zhang Y, Mai W, Yang S (2017) Carbon quantum dots as a visible light sensitizer to significantly increase the solar water splitting performance of bismuth vanadate photoanodes. Energ Environ Sci 10:772–779. https://doi.org/10.1039/C6EE03442J
91. Yin Y, Alivisatos AP (2005) Colloidal nanocrystal synthesis and the organic–inorganic interface. Nature 437:664. https://doi.org/10.1038/nature04165
92. Yuan M, Zhong R, Gao H, Li W, Yun X, Liu J, Zhao X, Zhao G, Zhang F (2015) One-step, green, and economic synthesis of water-soluble photoluminescent carbon dots by hydrothermal treatment of wheat straw, and their bio-applications in labeling, imaging, and sensing. Appl Surf Sci 355:1136–1144. https://doi.org/10.1016/j.apsusc.2015.07.095
93. Yuan M, Guo Y, Wei J, Li J, Long T, Liu Z (2017) Optically active blue-emitting carbon dots to specifically target the Golgi apparatus. RSC Adv 7:49931. https://doi.org/10.1039/C7RA09271G
94. Zhai Y, Zhu Z, Zhu C, Ren J, Wang E, Dong S (2014) Multifunctional water-soluble luminescent carbon dots for imaging and Hg^{2+} sensing. J Mater Chem B 2:6995. https://doi.org/10.1039/C4TB01035C
95. Zhang H, Ming H, Lian S, Huang H, Li H, Zhang L, Liu Y, Kang Z, Lee ST (2011) Fe_2O_3/carbon quantum dots complex photocatalysts and their enhanced photocatalytic activity under visible light. Dalton Trans 40:10822–10825. https://doi.org/10.1039/C1DT11147G
96. Zhang Z, Hao J, Zhang J, Zhang B, Tang J (2012) Protein as the source for synthesizing fluorescent carbon dots by a one-pot hydrothermal route. Rsc Adv 2:8599–8601. https://doi.org/10.1039/C2RA21217J
97. Zhang X, Zhang Y, Wang Y, Kalytchuk S, Kershaw SV, Wang Y, Wang P, Zhang T, Zhao Y, Zhang H, Cui T, Wang Y, Zhao J, Yu WW, Rogach AL (2013) Color-switchable electroluminescence of carbon dot light-emitting diodes. ACS Nano 7:11234–11241. https://doi.org/10.1021/nn405017q
98. Zhang J, Yu S-H (2016) Carbon dots: large-scale synthesis, sensing and bioimaging. Mater Today 19:382–393. https://doi.org/10.1016/j.mattod.2015.11.008
99. Zhang Y, Gao Z, Zhang W, Wang W, Chang J, Kai J (2018) Fluorescent carbon dots as nanoprobe for determination of lidocaine hydrochloride. Sensor Actuat B-Chem 262:928–937. https://doi.org/10.1016/j.snb.2018.02.079
100. Zhong DK, Choi S, Gamelin DR (2011) Near-complete suppression of surface recombination in solar photoelectrolysis by "Co-Pi" catalyst-modified W:$BiVO_4$. J Am Chem Soc 133:18370–18377. https://doi.org/10.1021/ja207348x
101. Zhou J, Sheng Z, Han H, Zou M, Li C (2012) Facile synthesis of fluorescent carbon dots using watermelon peel as a carbon source. Mater Let 66:222–224. https://doi.org/10.1016/j.matlet.2011.08.081
102. Zhou J, Liu Q, Feng W, Sun Y, Li F (2015) Upconversion luminescent materials: advances and applications. Chem Rev 115:395–465. https://doi.org/10.1021/cr400478f
103. Zhu Z, Ma J, Wang Z, Mu C, Fan Z, Du L, Bai Y, Fan L, Yan H, Phillips DL, Yang S (2014) Efficiency enhancement of perovskite solar cells through fast electron extraction: the role of graphene quantum dots. J Am Chem Soc 136:3760–3763. https://doi.org/10.1021/ja4132246
100. Zhu S, Meng Q, Wang L, Zhang J, Song Y, Jin H, Zhang K, Sun H, Wang H, Yang B (2013) Highly photoluminescent carbon dots for multicolor patterning, sensors, and bioimaging. Angew Chem 52:3953–3957. https://doi.org/10.1002/anie.201300519

105. Zhu H, Wang E, Li J, Wang J (2018) L-tyrosine methyl ester-stabilized carbon dots as fluorescent probes for the assays of biothiols. Anal Chim Act 1006:83–89. https://doi.org/10.1016/j.aca.2017.12.014
106. Zhang J, Yu S-H (2016) Carbon dots: large-scale synthesis, sensing and bioimaging. Mater Today 19:382–393
107. Liu S, Tian J, Wang L, Zhang Y, Qin X, Luo Y, et al (2012) Hydrothermal treatment of grass: a low-cost, green route to nitrogen-doped, carbon-rich, photoluminescent polymer nanodots as an effective fluorescent sensing platform for label-free detection of Cu(II)ions. Adv Mater 24(15):2037–41
108. Huang X, Zhang F, Zhu L, Choi KY, Guo N, Guo J, Tackett K, Anilkumar P, Liu G, Quan Q (2013) ACS Nano, 7:5684–5693
109. Li H, Liu R, Lian S, Liu Y, Huang H, Kang Z (2013) Near infrared light controlled photocatalytic activity of carbon quantum dots for highly selective oxidation reaction. Nanoscale, 5(8):3289–329
110. Yang YH, Cui JH, Zheng MT, Hu CF, Tan SZ, Xiao Y, et al (2012) One-step synthesis of amino-functionalized fluorescent carbon nanoparticles by hydrothermal carbonization of chitosan. Chem Commun 48:380–2
111. Zhang X, Zhang Y, Wang Y, Kalytchuk S, Kershaw SV, Wang Y, Wang P, Zhang T, Zhao Y, Zhang H, Cui T, Wang Y, Zhao J, Yu WW, Rogach AL (2013) Color-switchable electroluminescence of carbon dot light-emitting diodes. ACS Nano 7:11234–11241

Multifunctional Composite Aerogels—As Micropollutant Scavengers

Oumaima Mertah, Anina James, Masoumeh Zargar, Sushma Chauhan, Abdelhak Kherbeche, and Padmanaban Velayudhaperumal Chellam

Abstract Composite aerogels are low-density porous material with a high surface area, facilitating their application in wastewater treatment. The surface of these aerogels can be modified based on the ionic charge of the target pollutants. Improved properties like high gas holdup, a low mean free path of diffusion, high mechanical strength, and integrated 3D gel architecture make them an ideal matrix for several environmental applications. This review focuses on carbon, silica, metal-based, and biopolymer composite aerogels toward effluent treatment. The challenges in the synthesis of aerogels using various reduction methods and strategies for surface modification of aerogels having improved water treatment properties are compared. Composite aerogels' application for removing textile dyes, heavy metals, and pesticides, and the oil separation is discussed along with relevant reaction kinetics. The adsorptive and photocatalytic removal of micropollutants by aerogels are also compared. Though several novel composite aerogels have been experimented with in wastewater treatment, the toxicity implications limit their extensive scale application. The toxicity of precursor compounds such as polyacrylonitrile, resorcinol–formaldehyde, phenol, Tetramethyl orthosilicate (TMOS), and leaching of nanoparticles from aerogels are discussed. As a solution to these impediments, bio-aerogels' use and

---

O. Mertah et al.—All authors contributed equally.

O. Mertah · A. Kherbeche
Laboratory of Materials, Processes, Catalysis, and Environment (LPCME), Sidi Mohamed Ben Abdellah University, Fez, Morocco

A. James
Soil Microbial Ecology and Environmental Toxicology Laboratory, Department of Zoology, University of Delhi, Delhi, India

M. Zargar
School of Engineering, Edith Cowan University (ECU), Joondalup, Australia

S. Chauhan
Amity University Chhattisgarh Raipur, Raipur, India

P. V. Chellam (✉)
Department of Biotechnology, National Institute of Technology Andhra Pradesh, Tadepalligudem, Andhra Pradesh, India

E. Lichtfouse et al. (eds.), *Inorganic-Organic Composites for Water and Wastewater Treatment*, Environmental Footprints and Eco-design of Products and Processes,
https://doi.org/10.1007/978-981-16-5928-7_7

further improvement for water treatment are highly warranted. The possibility of using aerogels in large-scale applications in moving toward a sustainable circular economy is also emphasized.

Keywords Aerogels · Synthesis · Micropollutants · Reaction kinetics · Ecotoxicology

1 Introduction

Aerogels are one of the most promising and versatile material, as an outcome of advancements in the composite material science. The applications of aerogels depends on the type of composite backbone like silica, tungstic oxide, alumina, nickel tartrate, stannic oxide, gelatin, egg albumin, agar, nitrocellulose, and cellulose aerogels [1]. The early years of research were limited to and focused on silica-based aerogels. The basic plan of synthesis of aerogels, which are nearly dried gels, involves sol–gel reactions with specific drying processes applied to obtain pores and networks' functional structure. Attributes associated with its composition, texture, and surface area depends on the flexibility conferred by the sol–gel method, which also permits varying the constituents according to the utility [2]. Properties of aerogels, such as high porosities, improved surface area, and dielectric strengths, and low densities, acoustic impedance, thermal conductivities, and refractive index, has brought about monumental impact in different industrial sector. Since its origination, researchers have tried various permutations and combinations with several materials to make different kinds of aerogels complying with specific requirements. The last few decades of the twentieth century witnessed the development of silica aerogels based on metal oxides such as titanium, zirconium, aluminum [3], and synthetic polymers incorporating resorcinol–formaldehyde [4], polyurethane [5]. With respect to low thermal conductivity, polymer aerogels were superior to silica aerogels [6]. The advent of the twenty-first century witnessed the development of a new generation of composite aerogels that were carbon based, including carbon nanotubes, graphene [7, 8], semiconductor chalcogenides such as CdS, CdSe, PbTe [9, 10], biological aerogels or bio-aerogels such as cellulose and other polysaccharides including various proteins [11, 12] and, lately SiC-based aerogels [13, 14]. Bio-aerogels were quick to catch researchers' attention because compared to too fragile silica aerogels, and they demonstrated improved mechanical strength and stability [15, 16]. The preparation and utilization of bio-aerogels do not entail any toxic components; hence its use furthers the green chemistry revolution rendering it a mass appeal, particularly, but not limited to, life science industries [17, 18]. In recent years, the functionality of the aerogels has seen tremendous growth with the incorporation of nanomaterials in its making. Thermal insulation was the singular contribution of aerogels (silica-based) in its nascent years. Early studies highlighted their use in catalysis, which required high surface area and porosity [19, 20]. After a lull of few decades, by early 2000s, intensive research toward the development of novel aerogels toward environmental cleanup [21, 22], chemical sensors [23], filtering media [24] acoustic transducers [25], energy storage devices [26], metal casting molds [27], water repellant coatings [28], drug delivery and other biomedical and pharmaceutical applications [29], 30], extracting agents [31], and protective clothing [32].

The growing water pollution caused by widespread industrial and household effluents, fossil fuel use, radionuclides contamination, etc., is a grave concern. The pollutants' extreme toxicity, recalcitrant nature, and tendency to accumulate and transfer in the food chain and the ensuing prolonged catastrophic threat to different living ecosystems. Several strategies have been developed to mitigate water contamination, such as biological degradation [33], flocculation [34], electrochemical treatment [35], photocatalysis [36], advanced oxidation [37], nanofiltration [38], and adsorption [39].

Among these, adsorption has created a wide scope for research and implementation in industrial-scale applications. The preparation of low-cost novel functional materials toward sustainability with ease of operation toward the treatment with less secondary pollution has gained more research attention in recent years. Based on the type of pollutants, different types of adsorbents were used; As the separation of adsorbents and their composites require high-speed centrifugation and vacuum filtration [40] makes it more laborious to recycle them. Aerogels are the perfect candidates due to their high surface stable three-dimensional (3D) structures, intricate and extensive network of channels, and multidimensional mass transport pathways [41].

This review pivots on the various methods of fabrication and physiochemical properties of carbon, silica, metal-based, and biopolymer composite aerogels for water treatment, particularly for removing textile dyes, heavy metals and pesticides, and oil separation (Fig. 1). Though many types of composite aerogels have been used in wastewater treatment, there may have been an oversight regarding the toxicity of precursor compounds such as polyacrylonitrile, resorcinol–formaldehyde, phenol, TMOS, and leaching of nanoparticles, which may limit their large scale application. This leads the discussion to the practical, sustainable, and large-scale use and further improvement of bio-aerogels for water treatment to move toward a circular economy.

2 Methods for the Synthesis of Composites Aerogels

Aerogels' achievement is directly linked to its chemical compositions, crystal structures, pore structures, and surface morphologies. Scientists and engineers still desire aerogel composites with unique structures and superior performance. Generally, aerogels are made from different monomers (hydrophilic or hydrophobic) for particular uses and can be synthetic or natural polymers. Its low mechanical strength a slow degradation rate, and durability, which have to be balanced by optimal design [42, 43]. Besides pure aerogels, the fabrication of composites aerogels can be modified by different methods depending on their material compositions to improve their mechanical and thermal structures. In recent decades, nanoparticles and their composites have become exciting and challenging for scientists due to their extensive applications. However, the safety of nanoparticle technology inhibits its application. To overcome

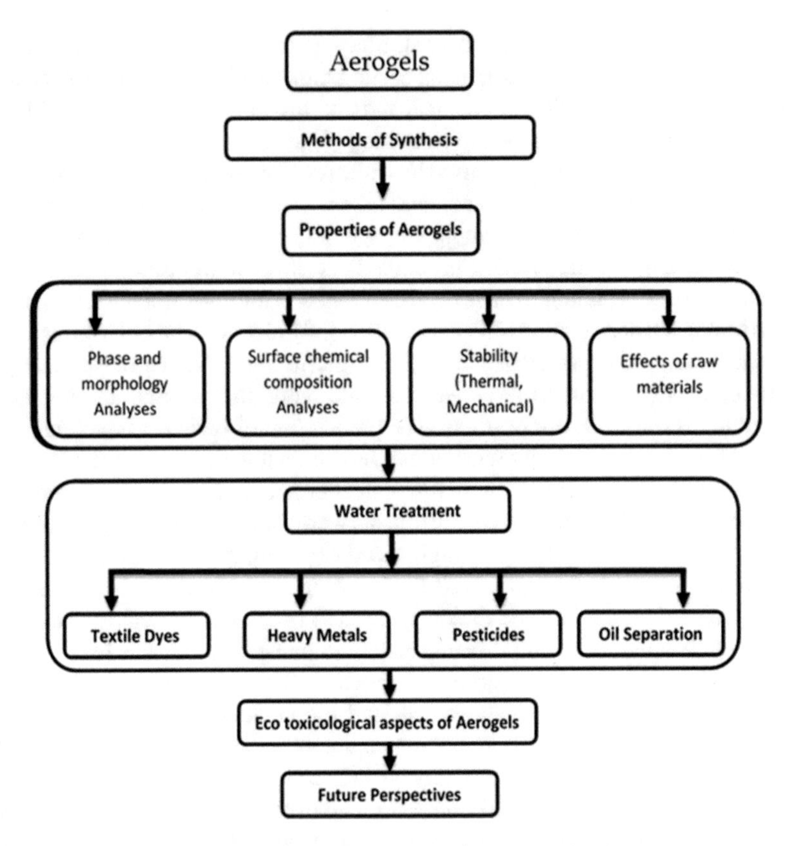

Fig. 1 Schematic outline of aerogels: synthesis, properties, and application

the hazards to human health and the environment, it is advantageous to combine them with gels and to improve their amalgamation properties at the same time [44]. The incorporation of nanoparticles into the gel network brings about a rearrangement and consequently, the creation of new materials. The gels can attain varied chemical and physical properties that favor their application in several fields [45]. Many nanoparticles have been introduced into polymeric networks to obtain composites aerogel such as carbon (graphene, carbon nano tubes, and carbon fibers), silica, and metal/metal oxide-based aerogels using different methods.

Carbon is the most abundant component in several nanostructures ranging from zero dimension (0D) to three dimensions (3D) with attractive properties, including high electric conductivity and ease of use, making it a promising material in various applications. Carbon-based nanoparticles like carbon nanotubes (CNTs), graphene, and carbon fibers are widely introduced into polymers to add functionalities to the polymer network using appropriate host–guest interactions [46, 47]. Polymer gelators and the carbon-based nanoparticles interact with each other in numerous forms to build composites, usually van der Waals forces over the aliphatic chains,

$\pi - \pi$ stacking among aromatic sites, dipolar and electrostatic interactions using polar moieties providing stability and flexibility [48]. The most recent studies proved that carbon-based nanoparticles incorporated aerogels exhibited strong properties, including biocompatibility, electrical conductivity, high porosity, efficient self-healing, and mechanical strength.

The intact interaction of silica nanoparticles with the polymer enhances the mechanical strength of the nanocomposite gel [49, 50]. These advantages have urged scientists to focus further on silica-based nanocomposite gels. The adsorption of poly (N, N_0-dimethyl acrylamide) (PDMA) chains on silica affect the formation of a physical network [51] that enhances the mechanical property of the gels network. Moreover, Luo et al. [52] studied the chemical crosslinking of trimethoxy silyl groups present in methyltrimethoxysilane (MTMS) solution with the hydroxyl groups of polyvinyl alcohol (PVA) to synthesize nanocomposite gels with the strong mechanical property. The addition of silica nanoparticles (Si NPs) in the polyacrylic acid and chitosan (PAA/CS) gel increased the compression strength and the fracture toughness [53]. Silica NPs increase the crosslinking density of gel, surface area, tunable biodegradability, and beneficial mechanical properties. Metal nanoparticles such as gold, silver, copper, nickel, and metal oxide nanoparticles based nanocomposites gel have garnered enormous attention due to their extensive application in the field of biomedicine toward sensor development.

Several researchers have studied gels incorporated with gold nanoparticles (Au NPs) due to their fluent surface functionalization, compatibility, facile synthesis, and high electrical and optical proprieties [54–56]. Song et al. [55] reported the aggregated systems of small Au-NPs in DNA gel using gel scaffold assembled using enzymatic ligating of X-shaped DNA. This modified gel finds its application for photo-thermal therapy and contrast CT imaging and radio-sensitization in disease diagnosis. Au-NPs-based nanocomposite gels have several useful properties, however, the price of gold limit their large-scale applications.

Silver nanoparticles (Ag NPs) were widely experimented with in various applications of the biomedical field because of their improved properties and characteristics toward diagnosis and treatment [57, 58]. Ag NPs impregnated nanocomposite gels increase gelling properties, electrical conductivity, optical effects, deformability, and antibacterial properties [59, 60]. Simultaneously, the in situ reduction of Ag^{+} ions could be obtained by robust reducing species formed during water radiolysis to enhance mechanical properties, elasticity, and antibacterial properties. Therefore, the synthesis of metal nanoparticles based nanocomposite gels results in promising new materials.

There is enormous research interest in the development of composite aerogels with specially designed functions. However, composites' preparation is still in the early stages requiring several slight adjustments in their synthesis for their applications in critical areas, especially the biomedical field (Table 1).

Table 1 Different types of aerogels with their synthesis methods and properties from reported literature

Type of nanoparticles-based composites aerogel		Methods	Properties	References
Aerogel	Fillers			
PPy nanotube hydrogels (PPyNHs)	Carbon Nanotubes (CNAs)	Bamboo-like N/S-codoped CNAs made using MnO_2-templated polymerization of pyrrole in an acidic environment	– High supercapacitive performances – Increased capability and capacity retention	[61]
Chitosan	Multiwalled carbon nanotubes (MWCNTs)	ZIF-8 or UiO66-NH_2 derived MOFs-based CNA (MPCA) In situ nucleation, enlargement of MOFs nanoparticles onto carbon nanotubes	– Good hydrophilia, compression, resilience, and thermostability – Reduced the risk of MOFs leakage and subsequent bioaccumulation	[62]
Polyimide (PI)	Carbon nanotubes (CNTs)	Lightweight and strong PI aerogels crosslinked with 4,4'-oxydianiline (ODA)-functionalized Carbon Nanotubes (CNTs) followed by chemical imidizing	– Increased mechanical properties – Increased yield strength – Improved thermal stability	[63]
Dopamine hydrochloride	Multiwalled carbon nanotubes (MWCNTs) and Graphene Oxide	Green synthesis of 3D graphene/polydopamine modified Multiwalled Carbon Nanotube (MWCNT-PDA) hybrid aerogels	– Improved structure stability – Prevents the stacking of GO sheets – Improved adsorption – Surface complexation, chelation improved	[64]

(continued)

Table 1 (continued)

Type of nanoparticles-based composites aerogel		Methods	Properties	References
Aerogel	Fillers			
Polydopamine (PDA)	Carbon Coaxial Nanotubes and Sb2S3	Sb_2S_3@nitrogen-doped carbon nanotubes encapsulated by porous graphene aerogel	– Improved electrical conductivity – Increased porosity – Excellent electrochemical properties	[65]
Konjac glucomannan (KGM)	Multiwalled carbon nanotube	The incorporation of mussel-inspired protein for preparation of KGM/PCCNT aerogels	– Potential applications targeted delivery of drugs via carrier	[66]
Nylon 66 (PA66) pellets	Carbon nanotube	CNT/PA66 aerogels have been fabricated by freeze-drying	– Good electromagnetic wave absorption properties	[67]
Nafion	Multi-walled carbon nanotubes and graphene screen-printed electrode and Gold NPs	Development of CAG based on MWCNT and Nafion frameworks called Nafion-enhanced CAG (NfCAG), which are modified on the surface of Gr-SPEs	– Capable of detecting dihydroxybenzene isomers in real water samples – Screening of environmental pollution	[68]
Chitosan	Carbon nanotube and montmorillonite	Preparation of chitosan/carbon nanotube composite aerogel (CCA)/montmorillonite using freeze-drying	– Exhibited self-extinguishing and excellent flame retardancy – Anti-fatigue performance – Thermal insulation – Good piezoresistivity	[69]

(continued)

Table 1 (continued)

Type of nanoparticles-based composites aerogel		Methods	Properties	References
Aerogel	Fillers			
CH_3-riched trimethylmethoxysilane (TMMS)	Silica aerogel	Monolithic silica aerogel is made by controlling the skeleton's crosslinking degree, and ambient pressure drying	– Flexibility – Porosity – Hydrophobicity – Good thermal stability	[70]
Silica aerogel	Tetraethoxysilane (TEOS)	Hydrophobization of silica alcogels resulted in the fabrication of superhydrophobic silica aerogel powders and granules	– High specific surface area – Good thermal conductivity – Highly transparent powders	[71]
Silica aerogel	IBPIF with partial opening cell	Layers of silica aerogel were added into IBPIF cells via an in situ growth process of Silica Sol (SS)	– High thermal insulation – Fire safe – Improved flame retardant quality compared to silica particles	[72]
Silica aerogel	Epoxy resin	Aerogel porosity maintained by choosing suitable epoxy viscosity	– Thermal conductivity – Compressive – Excellent for limited external loading	[73]
Silica aerogel	Carbon nanotube (CNT)	Silica aerogel was incorporated into the CNT-based composites	– Improved the stability in heat generation and changes in the electrical resistivity – Enhanced electrical properties	[74]

(continued)

Table 1 (continued)

Type of nanoparticles-based composites aerogel		Methods	Properties	References
Aerogel	Fillers			
Silica aerogel	APTES as an amine source	The silica aerogel microsphere was synthesized by a "ball drop" method	– Good sphericity – Enhanced CO_2 adsorption capacity	[75]
Silica aerogel	Polyacrylonitrile/Polyvinylidene fluoride (PAN/PVDF) membrane	Silica aerogel/Polyacrylonitrile/Polyvinylidene fluoride (SAPPF) webs made using electrospinning	– Increased porosity and thermal resistance – Enhanced the adsorption capacity	[76]
Silica aerogel	Graphite	Fabrication of SiC fiber through heating silica aerogel reinforced graphites	– Higher porosity – Thermal stability	[77]
Silica aerogel	Methyltriethoxysilane (MTES) and tetraethoxysilane (TEOS)	Dual-mesoporous silica aerogel made at atmospheric pressure drying	– Separate oil from oil-in-water emulsion using stabilized surfactant – Superhydrophobic – Thermally stable	[78]
Graphene-based aerogel	Gold nanoparticles and Graphene Oxide (GO)	Graphene-based aerogel with gold nanoparticles through a freeze-drying process	– Detection of H_2O_2 – Good repeatability – Low detection limit	[79]
TiO_2 aerogel	Gold nanoparticles	TiO_2–Au aerogels synthesized using a sol–gel method	– Higher degradation rates	[80]
Chitosan–silica aerogels	Gold nanoparticles	Photo-formation of gold nanoparticles in the solid monoliths of Au(III)-chitosan-silica aerogels with different Au/NH_2 molar ratios using photoacoustic spectroscopy	– Thermally thin	[81]

(continued)

Table 1 (continued)

Type of nanoparticles-based composites aerogel		Methods	Properties	References
Aerogel	Fillers			
Multiple Graphene Aerogel (MGA)	Gold nanocrystals	Synthesis of gold nanocrystals via reduction of $HAuCl_4$ with ascorbic acid	– Ultrahigh sensitivity – Detection of ctDNA with good specificity	[82]
Polyimide aerogels	Silver nanoparticles	Fabrication of crosslinked polyimide aerogels with Ag nanoparticles processed with supercritical CO_2 drying	– Highest mechanical properties – High-temperature microstructure stability	[83]
Chitosan	Silver nanoparticles	Chitosan modified using silver organosols via metal vapor synthesis	– Increased specific surface area	[84]
Silica aerogel	Silver nanoparticles	Supercritical deposition technique using laser irradiation to make Ag nanoparticles in silica aerogel	– Spherical AgNPs with a size of few nanometers can be synthesized fast and locally – The fluid molecules tightly envelop AgNPs to form a relatively dense coating	[85]
Carbon aerogels	Copper nanoparticles	Sol–gel-based preparation of Cu^{2+} embedded with resorcinol (R) and formaldehyde (F) followed by calcination at high temperature	– Impressive activity of electrochemical reduction of CO_2 to CO – High selectivity toward CO – Excellent stability	[86]

(continued)

Table 1 (continued)

Type of nanoparticles-based composites aerogel		Methods	Properties	References
Aerogel	Fillers			
Graphene aerogel	ZnO nanoparticles	Macroporous microrecycled zinc oxide nanoparticles (NPs) in three-dimensional (3D) graphene aerogel	– Detection of NO_2 gas – Rapid electron channels to ZnO	[87]
Carbon nanofibril aerogel	Fe_3O_4 nanoparticles	Necklace-like Fe_3O_4/carbon nanofibril aerogel (Fe_3O_4/CNF) by crosslinking followed by carbonization of the obtained ferric alginate aerogel	– High stability, – Electrical conductivity, – High specific capacity – Improved electrochemical storage devices	[88]
Cellulose	Copper nanoparticles and carbon nanotubes	Hybrid aerogel made from cellulose nanofiber embedded with transition metal divalent ions and strengthened by carbon nanotubes	– Deodorizer material for air pollution remediation – Eco-friendly and economical	[89]
Cellulose	Copper and nickel nanoparticles	Preparation of Bacterial Cellulose (BC) aerogels with metal nanoparticles (Cu and Ni) using swelling-induced adsorption process	– Excellent stability and reusability – Good catalytic performance	[90]
Alginate	Copper crystal and carbon dots	Copper-carbon dots complex binding with tigecycline (Cu/TGC@PDA) and alginate aerogel synthesized via the crosslinking method	– Antibacterial properties	[91]
Graphene aerogel	SnO_2 nanocrystals	3D N-doped graphene aerogel with ultra-small SnO_2 made by hydrothermal reaction	– Economical electrolyte infiltration and good structure stability – Excellent cyclability	[92]

(continued)

Table 1 (continued)

Type of nanoparticles-based composites aerogel		Methods	Properties	References
Aerogel	Fillers			
Graphene aerogel	α-Fe_2O_3	The robust 3D porous α-Fe_2O_3@3DrGO aerogel synthesized by a hydrothermal self-assembly process	– Remarkable cyclic life time – Fast rapid dynamic diffusion of lithium ions – Synergistic, boosted, ultra-fast pseudocapacitive kinetics	[93]
Alginate	TiO_2	GO and TiO_2 combined with sodium alginate to make reduced GO−TiO_2/sodium alginate (RGOT/SA) aerogel	– High synergy of adsorption and degradation – Efficient charge separation – Self-cleaning mechanism – Recyclable	[94]
Bacterial cellulose	Metal–organic framework nanoparticles	Bacteria cellulose base in which growth of MOF nanoparticles (zeolitic imidazolate framework-8 and UiO-66) induced	– High porosity – Mechanical flexibility – Low density – Superior adsorption performance	[95]
Titanium isopropoxide (TIP), ethanol, and nitric acid	TiO_2, Gold, and Silver nanoparticles	Porous nanocomposites of Au/Ag-TiO_2 aerogels by sol–gel process	– Higher specific surface area – Efficient salicylic acid degradation – Better performance in plasmonic photocatalysis	[96]

3 Properties of Aerogels

Aerogels consist of highly porous structures with high surface areas, and their structures are mainly defined by the process by which they are synthesized. The surface area of these aerogels, their pore volume, and size distribution can be tuned during their production to achieve aerogels of desired properties that can be applied for adsorption and catalytic processes during water treatment. When analyzing the aerogel's properties, some of the factors that mainly need discussion are its morphology, surface modifications to enhance their properties, and their stability characteristics under different environments. The density and microstructure of aerogels and their backbone materials and applied synthesis strategies greatly influence their properties [97]. Therefore, it is vital to conduct reliable characterizations to identify the correct structure–property relationships of aerogels and suitable materials for their production to target specific applications.

3.1 Morphological Analysis of Aerogels

Aerogels are low-density compounds whose three-dimensional macrostructure exert hydrophobicity characteristics. This hydrophobicity makes them suitable for removing oil-based contaminants from the environment and allows modifications in them to remove contaminants of different characteristics. Their porous structures contain several channels and confer good mass transfer properties. This homogenous porosity also makes them act as excellent catalysts for performing photocatalytic methods of water treatment. These aerogels exist in different morphological forms such as particulate, nonparticulate, or fibrous form and generally appear as a monolith, powder, or film [98]. The inner structures of these aerogels consist of micropores (<2 nm), mesopores (>2 to <50 nm), or mixed pores of both sizes. The inclusion of fillers or polymers in aerogels can improve their morphology by modifying their physical properties. The composition of aerogel composites can be beneficially modified by adding enhancers to them, and thus, improving their interfacial adhesion. A gelatin-based organic–inorganic composite aerogel was applied for removing multiple contaminants from complex wastewater treatment in which the amphiphilic structure and high porosity exerted by the aerogels showed excellent selectivity toward many pollutants such as dyes and heavy metals with good reusability properties [99]. In the case of starch-based aerogels, the changes in the structure of the synthesized aerogel, its properties and morphology depend on the nature of the raw material, composition of monomers like amylose and amylopectin [100].

The structure and morphology of aerogels, including their pore structure, pore size, pore connectivity, cell type, crystallinity, significantly contribute thermal and mechanical properties to them [101]. Morphological analysis can be performed using a wide range of techniques like X-ray Diffraction (XRD) spectroscopy, BET nitrogen adsorption/desorption analysis, helium and mercury porosimetry, Scanning Electron Microscopy (SEM). Synthesis and process conditions of aerogels, such as controlling their solvent solidification, freezing rate and method, and different organic or inorganic additives into the aerogel structures can significantly alter their microstructure [102, 103]. SEM-based morphological analysis of aerogels, which can assist with identifying aerogels cell size, homogeneity, and estimation of their pore size distribution (through subsequent image analysis). SEM gets more integral for the analysis of composite structures that show the material composition effects on the pore structure and material/pore size distribution [104, 105]. For instance, SEM imaging of bioaerogels with 2% nanofibrillated cellulose (NFC) formed through conventional (CFD) and spray (SFD) freezing techniques showed the significant contribution of the freezing step on the bioaerogels morphology; where bioaerogels generated through CFD show a 2D-sheet-like morphology while those made by SFD demonstrate a three-dimensional fibrillary skeleton structure (Fig. 2). This has been attributed to the freezing toward the variant structure of crystalline ice [102].

X-ray Diffraction (XRD) spectrometry is one of the direct approaches for analyzing crystal structure in aerogels (amorphous or crystalline). XRD estimates the aerogel's crystallinity according to the position of the most substantial diffraction peaks in their XRD pattern [104]. XRD is beneficial for analysis of the modified or hybrid aerogel structures to determine the nature of aerogel's crystallinity [104, 106]. For instance, Gong et al. studied the chitosan graphene oxide (CS GO) composite aerogels and, through XRD spectroscopy, the crystallinity of the chitosan was significantly improved upon graphene oxide integration (from 23.7 to 59.5%). This was attributed to the enhanced nucleation and hydrogen bond formation with chitosan amino groups [107, 108]. Brunauer–Emmett–Teller (BET) and Barrett–Joyner–Halena (BJH) models are useful approaches for the structural analysis of

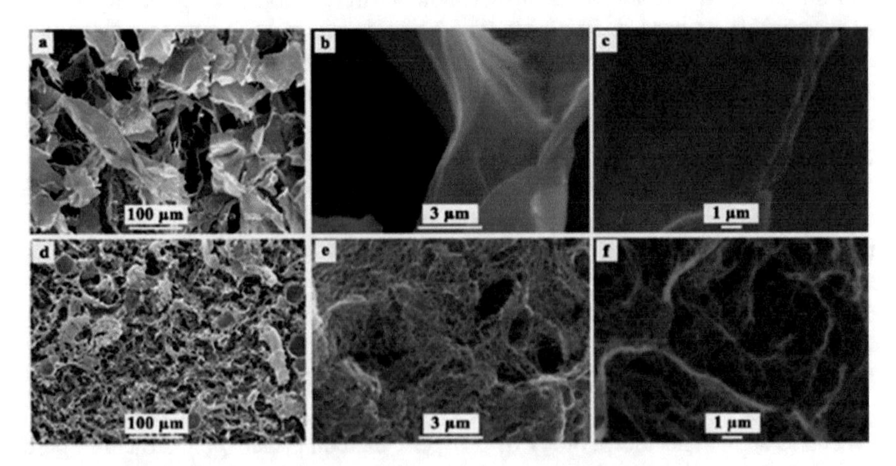

Fig. 2 SEM images showing the effect of freeze-drying strategy on the structure of aerogels prepared from 2% nanofibrillated cellulose, **a–c** conventional freeze-drying and **d–f** spray freeze-drying (reproduced from Jiménez-Saelices [97, 102] with permission from Elsevier)

aerogels [102]. To perform the BET-BJH analysis, the samples are first dried and degassed to remove physisorbed species at high vacuum at 100 °C for 24 h followed by nitrogen adsorption/desorption at −196 °C. The adsorbed/desorbed volume of nitrogen as well as the generated isotherms during the process identify the pore size, pore-volume, distribution, specific surface area of the aerogels [102]. Aerogels' porosity can be characterized using helium or mercury porosimetry [104, 109].

3.2 Surface Chemical Analysis

Efficient aerogels that can separate a wide range of pollutants from wastewater are achieved by modifying their surface using chemical modifiers containing different functional groups. These chemical modifiers form an intermediate reactive layer between the aerogel and water by using their functional groups and performing effective sorption. Polydopamine, tannic acid, and gallic acid are commonly used modifiers to perform the aerogels' surface functionalization [110]. The carboxyl, methyl, and amino groups based functional groups tailor the aerogel surface and provide strong affinities to adsorb the contaminants by increasing their pore size, density, and surface area [111]. A graphene-based aerogel with superhydrophobicity and superoleophilicity was fabricated for application in water treatment, and during their synthesis, functionalization using polydopamine and modification by fluoroalkyl silane promoted their selectivity and absorptivity toward contaminant removal providing high efficiency in oil/water separation processes [112]. A cellulose nanofibril aerogel was chemically modified through the oxidation–sulfonation process, and when applied for oil/water separation, the higher surface charge densities improved their separation efficiencies and the 〖 〖SO〗 _3〗 ^-groups on their hierarchical structure promoted the superoleophobic characteristics [113]. A multifunctional aerogel of cellulose nanofibrils decorated on its surface with different percentages of carbon nanotubes was fabricated for application in wastewater treatment. When applied, they exhibited antimicrobial and antioxidant properties proving them as potential candidates for industrial-level applications [114]. The highly amine-rich surface and a three-dimensional nanofiber-based aerogel's porosity showed excellent adsorption capacities toward heavy metal removal from wastewater. Its improved efficacy and versatility were achieved due to the aerogel's surface chemical modification [115]. Doping the surfaces of aerogels with doping materials can also help improvise their surface properties, and those which are metal doped can confer catalytic properties [116]. An Ag-doped carbon aerogel was fabricated and applied in drinking waters for removing halide ions. It was found that these modified surfaces had increased surface areas with improved porous structures and performed efficient chemisorption of the ions onto the aerogels [117].

The surface chemistry of the aerogels can be analyzed using a wide range of spectroscopic methods such as Fourier Transform Infrared (FTIR), ^{13}C Nuclear Magnetic Resonance (NMR), X-ray Photoelectron (XPS), and Energy Dispersive

Table 2 Fourier transform infrared spectral band assignments for composite polyisocyanurate (PIR) silica aerogels (reproduced from Zhao et al. [101] with permission from Elsevier)

Wave number (cm^{-1})	Functional groups
3369	– NH–stretching vibration absorption
2913	C–H stretching
2275	– N=C=O antisymmetric stretching
1712	C=O stretching vibration
1596	Benzene rings C=C stretching
1225	C–O asymmetric stretching
1068	C–O–C antisymmetric stretching in alcohol hydroxyl
814	=C–H variable angle vibration in benzene ring

X-Ray Analysis (EDX). FTIR spectroscopy is applied to identify the chemical structure of aerogels. It typically performs over around 100 scans within the range of 4000–400 cm^{-1}. A diverse range of chemical bonds has been identified for aerogels using FTIR depending on the aerogel's backbone materials and precursors, fabrication solvents, and final structures. For instance, the IR spectra band assignments for composite polyisocyanurate (PIR) silica aerogels are reported in Table 2, showing varied IR spectral peaks and the composition of the product aerogels [101].

The identified characteristic peaks determined by FTIR spectroscopy define the chemical structure of the developed aerogels. They indicate the extent of the precursors and additives (i.e., reacted, partially reacted, or non-reacted materials).

^{13}C NMR can also be used for spectral peak identification of the organic aerogels signposting the different species and chemical bonds present within the aerogel structure [118]. NMR can further identify the amorphous or crystalline structure of aerogels [103].

X-ray photoelectron spectroscopy analyzes aerogels' surface chemistry and its characterization [104]. It is usually conducted using a Kratos Axis Ultra instrument and has a nearly 15 nm analysis depth (from the surface) [119]. The spectra usually require deconvolution using commercial software such as CasaXPS to identify the exact chemical bonds associated with a single element and the composition of the original elements. This can clarify the materials' ratio and the relevant chemical interactions in aerogels [104]. The EDX detector of the SEM or Transmission Electron Microscopy (TEM) instruments can be used for the surface elemental analysis of the aerogels and the techniques mentioned above. EDX analysis can quickly assist with the identification of aerogels and modification efficiencies.

3.3 Thermal and Mechanical Stability

The mechanical strength of silica-based aerogels is low so combining them with other fibrous materials can improve their structural integrity. These aerogels' mechanical strength can be enhanced and tuned by combining them with polymers, by prolonging their aging step. Among these methods, the crosslinking of polymers with the silica backbone of the aerogel strengthen their network integrity and improve their strength [98]. Using chemical crosslinkers containing amine, phosphate, and carboxylic groups during aerogel synthesis can help modify and improve their mechanical properties. The addition of polyvinyl alcohol when synthesizing graphene oxide-based aerogels has been found to improve their stability, as the hydroxyl groups of polyvinyl alcohol crosslinks with the polar groups of graphene oxide layers; this enhances the number of vacant sites for adsorption on the aerogels for efficient adsorption of contaminants and can provide high structural stabilities [98]. Such polyvinyl alcohol crosslinked graphene oxide aerogels also have self-recovery properties by which they can maintain their original structure if they are subjected to extreme stress and strain conditions. The different proportions of polyvinyl alcohol and graphene oxide ratio can confer their different properties, making them a successful candidate for water treatment applications by providing high selective adsorption [120]. Another robust cellulose-based aerogel was synthesized by crosslinking polyethyleneimine (PEI) onto cellulose nanofibrils (CNF) by using 3-glycidyloxypropyl tri methoxy silane (GPTMS) as a crosslinking agent and improved mechanical. Adsorption capacities toward 〖 〖Cu〗 _2〗 ^+removal from waters were achieved when the increased mass ratios of PEI to GPTMS and increased amine content were maintained [121]. As maintaining the thermal strength of aerogels is vital for their widespread application, opacifiers can be added to improve their radiation properties to withstand high-temperature ranges and make them serve as excellent insulators. The proportion of opacifiers added may also affect their structural properties as they can affect their structural conformation, so appropriate proportions must be maintained [121]. When the aerogel's pore size is >68 nm, the heat conduction through these pores occurs at a higher rate, and their thermal conductivities are increased. Aerogels can have a range of thermal and mechanical stabilities depending on their pore structure, density, and material. For instance, silica aerogels have low mechanical stability due to their highly porous structure while being intrinsically thermal resistant [122, 123], while polyamide aerogels and cellulose/biochar aerogels have superior mechanical and relatively high thermal properties [118, 124]. Several techniques have been used to define the extent of aerogels' thermal and mechanical strength outlined here.

3.3.1 Mechanical Analysis of Aerogels

Aerogels' mechanical identity is usually classified through elasticity, brittleness, plasticity, reusability, compression, and expansion tests. The significant parameters of mechanical stability in terms of elastic and fracture properties of porous aerogels are their porosity and specific density level. These parameters are smaller for porous aerogels than non-porous ones [125]. The mechanical characterization techniques for aerogels can be categorized into static (three-point bending and uniaxial compression) or dynamic (ultrasonic, Brillouin scattering, dynamic mechanical analysis) techniques. The elastic properties of aerogels like Young's modulus, Poisson ratio, shear modulus, and internal friction are analyzed using dynamic techniques whereas elastic modulus and rupture strength are determined through static techniques [125–128]. Single Edge Notched Beam and double cleavage drilled compression tests are also commonly used to characterize the toughness and stress corrosion resistance of aerogels [125, 129]. Elasticity and deformation recovery of aerogels can be further tested by the execution of multiple squeezing and reabsorption cycles in a sample solution and measuring its adsorption capacity variation upon each cycle [106, 130, 131].

3.3.2 Thermal Analysis of Aerogels

The aerogels' thermal strength is characterized using Thermogravimetric Analysis (TGA) to identify the decomposition rates and maximum temperatures up to which the aerogels can tolerate without decomposition [101]. The experiment is usually conducted under dry air or nitrogen by increasing the chamber's temperature (loaded with a small amount of the aerogel sample (~5 mg)) from ~30 to 800 °C with designated gas flow and heating rates. Inorganic-based aerogels or organic aerogels incorporated with inorganic materials such as silica have, by nature, higher thermal resistance than pure organic aerogels. Aerogels' thermal conductivity is characterized using commercial or custom-made thermal conductivity analyzers that typically contain hot plates using nearly 25 cm × 25 cm × 2 cm aerogel specimens. The thermal conductivity of materials depends on their nature, density, cell size and structure, and pore size and connectivity [101]. For instance, silica aerogels have low thermal conductivity due to low density and high porosity [123].

3.4 Effect of Raw Materials

As noted before, several materials have been used to develop aerogels, which are generally classified into organic (e.g., cellulose, polyurethane, poly (vinyl alcohol) (PVA), polystyrene, polyimide, polysaccharide, etc.), inorganic (e.g., SiO_2 generated from various alkoxysilanes such as Al_2O_3, TiO_2, ZrO_2, SiC), carbon aerogels (pure carbon, carbon nanotubes (CNT), graphene) and other novel aerogels such as silicon, carbide, carbonitride, or composite/hybrid aerogels. All these have distinct properties that affect their applications and performance [7, 132–135]. Cellulose-based, polyimide-based, and derived carbon aerogels, have various properties, making them suitable for many applications [135]. Polymer aerogels have relatively high mechanical and environmental stability and have thermal conductivities in the same range as silica-based aerogels. They also have a high compressive modulus in the range of 1–5 MPa. The morphology of polymer aerogels can depend on the fabrication solvent and synthesis conditions and structural parameters (i.e., pore size, pore structure, ordering, etc.). The typical morphologies include but are not limited to colloidal-like nanoparticles, nano fibrillar/microfibrillar networks (e.g., globular superstructure nanofiber networks, homogeneous interwoven nanofiber networks), and sheet-like skeletons with a range of fiber sizes and alignment properties [134, 135]. Among polymers used for aerogels synthesis, cellulosic materials have typical physicochemical properties, high biodegradability, wide availability, and low cost and can be quickly processed and tailored in nanosized structures [134, 135]. Hence, cellulosic materials have recently gained significant attention to develop aerogels with applications ranging from sensors, energy storage, and thermal insulation [136–138] to mitigation of pollution [106, 109, 139–141]. Cellulosic aerogels typically have densities in the range of 10–105 kg/m^3, a high specific surface area up to 600 m^2/g, and can be fabricated with a wide range of morphological structures [15]. Cellulosic aerogels also have strong mechanical properties (i.e., high modulus up to 0.95 MPa and compression strength up to 150 MPa, and high toughness and energy absorption [134]. The challenges in the preparation of cellulosic aerogels, novel methods of preparation, and their properties are critically reviewed in the reported literature [134, 142, 143]. Resin-based aerogels such as polybenzoxazine typically have mesoporous structures and have relatively high mechanical strength and low shrinkage after polymerization as well as high density (around 300 kg/m^3) and high surface areas (384 m^2/g). These normally contain spherical polymer particles [135, 144, 145]. The polymer forming the organic aerogels can be later heated making carbon aerogels that have high electrical conductivity, high power and energy density [146], and high adsorption capacity (specific surface area up to 193 m^2/g for pure carbon aerogels and up to 998 m^2/g for graphene-based carbon aerogels). The mechanical and thermal stability of carbon aerogels is also relatively high [147, 148]. Carbon aerogels have the lowest density among all other types of aerogels (0.16 mg/cm^3) [149].

Silica aerogels are lightweight with a density of ~0.003–0.5 g/cm^3 and have highly porous structures (80–99.8%) with specific surface areas between 500 and 1200 m^2/g. The combination of low density and high porosity of regular silica-based aerogels makes them fragile with low mechanical properties [122, 123, 134]. The mechanical properties of silica aerogels (e.g., elasticity, flexibility) can be tailored to make reinforced silica aerogels by varying their processing conditions, the combination of silane precursors, or using additives such as organic polymers on their surface [122]. The highly porous structure of silica aerogels with low thermal conductivity increases the mechanical reinforcement due to their increased density [123, 150]. Figure 3 illustrates a silica aerogel sample's extremely low thermal conductivity protecting a delicate flower being damaged by heat [123].

4 Application of Aerogel on Environment

An increase in the scarcity of natural resources and their overexploitation has raised concerns about environmental pollution. For decades various efforts are being made to mitigate pollution by the application of green methods. The use of non-toxic, natural polymers and other biodegradable materials is one such method. Recently

Fig. 3 A silica aerogel slab protecting a delicate flower from excessive heat generated from a burner (https://commons.wikimedia.org/wiki/File:Aerogelflower_filtered.jpg)

aerogels have attracted much attention from researchers and governments for holding the potential of being eco-friendly. Klister developed aerogel in the 1930s, but its development was paced off due to a lack of technological development. However, recently much development in aerogel synthesis has taken place. Various types of aerogels are present, inorganic aerogels, synthetic polymer-based aerogel, carbon aerogels, and natural macromolecule-based aerogels [151]. Aerogels are prepared generally by a sol–gel process accompanied by drying. Aerogels were initially synthesized for thermal insulations in air space technology and building sectors. Recent yet substantial research shows aerogels are mutifunctional. Aerogels could be a promising solution to environmental pollution. For environment cleanup purposes, aerogels used are fabricated from carbon, silica, and natural polymer, (for example, cellulose) [152]. For instance, pesticides and herbicides are widely used in agriculture to protect the crop from invading pests and weeds. However, they are a significant threat to the environment. Cotet and coworkers have reported the use of carbon aerogel for pesticide adsorption. They prepared carbon aerogel along with xerogel and investigated their adsorption property against the pesticide alpha-cypermethrin. Their study showed that carbon aerogels were better adsorbents when compared to xerogels. The observed adsorption capacity is found to be 28.44 mg/g. Also, the grain size affected the adsorption capability. A decrease of 6.45% in adsorption was observed in increasing grain size attributed to internal diffusion limitation. Temperature inductive study suggests an increase of about 14% in adsorption capacity compared with xerogel. The study indicates aerogels as promising adsorbents for environmental cleanup [153]. In context with the above study, the applications of aerogels for environmental cleanup are discussed.

4.1 Aerogel for Heavy Metal Cleanup

Heavy metals are high-density metals present in the environment and are hazardous to human health. Heavy metals get dumped into the environment from industries like metal processing, chemical production industries, etc. They get incorporated as part of the food chain and get accumulated in living forms. They can be carcinogenic upon long exposure. Many methods are employed to remove heavy metals from the environment, like chemical precipitation and coagulation-flocculation. But these methods are inefficient as heavy metals are present in minimal quantities. Aerogels, due to their excellent adsorbing property, can be used for heavy metal removal. Electro-sorption processes are emerging as an efficient way to remove transition heavy metals from water. Many reports have shown that carbon aerogels are useful in this aspect. Many Southeast and far east Asia countries use cadmium-contaminated water for agricultural land irrigation. Cadmium is associated with cardiovascular and kidney diseases. Carbon electrode-based electrosorption has been attempted for its removal. When carbon aerogel was used as electrode material for electrosorption, up to 97.5% adsorption was observed. This value is higher than any other material used for the same purpose.

In chromium electrosorption, carbon aerogel showed about 71% adsorption. Aerogels have been reported to adsorb copper ions also. Most of the electrosorption occurs around 1.2 V. It is also suggested that altering the carbon aerogel material like doping, can enhance the aerogel's adsorption capacity. Graphene aerogel also could remove lead when doped with nitrogen. Up to 42% of lead was removed by nitrogen-doped graphene aerogel. However, graphene doped with sulfur showed better electrosorption [154].

Graphene nanosheets have recently attracted researchers for their use in heavy metal adsorption. Developing 3D graphene nanosheets is reported to be beneficial. Some of the useful properties found in graphene 3D aerogels are the following:

- Peculiar characteristics based on the formation of graphene building blocks.
- Unique 3D porous network with micro-, meso-, and macropores facilitates the diffusion of pollutants.
- Fully exposed active sites enhance the adsorption of pollutant molecules.
- The integrated morphology of the graphene promotes recycle of water treatment strategies.

To increase the removal capacity, a multilayered system could be beneficial such as that provided by MnO_2 nanostructure. It has a high surface area and is environmentally friendly. Thus, Liu and coworkers attempted to design new material from graphene and MnO_2; they prepared a hybrid aerogel, and tested it for heavy metal removal. Their results indicated that pH plays a vital role in the sorption of heavy metals like lead, cadmium, and copper. The aerogels so created showed a good deal of adsorption and reusability. The adsorption kinetics was also found to be enhanced [155].

4.2 Aerogels in Dye Removal

Dyes are cationic or anionic colored chemicals widely used in the textile, printing, plastic, rubber, cosmetics, leather tanning, paper, and food processing industries. However, their release into the environment is a serious concern. They have been shown to reduce photosynthesis in aquatic life. They may cause eye burns, skin irritation, affect the gastrointestinal tract, and may also cause cancer. However, various methods are employed to remove the dye, like adsorption, chemical oxidation or reduction, microbial treatment, nanofiltration, etc. Adsorption is a preferred method due to its simplicity and low cost. Aerogels made from various materials are used for this purpose. Among them are carbon nanotubes (CNTs), graphene, and nanocellulose that have caught the attention because they have the potential to be used as high-efficiency adsorbents. Aerogels prepared from hybrids of cellulose and graphene have been reported for the removal of dyes. These hydrogels showed adsorption of methylene blue and congo red up to 1166 mg/g and 507 mg/g, respectively [156]. Functionalization of graphene is essential for enhancing the adsorption property. Oxidation is the most common method. It helps in the formation of moieties like COO, COOH, CO, and OH. The aerogels of graphene are widely used to absorb organic dyes [157]. Aerogels prepared using graphene oxide and graphene nanosheets impregnated with cellulose from the waste newspaper are synthesized. Such aerogels were reported to be prepared by the sol–gel and freeze-drying process.

Structural analysis and characterization of aerogel developed from graphene oxide/cellulose show that it has a foam-like structure with adequate adsorption capacity. Various interactions were reported between aerogel and dyes, like electrostatic interaction, charge repulsion, and pi–pi interaction [158]. Graphene oxide is also reported to be modified with silk fibroin. Structural characterization by SEM, XPS, XRD, Raman, and TG showed a well-developed porous structure. It showed the adsorption capacity of organic dye up to 1322 mg/g [159]. CuS/graphene aerogels have also been reported for anionic and cationic dye degradation due to their unique network structure and photoelectric properties [160].

Graphene oxide with agar is also used to make composite aerogel for dye removal and can be recycled [161]. Apart from graphene, cellulose nanofiber also has gained attention to be used as aerogel for dye removal. It is eco-friendly, sustainable, and biodegradable, making it a potent candidate for the purpose. Carbon nanofibers are mostly hydrophilic, restricting their water treatment usage; however, if their surface can be modified to hydrophobicity, they are beneficial. Zhou and Hsieh reported silane-modified carbon nanofibers [138]. Hasan and coworkers also reported their potential for the adsorption of dye from the water of up to 150 mg/g. It also showed low thermal conductivity, therefore, it could be used in building insulators [162].

4.3 Aerogel for Oil Removal

Many methods are used for the removal of oil spillage from water bodies and can be classified into three types: (1) biological method; (2) physical method; and (3) chemical method. The microbial method of removal is fair but requires a huge time to process. Absorbents are also a useful method of oil removal. They are categorized as inorganic mineral, natural organic material, and synthetic organic. However, most of the absorbents used have a low absorbing capacity and absorbs water. Recycled cellulose from waste can be used for making aerogels for oil removal because of its highly porous structure. Cellulose is hydrophilic, and to introduce hydrophobicity in it, methyltrimethoxysilane is used. The hydrophobic recycled cellulose aerogel showed an absorption capacity of up to 20.5 g/g. The absorption took place best around 40 °C [163]. Silica aerogels developed by the sol–gel process are also utilized for oil removal [164, 165].

Graphene-based aerogels with superhydrophobic/super oleophilic attributes can be used for water treatment. Graphene nanosheets have an intrinsic hydrophobic property that makes them suitable for oil removal from water. In a study, graphene was modified with fluoroalkyl silane, which increased its water contact and increased its ability to absorb oil and removed other pollutants as well [159]. Highly porous monolithic aerogels based on ZnO photocatalyst and polystyrene were reported to show the ability to remove organic pollutants from water [166].

4.4 Aerogel for Water Treatment

Aerogels are used in water treatment and purification because of their attractant properties. Heavy metals are highly hazardous and very difficult to remove from the contaminated environment, causing massive damage to humans' internal organs. The heavy metal contaminated water is efficiently treated and removed by the aerogel. Amino-functionalized resorcinol–formaldehyde aerogels were reported to adsorb heavy metal ions of Cd (II), Hg (II), and Pb (II) from solution [167]. The adsorptions of different heavy metals were optimized with several parameters such as pH, temperature, contact time, the capacity of heavy metal sorption, and sorbent concentration. The modified amino group adsorbs the heavy metals on the aerogel surface. Such aerogels could be used to eliminate heavy metals from contaminated water. Aerogel can be synthesized using polybenzoxazine, amines, phenol, formaldehyde as base material by sol–gel reaction [144]. Polybenzoxazine acts as a potential polymer to chelate heavy metals during the treatment of wastewater. They explained that the high capacity and strong coordination of polybenzoxazine with Sn^{2+} was by the Van der Waals force and Irving–Williams rule on aerogel's surface area. Chen et al. [168] prepared the cheapest biomass-based aerogels of carbon (CDPC) and carbon oxide (CDPCO) using cotton with alkaline engraving methods to eliminate heavy metals from the solution [168]. The experiments proved that the ions of heavy metals depend on the interaction of heavy metals with the active groups of –COOH and –OH on the surface of CDPCO. This aerogel adsorption of metal ions profile was well adopted with the Langmuir model. Carbogel synthesized by the basis of protein-doped cellulose through hydrothermal carbonization of cellulose performed at low pH with the occurrence of glycoprotein to enhance both types of anionic and cationic heavy metals attached with carbo gel treatment of contaminated water [169].

4.5 *Removal of Pesticides by Aerogels*

Chemical pesticides cause toxicity and lead to cancer and neurological diseases among other morbidities in human beings. Carbon aerogel was analyzed for the efficiency of pesticide removal. The exclusive properties of aerogel such as manageable porosity, high surface area with large volume pores [170], thermal stability are advantageous for removing pesticides from the contaminated water. α-Cypermethrin is a pesticide classified as pyrethroid, commonly used against insect pests and arachnid-like ticks and mites. It could be absorbed and removed by carbon aerogel from the polluted water [153]. Recently, Kien et al. prepared and proved carbon aerogel's competence for removing pesticides such as DDT and cypermethrin up to 95–99% from polluted river water samples in Cuu Long Delta [171]. Modified metal–organic frameworks (MOFs) composite with carbon nanoparticles aerogels have been synthesized for removing the toxic pesticides polluting water [62] (Table 3).

5 Eco-Toxicological Aspects of Aerogels

In the twenty-first century, no other human-made material has piqued researchers' curiosity worldwide, as have the aerogels. Vast numbers of research articles are published extolling the potential as well as useful applications of aerogels. Remediation of water pollution through adsorption has been highlighted as one of the aerogels' applications. It averts the impracticalities and lack of sustainability due to toxicity and high cost of techniques such as flocculation, electrochemical treatment, photocatalysis, advanced oxidation, nanofiltration, and biological degradation. However, there may be an oversight regarding the toxicity of chemicals used to synthesize the aerogels, particularly those used for water treatment.

Graphene oxide (GO), at dosage 50 mg/L, has been reported to induce in zebrafish embryos, minor cell growth hindrance and deferral in the hatching, and multiwalled carbon nanotubes, even at relatively low concentration of 25 mg/L, exhibit severe toxicity leading to potent cell growth hindrance and acute morphological defects in growing embryos [172]. Resorcinol–formaldehyde aerogels have been shown to absorb several heavy metals [167, 173], however, the synthesis of these aerogels involves toxic components [16]. GO polypyrrole composite, GO poly (acrylic

Table 3 Types of aerogel with their porous compound and its application [152]

Type of composite aerogel	Porous compound	Application
Cellulose aerogel	Waste engine oil	Biomedical applications such as thermal insulation, drug delivery system
Starch-based carbon aerogel	Crystal violet Methylene blue Methyl violet	Removal of azo dye Oil spill cleanup
Carbon aerogel	CI Reactive Red 2 dye Toluene	Hydrogen and electrical energy storage, desalination, and electrocatalysis
Carbon micro belt aerogel	Oils	Oil remediation
Graphene aerogel	Organic solvent and Oils	Removal of organic solvent and oil remediation
Magnetic graphene aerogel	Organic solvent and dyes Motor oils	Removal of organic solvent and oil remediation
Magnetic cellulose aerogel	Oils	Oil remediation
Graphene carbon nanotube aerogel	Petroleum products, organic solvent, and fats	Oil and dye as well as water purification and soil remediation
Hydrocarbon silica xerogel's Hydrocarbon silica aerogel	Dieldrin RhB	Used as catalysis, as a template to metal oxide
Hydrocarbon granular silica aerogel	Phenol	Removal of the toxic compound
Hydrophobic silica-based aerogel and xerogels	Toxic organic solvents, oils	Oil spill cleanup
Particulate hydrocarbon silica aerogel	Liquid oils	Oil spill cleanup
Poly (alkoxysilane) organogels	Oils and crude oils	Oil spill cleanup
Titana aerogel	Azo-Dye Orange II	Used as photocatalysts water splitting for hydrogen production
Montmorillonite clay-polymer composite aerogel-hydrophobized with TMOS	Motor oils, Dodecane	Industrial application
Sodium silicate-based aerogel	Phenol removal	Removal of the toxic compound
Silica aerogel-activated carbon nanocomposite	Heavy metal (Pb^{2+}, Cu^{2+})	Removal of heavy metals
Alginate aerogel	Heavy metal (Pb^{2+}, Cu^{2+})	Used as a drug delivery system
Calcium alginate carbon aerogel	Organic compound	Removal of lead, copper, cadmium ions
Polysaccharide based aerogel	Organic compound	Pharmaceuticals application as drug delivery system

acid) hydrogel, G.O./poly(amidoamine) nanocomposite, and several such aerogles created from GO sheets to which polymers have been incorporated using crosslinking agents increases their efficiency; but, these polymers are obtained from toxic, non-renewable sources [174]. Although chemical crosslinking is an efficient method to improve aerogels' functionality, chemicals used as crosslinkers have potential toxicity. Further, undesired side product formation due to any non-selective activity of the chemicals also needs attention. Toxicity of leachates from polymers of acrylonitrile, styrene, epoxy was reported in *Daphnia magna* [175]. Kwon et al. [176] reported the marine pollution caused by styrene oligomers and the need to monitor it. Reproductive toxicity was observed in mice orally administered styrene [176]. Similarly, oxidative stress and membrane destruction were seen in cyanobacteria exposed to polystyrene nanoparticles [177]. The rapid expansion in nanoparticles' usage and studies on their harmful effect on humans, other animals, and the environment has raised legitimate concerns as their minimal size potentiates their accumulation in the body [178, 179]. Schrand et al. [180] reviewed the toxicity of metal-based nanoparticles on mammalian cells and emphasized the need for a comprehensive database on nanoparticles' health, safety, and environmental impact. Wang et al. [66, 112, 159, 181] highlighted that conventional aerogels' raw materials come from toxic, non-renewable inorganic or petrochemical-based materials.

Concerns about the harmful environmental impact of aerogels and their precursor chemicals have led researchers worldwide to focus on developing non-toxic alternatives such as polysaccharide-based aerogels or bio-aerogels. These aerogels are formulated from natural ingredients that are biodegradable and hence eco-friendly. Cellulose as a precursor to bioaerogels has legitimately garnered most of the attention due to its widespread presence in plant parts such as cotton [182], coconut husk [183], and properties like biocompatibility, sustainability, low toxicity, renewability make it an excellent candidate for several industries [66] as well as for water treatment. [184] used cellulose in polymer-graphene composites to boost oil–water separation. Alatalo et al. [169] fabricated a cellulose-based carbon aerogel to remove ionic metal pollutants from water. Huang and Wang [185] suggested that treatments such as chemical precipitation, oxidation/reduction, membrane separation are adequate for removal of high concentrations of heavy metals but may not be effective in treating water with low concentration; however, cheap adsorbents such as alginate, clay, etc., could be utilized for efficacious removal of low concentration of heavy metals from water. They fabricated an aerogel constituting polysaccharide alginate and calcium by the freeze-drying technique to decontaminate Pb^{2+} and Cu^{2+} from water.

Similarly, Yan et al. [174] synthesized chitosan crosslinked Grapheneoxide lignosulfonate aerogel (GLCA) for dye removal; lignosulfonate (LS), an integral derivative of naturally occurring lignin polymer, is a byproduct of paper and pulping factories, hence easily procurable and has been abundantly utilized for adsorption of pollutants in aqueous solutions. Chitosan (CS) is also economical, plentiful biomaterial with considerable biocompatibility that aids in circumventing limitations of reduced selectivity and efficiency, and elevated energy consumption and use of polluting chemicals precursors. Of the several biopolymeric materials such as pectin, guar gum, etc., studied by researchers for the creation of composite materials [186], starch, found in abundance in nature, is a material with a vast number of potential applications. It is easily extractable from plants, has high molecular weight and large surface area, is non-toxic, inexpensive, and biodegradable [187], making it a good adsorbent for decontamination of harmful toxins from water [188, 189]. The presence of multiple hydroxyl groups in the structure of starch accentuates its utility because these groups undergo chemical alteration easily, forming ethers, esters, hydrogen bonds, etc. Furthermore, the functionality of starch is enhanced by its derivatization and crosslinking [190, 191]. Naushad et al. [192] fabricated starch/SnO_2 nanocomposite that could remove toxic mercury ions from an aqueous medium, establishing starch's application as a potential adsorbent for remediation of water contaminated with heavy metals. Karaki et al. [193] emphasize the potential of enzymatic modification of polysaccharides to improve their specificity and selectivity properties and reduce the input energy and the negative environmental impact, making it attractive alternatives to toxic and non-specific chemical approaches and enlarging the field of their potential applications. With the boom in the research on aerogels, it is imperative to focus on alternatives that are substantially less costly to the environment, bioaerogels are excellent alternatives that have applications in several fields, including wastewater remediation.

6 Conclusion

Aerogels are one of the most promising materials of the twenty-first century. Its potential to be structurally altered to suit the incumbent needs is one of its trademark characteristics. Various materials have been successfully used in the creation of different types of aerogels, but in the present scenario when we are grappling with pollution the world over, aerogels made of biopolymers seem to be highly encouraging. Aerogels can play a vital role in environmental cleanup. It has shown its ability to remove pollutants from various scapes of nature particularly water, where it can remediate oil spills and dyes to heavy metals and pesticides. Owing to its unique structural properties, pollution mitigating potential, eco-friendliness, and biodegradability, aerogels demand much more insight and research to boost their usage.

References

1. Kistler SS (1931) Coherent expanded Aerogels and Jellies. Nature 127(3211):3211
2. Suh DJ (2004) Catalytic applications of composite aerogels. J Non-cryst Solids 350:314–319. https://doi.org/10.1016/j.jnoncrysol.2004.08.230
3. Teichner SJ (1986). Aerogels of inorganic oxides. https://doi.org/10.1007/978-3-642-93313-4_2
4. Pekala RW (1989) Organic aerogels from the polycondensation of resorcinol with formaldehyde. J Mater Sci 24(9):3221–3227

5. Biesmans G, Randall D, Francais E, Perrut M (1998) Polyurethane-based organic aerogels' thermal performance. J Non-cryst Solids 225(1–3):36–40. https://doi.org/10.1016/S0022-3093(98)00103-3
6. Aegerter MA, Leventis N, Koebel MM (2011) Aerogels handbook. Aerogels Handbook. https://doi.org/10.1007/978-1-4419-7589-8
7. Worsley MA, Pauzauskie PJ, Olson TY, Biener J, Satcher JH, Baumann TF (2010) Synthesis of graphene aerogel with high electrical conductivity. J Am Chem Soc 132(40):14067–14069. https://doi.org/10.1021/ja1072299
8. Worsley MA, Satcher JH, Baumann TF (2008) Synthesis and characterization of monolithic carbon aerogel nanocomposites containing double-walled carbon nanotubes. Langmuir 24(17):9763–9766. https://doi.org/10.1021/la8011684
9. Brock SL, Yu H (2011) Chalcogenide Aerogels. Aerogels Handb. https://doi.org/10.1007/978-1-4419-7589-8_17
10. Mohanan JL, Arachchige IU, Brock SL (2005) Porous semiconductor chalcogenide aerogels. Science (New York, NY), 307(5708):397–400. https://doi.org/10.1126/science.1106525
11. Betz M, García-González CA, Subrahmanyam RP, Smirnova I, Kulozik U (2012) Preparation of novel whey protein-based aerogels as drug carriers for life science applications. J Supercrit Fluids 72:111–119. https://doi.org/10.1016/j.supflu.2012.08.019
12. Ratke L (2011) Monoliths and fibrous cellulose aerogels. Aerogels Handb. https://doi.org/10.1007/978-1-4419-7589-8_9
13. Kong Y, Shen X, Cui S, Fan M (2014) Preparation of monolith SiC aerogel with high surface area and large pore volume and the structural evolution during the preparation. Ceram Int 40(6):8265–8271. https://doi.org/10.1016/j.ceramint.2014.01.025
14. Kong Y, Zhong Y, Shen X, Gu L, Cui S, Yang M (2013) Synthesis of monolithic mesoporous silicon carbide from resorcinol-formaldehyde/silica composites. Mater Lett 99:108–110. https://doi.org/10.1016/j.matlet.2013.02.047
15. Pircher N, Carbajal L, Schimper C, Bacher M, Rennhofer H, Nedelec JM, … Liebner F (2016) Impact of selected solvent systems on the pore and solid structure of cellulose aerogels. Cellulose 23(3):1949–1966. https://doi.org/10.1007/s10570-016-0896-z
16. Rudaz C, Courson R, Bonnet L, Calas-Etienne S, Sallée H, Budtova T (2014) Aeropectin: fully biomass-based mechanically strong and thermal superinsulating aerogel. Biomacromol 15(6):2188–2195. https://doi.org/10.1021/bm500345u
17. García-González CA, Alnaief M, Smirnova I (2011) Polysaccharide-based aerogels—promising biodegradable carriers for drug delivery systems. Carbohyd Polym 86(4):1425–1438. https://doi.org/10.1016/j.carbpol.2011.06.066
18. Veronovski A, Tkalec G, Knez Z, Novak Z (2014) Characterisation of biodegradable pectin aerogels and their potential use as drug carriers. Carbohyd Polym 113:272–278. https://doi.org/10.1016/j.carbpol.2014.06.054
19. Kistler SS, Swann S, Appel EG (1934) Aërogel catalysts: Thoria: preparation of catalyst and conversions of organic acids to Ketones. Ind Eng Chem 26(4):388–391. https://doi.org/10.1021/ie50292a007
20. Swann S Jr, Appel EG, Kistler SS (1934) Thoria aerogel catalyst: aliphatic. Ind Eng Chem 26(9):1934
21. Adebajo MO, Frost RL, Kloprogge JT, Carmody O, Kokot S (2003) Porous materials for oil spill cleanup: a review of synthesis and absorbing properties. J Porous Mater 10(3):159–170. https://doi.org/10.1023/A:1027484117065
22. Reynolds JG, Coronado PR, Hrubesh LW (2001b) Hydrophobic aerogels for oil-spill cleanup—intrinsic absorbing properties. Energy Sourc 23(9):831–843. https://doi.org/10.1080/009083101316931906
23. Plata DL, Briones YJ, Wolfe RL, Carroll MK, Bakrania SD, Mandel SG, Anderson A (2004) Aerogel platform optical sensors for oxygen gas. J Non-cryst Solids 350:326–335
24. Pakowski Z, Maciszewska K (2004) Permeability of nonwoven glass fiber filters with hydrophobic silica aerogel layer. Inz Chem Procesowa 25(3):1435–1441

25. Nagahara H, Suginouchi T, Hashimoto M (2006) Acoustic properties of nanofoam and its applied air-borne ultrasonic transducers. Proc—IEEE Ultrasonics Symp 1:1541–1544. https://doi.org/10.1109/ULTSYM.2006.391
26. Long JW, Fischer AE, McEvoy TM, Bourg ME, Lytle JC, Rolison DR (2008) Self-limiting electropolymerization en route to ultrathin, conformal polymer coatings for energy-storage applications. ACS National Meeting Book of Abstracts
27. Steinbach S, Ratke L (2007) The microstructure response to fluid flow fields in Al-cast alloys. Trans Indian Inst Met 60(2–3):167–171
28. Latthe SS, Nadargi DY, Venkateswara Rao A (2009) TMOS based water repellent silica thin films by co-precursor method using TMES as a hydrophobic agent. Appl Surf Sci 255(6):3600–3604. https://doi.org/10.1016/j.apsusc.2008.10.005
29. Smirnova I, Suttiruengwong S, Arlt W (2004) Feasibility study of hydrophilic and hydrophobic silica aerogels as drug delivery systems. J Non-cryst Solids 350:54–60. https://doi.org/10.1016/j.jnoncrysol.2004.06.031
30. Smirnova I (2011) Pharmaceutical applications of Aerogels. Aerogels Handb. https://doi.org/10.1007/978-1-4419-7589-8_31
31. Mora MF, Jones SM, Creamer J, Willis PA (2018) Extraction of amino acids from aerogel for analysis by capillary electrophoresis. Implications for a mission concept to Enceladus' Plume. Electrophoresis 39(4):620–625. https://doi.org/10.1002/elps.201700323
32. Qi Z, Huang D, He S, Yang H, Hu Y, Li L, Zhang H (2013) Thermal protective performance of aerogel embedded firefighter's protective clothing. J Eng Fibers Fabr 8(2):134–139. https://doi.org/10.1177/155892501300800216
33. Hsu CA, Wen TN, Su YC, Jiang ZB, Chen CW, Shyur LF (2012) Biological degradation of anthroquinone and azo dyes by a novel laccase from Lentinus sp. Environ Sci Technol 46(9):5109–5117. https://doi.org/10.1021/es2047014
34. Yang Z, Yang H, Jiang Z, Cai T, Li H, Li H, … Cheng R (2013) Flocculation of both anionic and cationic dyes in aqueous solutions by the amphoteric grafting flocculant carboxymethyl chitosan-graft-polyacrylamide. J Hazard Mater 254–255(1):36–45. https://doi.org/10.1016/j.jhazmat.2013.03.053
35. Bai H, He P, Chen J, Liu K, Lei H, Dong F, … Li H (2017) Fabrication of Sc2O3-magneli phase titanium composite electrode and its application in efficient electrocatalytic degradation of methyl orange. Appl Surf Sci 401:218–224. https://doi.org/10.1016/j.apsusc.2017.01.019
36. Wen J, Xie J, Chen X, Li X (2017) A review on g-C_3N_4-based photocatalysts. Appl Surf Sci 391:72–123. https://doi.org/10.1016/j.apsusc.2016.07.030
37. Laxman K, Al Rashdi M, Al Sabahi J, Al Abri M, Dutta J (2017) Supported versus colloidal zinc oxide for advanced oxidation processes. Appl Surf Sci 411:285–290. https://doi.org/10.1016/j.apsusc.2017.03.139
38. Nikooe N, Saljoughi E (2017) Preparation and characterization of novel PVDF nanofiltration membranes with hydrophilic property for filtration of dye aqueous solution. Appl Surf Sci 413:41–49. https://doi.org/10.1016/j.apsusc.2017.04.029
39. Li Y, Liu FT, Zhang HX, Li X, Dong XF, Wang CW (2019) DMF-treated strategy of carbon nanospheres for high-efficient and selective removal of organic dyes. Appl Surf Sci 484:144–151. https://doi.org/10.1016/j.apsusc.2019.04.080
40. Zhu W, Lin Y, Kang W, Quan H, Zhang Y, Chang M, … Hu H (2020) An aerogel adsorbent with bio-inspired interfacial adhesion between graphene and MoS_2 sheets for water treatment. Appl Surf Sci 512. https://doi.org/10.1016/j.apsusc.2020.145717
41. Lu KQ, Xin X, Zhang N, Tang ZR, Xu YJ (2018) Photoredox catalysis over graphene aerogel-supported composites. J Mater Chem A 6(11):4590–4604. https://doi.org/10.1039/c8ta00728d
42. Nguyen TD (2018) Hydrogel: Preparation, characterization, and applications: a review. J Wood Sci 64(3):105–121. https://doi.org/10.1515/hf-2012-0181
43. Xiang J, Shen L, Hong Y (2020) Status and future scope of hydrogels in wound healing: synthesis, materials and evaluation. Eur Polym J 130. https://doi.org/10.1016/j.eurpolymj.2020.109609

44. Thoniyot P, Tan MJ, Karim AA, Young DJ, Loh XJ (2015) Nanoparticle–hydrogel composites: concept, design, and applications of these promising, multi-functional materials. Adv Sci 2(1–2). https://doi.org/10.1002/advs.201400010
45. Wahid F, Zhao XJ, Jia SR, Bai H, Zhong C (2020) Nanocomposite hydrogels as multifunctional systems for biomedical applications: current state and perspectives. Compos Part B: Eng 200. https://doi.org/10.1016/j.compositesb.2020.108208
46. Alam A, Zhang Y, Kuan HC, Lee SH, Ma J (2018) Polymer composite hydrogels containing carbon nanomaterials—morphology and mechanical and functional performance. Prog Polym Sci 77:1–18. https://doi.org/10.1016/j.progpolymsci.2017.09.001
47. Dhanjai Sinha A, Kalambate PK, Mugo SM, Kamau P, Chen J, Jain R (2019) Polymer hydrogel interfaces in electrochemical sensing strategies: a review. TrAC—Trends Analytical Chem 118:488–501. https://doi.org/10.1016/j.trac.2019.06.014
48. Bhattacharya S, Samanta SK (2016) Soft-nanocomposites of nanoparticles and nanocarbons with supramolecular and polymer gels and their applications. Chem Rev 116(19):11967–12028. https://doi.org/10.1021/acs.chemrev.6b00221
49. Maharjan B, Kumar D, Awasthi GP, Bhattarai DP, Kim JY, Park CH, Kim CS (2019) Synthesis and characterization of gold/silica hybrid nanoparticles incorporated gelatin methacrylate conductive hydrogels for H9C2 cardiac cell compatibility study. Compos Part B: Eng 177. https://doi.org/10.1016/j.compositesb.2019.107415
50. Xu X, Lü S, Gao C, Wang X, Bai X, Gao N, Liu M (2014) One-pot facile synthesis of silica reinforced double network hydrogels based on triple interactions. Chem Eng J 240:331–337. https://doi.org/10.1016/j.cej.2013.11.084
51. Petit L, Carlsson L, Rose S, Marcellan A, Narita T, Hourdet D (2014) Design and viscoelastic properties of PDMA/silica assemblies in aqueous media. Macromol Symp 337(1):58–73. https://doi.org/10.1002/masy.201450307
52. Luo X, Akram MY, Yuan Y, Nie J, Zhu X (2019) Silicon dioxide/poly(vinyl alcohol) composite hydrogels with high mechanical properties and low swellability. J Appl Polym Sci 136(1). https://doi.org/10.1002/app.46895
53. Lin YJ, Hsu FC, Chou CW, Wu TH, Lin HR (2014) Poly(acrylic acid)-chitosan-silica hydrogels carrying platelet gels for bone defect repair. J Mater Chem B 2(47):8329–8337. https://doi.org/10.1039/c4tb01356e
54. Manickam P, Vashist A, Madhu S, Sadasivam M, Sakthivel A, Kaushik A, Nair M (2020) Gold nanocubes embedded biocompatible hybrid hydrogels for electrochemical detection of H_2O_2. Bioelectrochemistry 131. https://doi.org/10.1016/j.bioelechem.2019.107373
55. Song J, Hwang S, Im K, Hur J, Nam J, Hwang S, … Park N (2015) Light-responsible DNA hydrogel-gold nanoparticle assembly for synergistic cancer therapy. J Mater Chem B 3(8):1537–1543. https://doi.org/10.1039/c4tb01519c
56. Villalobos LF, Neelakanda P, Karunakaran M, Cha D, Peinemann KV (2014) Poly-thiosemicarbazide/gold nanoparticles catalytic membrane: In-situ growth of well-dispersed, uniform and stable gold nanoparticles in a polymeric membrane. Cataly Today 236(Part A):92–97. https://doi.org/10.1016/j.cattod.2013.10.067
57. Alhokbany N, Ahama T, Ruksana NM, Alshehri SM (2019) AgNPs embedded N- doped highly porous carbon derived from chitosan based hydrogel as catalysts for the reduction of 4-nitrophenol. Comp Part B: Eng 173. https://doi.org/10.1016/j.compositesb.2019.106950
58. Zhang XF, Liu ZG, Shen W, Gurunathan S (2016) Silver nanoparticles: Synthesis, characterization, properties, applications, and therapeutic approaches. Int J Molecul Sci 17(9). https://doi.org/10.3390/ijms17091534
59. Murali Mohan Y, Vimala K, Thomas V, Varaprasad K, Sreedhar B, Bajpai SK, Mohana Raju K (2010) Controlling of silver nanoparticles structure by hydrogel networks. J Colloid Interf Sci 342(1):73–82. https://doi.org/10.1016/j.jcis.2009.10.008
60. Prusty K, Biswal A, Biswal SB, Swain SK (2019) Synthesis of soy protein/polyacrylamide nanocomposite hydrogels for delivery of ciprofloxacin drug. Mater Chem Phys 234:378–389. https://doi.org/10.1016/j.matchemphys.2019.05.038

61. Yang C, Pan Q, Jia Q, Qi W, Jiang W, Wei H, … Cao B (2020) Bamboo-like N/S-codoped carbon nanotube aerogels for high-power and high-energy supercapacitors. J Alloys Comp. https://doi.org/10.1016/j.jallcom.2020.157946
62. Liang W, Wang B, Cheng J, Xiao D, Xie Z, Zhao J (2021) 3D, eco-friendly metal-organic frameworks@carbon nanotube aerogels composite materials for removal of pesticides in water. J Hazardous Mater 401. https://doi.org/10.1016/j.jhazmat.2020.123718
63. Zhu Z, Yao H, Dong J, Qian Z, Dong W, Long D (2019) High-mechanical-strength polyimide aerogels crosslinked with 4, 4′-oxydianiline-functionalized carbon nanotubes. Carbon 144:24–31. https://doi.org/10.1016/j.carbon.2018.11.057
64. Zhan W, Gao L, Fu X, Siyal SH, Sui G, Yang X (2019) Green synthesis of amino-functionalized carbon nanotube-graphene hybrid aerogels for high performance heavy metal ions removal. Appl Surf Sci 467–468:1122–1133. https://doi.org/10.1016/j.apsusc.2018.10.248
65. Zhan W, Zhu M, Lan J, Wang H, Yuan H, Yang X, Sui G (2021) 1D Sb2S3@Nitrogen-doped carbon coaxial nanotubes uniformly encapsulated within 3D porous graphene aerogel for fast and stable sodium storage. Chem Eng J 408
66. Wang Y, Su Y, Wang W, Fang Y, Riffat SB, Jiang F (2019) The advances of polysaccharide-based aerogels: preparation and potential application. Carbohyd Polym 226. https://doi.org/10.1016/j.carbpol.2019.115242
67. Fu X, Guo Y, Guan L, Liu J (2019) Three dimensional nylon66@carbon nanotube aerogel: a platform for high-performance electromagnetic wave absorbing composites. Mater Lett 247:147–150. https://doi.org/10.1016/j.matlet.2019.03.116
68. Avan AA, Filik H (2020) Simultaneous electrochemical sensing of dihydroxybenzene isomers at multi-walled carbon nanotubes aerogel/gold nanoparticles modified graphene screen-printed electrode. J Electroanalytical Chem 878. https://doi.org/10.1016/j.jelechem.2020.114682
69. Chen J, Xie H, Lai X, Li H, Gao J, Zeng X (2020) An ultrasensitive fire-warning chitosan/montmorillonite/carbon nanotube composite aerogel with high fire-resistance. Chem Eng J 399. https://doi.org/10.1016/j.cej.2020.125729
70. Gao H, Bo L, Liu P, Chen D, Li A, Ou Y, … Wang G (2019) Ambient pressure dried flexible silica aerogel for construction of monolithic shape-stabilized phase change materials. Solar Energy Mater Solar Cells 201. https://doi.org/10.1016/j.solmat.2019.110122
71. Liu R, Wang J, Du Y, Liao J, Zhang X (2019) Phase-separation induced synthesis of superhydrophobic silica aerogel powders and granules. J Solid State Chem 279. https://doi.org/10.1016/j.jssc.2019.120971
72. Sun G, Duan T, Liu C, Zhang L, Chen R, Wang J, Han S (2020) Fabrication of flame-retardant and smoke-suppressant isocyanate-based polyimide foam modified by silica aerogel thermal insulation and flame protection layers. Polym Test 91. https://doi.org/10.1016/j.polymertesting.2020.106738
73. Kucharek M, MacRae W, Yang L (2020) Investigation of the effects of silica aerogel particles on thermal and mechanical properties of epoxy composites. Compos Part A: Appl Sci Manuf 139. https://doi.org/10.1016/j.compositesa.2020.106108
74. Jang D, Yoon HN, Seo J, Lee HK, Kim GM (2021) Effects of silica aerogel inclusion on the stability of heat generation and heat-dependent electrical characteristics of cementitious composites with CNT. Cement Concrete Compos 115. https://doi.org/10.1016/j.cemconcomp.2020.103861
75. Jiang X, Ren J, Kong Y, Zhao Z, Shen X, Fan M (2020) Shape-tailorable amine grafted silica aerogel microsphere for CO_2 capture. Green Chem Eng. https://doi.org/10.1016/j.gce.2020.11.010
76. Mahmoodi NM, Mokhtari-Shourijeh Z, Langari S, Naeimi A, Hayatti B, Jalili M, Seifpanahi-Shabani K (2020) Silica aerogel/polyacrylonitrile/polyvinylidene fluoride nanofiber and its ability for treatment of colored wastewater. J Mol Struct. https://doi.org/10.1016/j.molstruc.2020.129418
77. Huang Y, Peng X, Liu X, Chen C, Han X (2021) Development of SiC fiber through heat treatment of silica aerogel by in-situ curing. Mater Lett 283. https://doi.org/10.1016/j.matlet.2020.128797

78. Zhang Y, Xiang L, Shen Q, Li X, Wu T, Zhang J, Nie C (2020) Rapid synthesis of dual-mesoporous silica aerogel with excellent adsorption capacity and ultra-low thermal conductivity. J Non-cryst Solids. https://doi.org/10.1016/j.jnoncrysol.2020.120547
79. Liu X, Shen T, Zhao Z, Qin Y, Zhang P, Luo H, Guo ZX (2018) Graphene/gold nanoparticle aerogel electrode for electrochemical sensing of hydrogen peroxide. Mater Lett 229:368–371. https://doi.org/10.1016/j.matlet.2018.07.024
80. Pap Z, Radu A, Hidi IJ, Melinte G, Diamandescu L, Popescu T, … Baia M (2013) Behavior of gold nanoparticles in a titania aerogel matrix: photocatalytic activity assessment and structure investigations. Cuihua Xuebao/Chinese J Catalysis 34(4):734–740. https://doi.org/10.1016/s1872-2067(11)60500-7
81. Kuthirummal N, Dean A, Yao C, Risen W (2008) Photo-formation of gold nanoparticles: Photoacoustic studies on solid monoliths of Au(III)-chitosan-silica aerogels. Spectrochimica Acta—Part A: Molecul Biomolecul Spectrosc 70(3):700–703. https://doi.org/10.1016/j.saa.2007.09.011
82. Yuanfeng P, Ruiyi L, Xiulan S, Guangli W, Zaijun L (2020) Highly sensitive electrochemical detection of circulating tumor DNA in human blood based on urchin-like gold nanocrystal-multiple graphene aerogel and target DNA-induced recycling double amplification strategy. Anal Chim Acta 1121:17–25. https://doi.org/10.1016/j.aca.2020.04.077
83. Zhang T, Zhao Y, Muhetaer M, Wang K (2020) Silver nanoparticles cross-linked polyimide aerogels with improved high temperature microstructure stabilities and high mechanical performances. Microporous Mesoporous Mater 297. https://doi.org/10.1016/j.micromeso.2020.110035
84. Rubina MS, Elmanovich IV, Shulenina AV, Peters GS, Svetogorov RD, Egorov AA, … Vasil'kov AY (2020) Chitosan aerogel containing silver nanoparticles: from metal-chitosan powder to porous material. Polym Test 86. https://doi.org/10.1016/j.polymertesting.2020.106481
85. Arakcheev V, Bagratashvili V, Bekin A, Khmelenin D, Minaev N, Morozov V, Rybaltovsky A (2017) Laser assisted synthesis of silver nanoparticles in silica aerogel by supercritical deposition technique. J Supercrit Fluids 127:176–181. https://doi.org/10.1016/j.supflu.2017.03.028
86. Xiao X, Xu Y, Lv X, Xie J, Liu J, Yu C (2019). Electrochemical CO_2 reduction on copper nanoparticles-dispersed carbon aerogels. J Colloid Interf Sci 545:1–7
87. Hassan K, Hossain R, Sahajwalla V (2020) Novel microrecycled ZnO nanoparticles decorated macroporous 3D graphene hybrid aerogel for efficient detection of NO_2 at room temperature. Sens Actuat B: Chem. https://doi.org/10.1016/j.snb.2020.129278
88. Liu Y, Chen J, Liu Z, Xu H, Shi Z, Yang Q, … Xiong C (2020) Necklace-like ferroferric oxide (Fe_3O_4) nanoparticle/carbon nanofibril aerogels with enhanced lithium storage by carbonization of ferric alginate. J Colloid Interf Sci 576:119–126. https://doi.org/10.1016/j.jcis.2020.04.128
89. Adavan Kiliyankil V, Fugetsu B, Sakata I, Wang Z, Endo M (2021) Aerogels from copper (II)-cellulose nanofibers and carbon nanotubes as absorbents for the elimination of toxic gases from air. J Colloid Interf Sci 582:950–960. https://doi.org/10.1016/j.jcis.2020.08.100
90. Song L, Shu L, Wang Y, Zhang XF, Wang Z, Feng Y, Yao J (2020) Metal nanoparticle-embedded bacterial cellulose aerogels via swelling-induced adsorption for nitrophenol reduction. Int J Biol Macromol 143:922–927. https://doi.org/10.1016/j.ijbiomac.2019.09.152
91. Wu XX, Zhang Y, Hu T, Li WX, Li ZL, Hu HJ, … Jiang GB (2020) Long-term antibacterial composite via alginate aerogel sustained release of antibiotics and Cu used for bone tissue bacteria infection. Int J Biol Macromolecules https://doi.org/10.1016/j.ijbiomac.2020.11.075
92. Song D, Wang S, Liu R, Jiang J, Jiang Y, Huang S, … Zhao B (2019) Ultra-small SnO_2 nanoparticles decorated on three-dimensional nitrogen-doped graphene aerogel for high-performance bind-free anode material. Appl Surf Sci 478:290–298. https://doi.org/10.1016/j.apsusc.2019.01.143

93. Wu C, Xu Y, Ao L, Jiang K, Shang L, Li Y, … Chu J (2020) Robust three-dimensional porous rGO aerogel anchored with ultra-fine α-Fe_2O_3 nanoparticles exhibit dominated pseudocapacitance behavior for superior lithium storage. J Alloys Comp 816. https://doi.org/10.1016/j.jallcom.2019.152627
94. Nawaz M, Khan AA, Hussain A, Jang J, Jung HY, Lee DS (2020) Reduced graphene oxide−TiO_2/sodium alginate 3-dimensional structure aerogel for enhanced photocatalytic degradation of ibuprofen and sulfamethoxazole. Chemosphere 261. https://doi.org/10.1016/j.chemosphere.2020.127702
95. Ma X, Lou Y, Chen XB, Shi Z, Xu Y (2019) Multifunctional flexible composite aerogels constructed through in-situ growth of metal-organic framework nanoparticles on bacterial cellulose. Chem Eng J 356:227–235. https://doi.org/10.1016/j.cej.2018.09.034
96. Sadrieyeh S, Malekfar R (2018) Photocatalytic performance of plasmonic Au/Ag-TiO_2 aerogel nanocomposites. J Non-cryst Solids 489:33–39. https://doi.org/10.1016/j.jnoncrysol.2018.03.020
97. Jiménez-Saelices C, Seantier B, Cathala B, Grohens Y (2017b) Spray freeze-dried nanofibrillated cellulose aerogels with thermal superinsulating properties. Carbohydrate Polym 157:105–113. https://doi.org/10.1016/j.carbpol.2016.09.068
98. Karamikamkar S, Naguib HE, Park CB (2020) Advances in precursor system for silica-based aerogel production toward improved mechanical properties, customized morphology, and multifunctionality: a review. Adv Colloid Interf Sci 276. https://doi.org/10.1016/j.cis.2020.102101
99. Jiang J, Zhang Q, Zhan X, Chen F (2019) A multifunctional gelatin-based aerogel with superior pollutants adsorption, oil/water separation and photocatalytic properties. Chem Eng J 358:1539–1551. https://doi.org/10.1016/j.cej.2018.10.144
100. Zou F, Budtova T (2020) Tailoring the morphology and properties of starch aerogels and cryogels via starch source and process parameter. Carbohyd Polym. https://doi.org/10.1016/j.carbpol.2020.117344
101. Zhao C, Yan Y, Hu Z, Li L, Fan X (2015) Preparation and characterization of granular silica aerogel/polyisocyanurate rigid foam composites. Constr Build Mater 93:309–316. https://doi.org/10.1016/j.conbuildmat.2015.05.129
102. Jiménez-Saelices C, Seantier B, Cathala B, Grohens Y (2017a) Effect of freeze-drying parameters on the microstructure and thermal insulating properties of nanofibrillated cellulose aerogels. J Sol-Gel Sci Technol 84(3):475–485. https://doi.org/10.1007/s10971-017-4451-7
103. Revin VV, Pestov NA, Shchankin MV, Mishkin VP, Platonov VI, Uglanov DA (2019) A study of the physical and mechanical properties of aerogels obtained from bacterial cellulose. Biomacromol 20(3):1401–1411. https://doi.org/10.1021/acs.biomac.8b01816
104. Gong Y, Yu Y, Kang H, Chen X, Liu H, Zhang Y, … Song H (2019) Synthesis and characterization of graphene oxide/chitosan composite aerogels with high mechanical performance. Polymers 11(5). https://doi.org/10.3390/polym11050777
105. Zargar M, Hartanto Y, Jin B, Dai S (2017) Understanding functionalized silica nanoparticles incorporation in thin film composite membranes: interactions and desalination performance. J Membr Sci 521:53–64. https://doi.org/10.1016/j.memsci.2016.08.069
106. Xu X, Dong F, Yang X, Liu H, Guo L, Qian Y, … Luo J (2019) Preparation and characterization of cellulose grafted with epoxidized soybean oil aerogels for oil-absorbing materials. J Agricult Food Chem 67(2):637–643. https://doi.org/10.1021/acs.jafc.8b05161
107. Mondal T, Ashkar R, Butler P, Bhowmick AK, Krishnamoorti R (2016) Graphene nanocomposites with high molecular weight poly(ε-caprolactone) grafts: controlled synthesis and accelerated crystallization. ACS Macro Lett 5(3):278–282. https://doi.org/10.1021/acsmacrolett.5b00930
108. Mourya VK, Inamdar NN (2008) Chitosan-modifications and applications: opportunities galore. React Funct Polym 68(6):1013–1051. https://doi.org/10.1016/j.reactfunctpolym.2008.03.002
109. Patil PG, Sharma RP, Chakrabarty A (2020) Synthesis and characterization of highly hydrophobic, oil-absorbing aerogels for oil spill applications; A review. J Phys: Conf Ser 1644(1). https://doi.org/10.1088/1742-6596/1644/1/012047

110. Ji Y, Wen Y, Wang Z, Zhang S, Guo M (2020) Eco-friendly fabrication of a cost-effective cellulose nanofiber-based aerogel for multifunctional applications in Cu(II) and organic pollutants removal. J Clean Prod 255. https://doi.org/10.1016/j.jclepro.2020.120276
111. Lamy-Mendes A, Torres RB, Vareda JP, Lopes D, Ferreira M, Valente V, … Durães L (2019). Amine modification of silica aerogels/xerogels for removal of relevant environmental pollutants. Molecules 24(20). https://doi.org/10.3390/molecules24203701
112. Wang H, Wang C, Liu S, Chen L, Yang S (2019) Superhydrophobic and superoleophilic graphene aerogel for adsorption of oil pollutants from water. RSC Adv 9(15):8569–8574. https://doi.org/10.1039/c9ra00279k
113. Sun F, Liu W, Dong Z, Deng Y (2017) Underwater superoleophobicity cellulose nanofibril aerogel through regioselective sulfonation for oil/water separation. Chem Eng J 330:774–782. https://doi.org/10.1016/j.cej.2017.07.142
114. Gopi S, Balakrishnan P, Divya C, Valic S, Govorcin Bajsic E, Pius A, Thomas S (2017) Facile synthesis of chitin nanocrystals decorated on 3D cellulose aerogels as a new multi-functional material for waste water treatment with enhanced anti-bacterial and anti-oxidant properties. New J Chem 41(21):12746–12755. https://doi.org/10.1039/c7nj02392h
115. Roy S, Maji PK, Goh KL (2020) Sustainable design of flexible 3D aerogel from waste PET bottle for wastewater treatment to energy harvesting device. Chem Eng J. https://doi.org/10.1016/j.cej.2020.127409
116. Moreno-Castilla C, Maldonado-Hódar FJ (2005) Carbon aerogels for catalysis applications: an overview. Carbon 43(3):455–465. https://doi.org/10.1016/j.carbon.2004.10.022
117. Sánchez-Polo M, Rivera-Utrilla J, Salhi E, von Gunten U (2007) Ag-doped carbon aerogels for removing halide ions in water treatment. Water Res 41(5):1031–1037. https://doi.org/10.1016/j.watres.2006.07.009
118. Williams JC, Nguyen BN, McCorkle L, Scheiman D, Griffin JS, Steiner SA, Meador MAB (2017) Highly porous, rigid-rod polyamide aerogels with superior mechanical properties and unusually high thermal conductivity. ACS Appl Mater Interf 9(2):1801–1809. https://doi.org/10.1021/acsami.6b13100
119. Zargar M, Hartanto Y, Jin B, Dai S (2016) Hollow mesoporous silica nanoparticles: a peculiar structure for thin film nanocomposite membranes. J Membr Sci 519:1–10. https://doi.org/10.1016/j.memsci.2016.07.052
120. Ye S, Liu Y, Feng J (2017) Low-density, mechanical compressible, water-induced self-recoverable graphene aerogels for water treatment. ACS Appl Mater Interf 9(27):22456–22464. https://doi.org/10.1021/acsami.7b04536
121. Tang C, Brodie P, Li Y, Grishkewich NJ, Brunsting M, Tam KC (2020) Shape recoverable and mechanically robust cellulose aerogel beads for efficient removal of copper ions. Chem Eng J 392. https://doi.org/10.1016/j.cej.2020.124821
122. Maleki H, Durães L, Portugal A (2014) An overview on silica aerogels synthesis and different mechanical reinforcing strategies. J Non-cryst Solids 385:55–74. https://doi.org/10.1016/j.jnoncrysol.2013.10.017
123. Yeo J, Liu Z, Ng TY (2020) Silica aerogels: a review of molecular dynamics modelling and characterization of the structural, thermal, and mechanical properties. Handb Mater Model 1575–1595. https://doi.org/10.1007/978-3-319-44680-6_83
124. Lazzari LK, Perondi D, Zampieri VB, Zattera AJ, Santana RMC (2019) Cellulose/biochar aerogels with excellent mechanical and thermal insulation properties. Cellulose 26(17):9071–9083. https://doi.org/10.1007/s10570-019-02696-3
125. Woignier T, Primera J, Alaoui A, Despetis F, Calas-Etienne S, Faivre A, … Etienne P (2020) Techniques for characterizing the mechanical properties of aerogels. J Sol-Gel Sci Technol 93(1):6–27. https://doi.org/10.1007/s10971-019-05173-2
126. Dong W, Faltens T, Pantell M, Simon D, Thompson T, Dong W (2009) Acoustic properties of organic/inorganic composite aerogels. Mater Res Soc Symp Proc 1188:219–229. https://doi.org/10.1557/proc-1188-ll07-02
127. Perin L, Faivre A, Calas-Etienne S, Woignier T (2004) Nanostructural damage associated with isostatic compression of silica aerogels. J Non-cryst Solids 333(1):68–73. https://doi.org/10.1016/j.jnoncrysol.2003.09.046

128. Sussner H, Vacher R (1979) High-precision measurements of Brillouin scattering frequencies. Appl Opt 18(22):3815. https://doi.org/10.1364/ao.18.003815
129. Alaoui AH, Woignier T, Scherer GW, Phalippou J (2008) Comparison between flexural and uniaxial compression tests to measure the elastic modulus of silica aerogel. J Non-cryst Solids 354(40–41):4556–4561. https://doi.org/10.1016/j.jnoncrysol.2008.06.014
130. Hartanto Y, Zargar M, Cui X, Jin B, Dai S (2019) Non-ionic copolymer microgels as high-performance draw materials for forward osmosis desalination. J Membr Sci 572:480–488. https://doi.org/10.1016/j.memsci.2018.11.042
131. Hartanto Y, Zargar M, Cui X, Shen Y, Jin B, Dai S (2016) Thermoresponsive cationic copolymer microgels as high performance draw agents in forward osmosis desalination. J Membr Sci 518:273–281. https://doi.org/10.1016/j.memsci.2016.07.018
132. Anderson AM, Carroll MK (2011) Hydrophobic silica aerogels: review of synthesis, properties and applications. Aerogels Handb 47–77. https://doi.org/10.1007/978-1-4419-7589-8_3
133. Horikawa T, Hayashi J, Muroyama K (2004) Size control and characterization of spherical carbon aerogel particles from resorcinol-formaldehyde resin. Carbon 42(1):169–175. https://doi.org/10.1016/j.carbon.2003.10.007
134. Zaman A, Huang F, Jiang M, Wei W, Zhou Z (2020) Preparation, properties, and applications of natural cellulosic aerogels: a review. Energy Built Environ 1(1):60–76. https://doi.org/10.1016/j.enbenv.2019.09.002
135. Zuo L, Zhang Y, Zhang L, Miao YE, Fan W, Liu T (2015) Polymer/Carbon-based hybrid aerogels: preparation, properties and applications. Materials 8(10):6806–6848. https://doi.org/10.3390/ma8105343
136. Li D, Wang Y, Sun Y, Lu Y, Chen S, Wang B, … Yang D (2018) Turning gelidium amansii residue into nitrogen-doped carbon nanofiber aerogel for enhanced multiple energy storage. Carbon 137:31–40. https://doi.org/10.1016/j.carbon.2018.05.011
137. Qin J, Chen L, Zhao C, Lin Q, Chen S (2017) Cellulose nanofiber/cationic conjugated polymer hybrid aerogel sensor for nitroaromatic vapors detection. J Mater Sci 52(14):8455–8464. https://doi.org/10.1007/s10853-017-1065-y
138. Zhou J, Hsieh YL (2018) Conductive polymer protonated nanocellulose aerogels for tunable and linearly responsive strain sensors. ACS Appl Mater Interf 10(33):27902–27910. https://doi.org/10.1021/acsami.8b10239
139. Jiang F, Hsieh YL (2014) Amphiphilic superabsorbent cellulose nanofibril aerogels. Journal of Materials Chemistry A 2(18):6337–6342. https://doi.org/10.1039/c4ta00743c
140. Liu H, Geng B, Chen Y, Wang H (2017) Review on the aerogel-type oil sorbents derived from nanocellulose. ACS Sustain Chem Eng 5(1):49–66. https://doi.org/10.1021/acssuschemeng.6b02301
141. Rafieian F, Hosseini M, Jonoobi M, Yu Q (2018) Development of hydrophobic nanocellulose-based aerogel via chemical vapor deposition for oil separation for water treatment. Cellulose 25(8):4695–4710. https://doi.org/10.1007/s10570-018-1867-3
142. Budtova T (2019) Cellulose II aerogels: a review. Cellulose 26(1):81–121. https://doi.org/10.1007/s10570-018-2189-1
143. Liebner F, Pircher N, Schimper C, Haimer E, Rosenau T (2016) Aerogels: cellulose-based. Encyclopedia Biomed Polym Polymeric Biomater 37–75. https://doi.org/10.1081/e-ebpp-120051062
144. Chaisuwan T, Komalwanich T, Luangsukrerk S, Wongkasemjit S (2010) Removal of heavy metals from model wastewater by using polybenzoxazine aerogel. Desalination 256(1–3):108–114. https://doi.org/10.1016/j.desal.2010.02.005
145. Lorjai P, Wongkasemjit S, Chaisuwan T, Jamieson AM (2011) Significant enhancement of thermal stability in the non-oxidative thermal degradation of bisphenol-A/aniline based polybenzoxazine aerogel. Polym Degrad Stab 96(4):708–718. https://doi.org/10.1016/j.polymdegradstab.2010.12.005
146. Frackowiak E, Béguin F (2001) Carbon materials for the electrochemical storage of energy in capacitors. Carbon 39(6):937–950. https://doi.org/10.1016/S0008-6223(00)00183-4

147. Biener J, Stadermann M, Suss M, Worsley MA, Biener MM, Rose KA, Baumann TF (2011) Advanced carbon aerogels for energy applications. Energy Environ Sci 4(3):656–667. https://doi.org/10.1039/c0ee00627k
148. White RJ, Brun N, Budarin VL, Clark JH, Titirici MM (2014) Always look on the "light" side of life: sustainable carbon aerogels. Chemsuschem 7(3):670–689. https://doi.org/10.1002/cssc.201300961
149. Sun H, Xu Z, Gao C (2013) Multifunctional, ultra-flyweight, synergistically assembled carbon aerogels. Adv Mater 25(18):2554–2560. https://doi.org/10.1002/adma.201204576
150. Groß J, Fricke J (1995) Scaling of elastic properties in highly porous nanostructured aerogels. Nanostruct Mater 6(5–8):905–908. https://doi.org/10.1016/0965-9773(95)00206-5
151. Long LY, Weng YX, Wang YZ (2018) Cellulose aerogels: synthesis, applications, and prospects. Polymers 8(6). https://doi.org/10.3390/polym10060623
152. Maleki H (2016) Recent advances in aerogels for environmental remediation applications: a review. Chem Eng J 300:98–118. https://doi.org/10.1016/j.cej.2016.04.098
153. Coteţ LC, Măicăneanu A, Forţ CI, Danciu V (2013) Alpha-cypermethrin pesticide adsorption on carbon aerogel and xerogel. Separat Sci Technol (Philadelphia) 48(17):2649–2658. https://doi.org/10.1080/01496395.2013.805782
154. Chen R, Sheehan T, Ng JL, Brucks M, Su X (2020) Capacitive deionization and electrosorption for heavy metal removal. Environ Sci: Water Res Technol 6(2):258–282. https://doi.org/10.1039/c9ew00945k
155. Liu J, Ge X, Ye X, Wang G, Zhang H, Zhou H, … Zhao H (2016) 3D graphene/δ-MnO_2 aerogels for highly efficient and reversible removal of heavy metal ions. J Mater Chem A 4(5):1970–1979. https://doi.org/10.1039/c5ta08106h
156. Yu Z, Hu C, Dichiara AB, Jiang W, Gu J (2020) Cellulose nanofibril/carbon nanomaterial hybrid aerogels for adsorption removal of cationic and anionic organic dyes. Nanomaterials 10(1). https://doi.org/10.3390/nano10010169
157. Fraga TJM, Carvalho MN, Ghislandi MG, Da Motta Sobrinho MA (2019) Functionalized graphene-based materials as innovative adsorbents of organic pollutants: a concise overview. Braz J Chem Eng 36(1):1–31. https://doi.org/10.1590/0104-6632.20190361s20180283
158. Feng C, Ren P, Li Z, Tan W, Zhang H, Jin Y, Ren F (2020) Graphene/waste-newspaper cellulose composite aerogels with selective adsorption of organic dyes: preparation, characterization, and adsorption mechanism. New J Chem 44(6):2256–2267. https://doi.org/10.1039/c9nj05346h
159. Wang S, Ning H, Hu N, Huang K, Weng S, Wu X, … Alamusi (2019) Preparation and characterization of graphene oxide/silk fibroin hybrid aerogel for dye and heavy metal adsorption. Compos Part B: Eng 163:716–722. https://doi.org/10.1016/j.compositesb.2018.12.140
160. Cai T, Ding Y, Xu L (2019) Synthesis of flower-like CuS/graphene aerogels for dye wastewater treatment. Funct Mater Lett 12(2). https://doi.org/10.1142/S1793604719500024
161. Chen L, Li Y, Du Q, Wang Z, Xia Y, Yedinak E, … Ci L (2017) High performance agar/graphene oxide composite aerogel for methylene blue removal. Carbohydrate Polym 155:345–353. https://doi.org/10.1016/j.carbpol.2016.0
162. Hasan M, Gopakumar DA, Arumughan V, Pottathara YB, Sisanth KS, Pasquini D, … Abdul Khalil HPS (2019) Robust superhydrophobic cellulose nanofiber aerogel for multifunctional environmental applications. Polymers 11(3). https://doi.org/10.3390/polym11030495
163. Nguyen ST, Feng J, Le NT, Le ATT, Hoang N, Tan VBC, Duong HM (2013) Cellulose aerogel from paper waste for crude oil spill cleaning. Ind Eng Chem Res 52(51):18386–18391. https://doi.org/10.1021/ie4032567
164. Olalekan AP, Dada AO, Adesina OA (2014) Review: Silica Aerogel as a viable absorbent for oil spill remediation. J Encapsulation Adsorption Sci 04(04):122–131. https://doi.org/10.4236/jeas.2014.44013
165. Reynolds JG, Coronado PR, Hrubesh LW (2001a) Hydrophobic aerogels for oil-spill clean up—synthesis and characterization. J Non-crystalline Solids 292(1–3):127–137. https://doi.org/10.1016/S0022-3093(01)00882-1

166. Sacco O, Vaiano V, Daniel C, Navarra W, Venditto V (2019) Highly robust and selective system for water pollutants removal: how to transform a traditional photocatalyst into a highly robust and selective system for water pollutants removal. Nanomaterials 9(11). https://doi.org/10.3390/nano9111509
167. Motahari S, Nodeh M, Maghsoudi K (2016) Absorption of heavy metals using resorcinol formaldehyde aerogel modified with amine groups. Desalin Water Treat 57(36):16886–16897. https://doi.org/10.1080/19443994.2015.1082506
168. Chen H, Wang X, Li J, Wang X (2015) Cotton derived carbonaceous aerogels for the efficient removal of organic pollutants and heavy metal ions. J Mater Chem A 3(11):6073–6081. https://doi.org/10.1039/c5ta00299k
169. Alatalo SM, Pileidis F, Mäkilä E, Sevilla M, Repo E, Salonen J, … Titirici MM (2015) Versatile cellulose-based carbon aerogel for the removal of both cationic and anionic metal contaminants from water. ACS Appl Mater Interf 7(46):25875–25883. https://doi.org/10.1021/acsami.5b08287
170. Coteť LC, Danciu V, Cošoveanu V, Popescu IC, Roig A, Molins E (2007) Synthesis of meso- and macroporous carbon aerogels. Rev Roum Chim 52(11):1077–1081
171. Le KA, Pham NQ, Pham MC (2020) Effect of synthesis conditions on carbon aerogels material to remove pesticide in cuu long delta rivers. Chem Eng Trans 78:571–576. https://doi.org/10.3303/CET2078096
172. Chen LQ, Hu PP, Zhang L, Huang SZ, Luo LF, Huang CZ (2012) Toxicity of graphene oxide and multi-walled carbon nanotubes against human cells and zebrafish. Sci China Chem 55(10):2209–2216. https://doi.org/10.1007/s11426-012-4620-z
173. Lin YF, Chen JL (2014) Magnetic mesoporous Fe/carbon aerogel structures with enhanced arsenic removal efficiency. J Colloid Interf Sci 420:74–79. https://doi.org/10.1016/j.jcis.2014.01.008
174. Yan M, Huang W, Li Z (2019) Chitosan cross-linked graphene oxide/lignosulfonate composite aerogel for enhanced adsorption of methylene blue in water. Int J Biol Macromol 136:927–935. https://doi.org/10.1016/j.ijbiomac.2019.06.144
175. Lithner D, Nordensvan I, Dave G (2012) Comparative acute toxicity of leachates from plastic products made of polypropylene, polyethylene, PVC, acrylonitrile-butadiene-styrene, and epoxy to Daphnia magna. Environ Sci Pollut Res 19(5):1763–1772. https://doi.org/10.1007/s11356-011-0663-5
176. Kwon BG, Koizumi K, Chung SY, Kodera Y, Kim JO, Saido K (2015) Global styrene oligomers monitoring as new chemical contamination from polystyrene plastic marine pollution. J Hazard Mater 300:359–367. https://doi.org/10.1016/j.jhazmat.2015.07.039
177. Feng LJ, Li JW, Xu EG, Sun XD, Zhu FP, Ding Z, … Yuan XZ (2019) Short-term exposure to positively charged polystyrene nanoparticles causes oxidative stress and membrane destruction in cyanobacteria. Environ Sci: Nano 6(10):3072–3079. https://doi.org/10.1039/c9en00807a
178. Sapsford KE, Algar WR, Berti L, Gemmill KB, Casey BJ, Oh E, … Medintz IL (2013) Functionalizing nanoparticles with biological molecules: developing chemistries that facilitate nanotechnology. Chem Rev 113(3):1904–2074. https://doi.org/10.1021/cr300143v
179. Yang Z, Liu ZW, Allaker RP, Reip P, Oxford J, Ahmad Z, Reng G (2013) A review of nanoparticle functionality and toxicity on the central nervous system. Nanotechnol Brain Fut 313–332. https://doi.org/10.1007/978-94-007-1787-9_18
180. Schrand A, Rahman M, Hussain S, Schlager J, Smith D, Syed A (2010) Metal-based nanoparticles and their toxicity assessment. Wiley Interdisc Rev: Nanomed Nanobiotechnol 2(5):544–568. https://doi.org/10.1002/wnan.v2:5
181. Wang L, Mu RJ, Lin L, Chen X, Lin S, Ye Q, Pang J (2019) Bioinspired aerogel based on konjac glucomannan and functionalized carbon nanotube for controlled drug release. Int J Biol Macromolecules 133:693–701. https://doi.org/10.1016/j.ijbiomac.2019.04.148
182. Cheng H, Gu B, Pennefather MP, Nguyen TX, Phan-Thien N, Duong HM (2017) Cotton aerogels and cotton-cellulose aerogels from environmental waste for oil spillage cleanup. Mater Des 130:452–458. https://doi.org/10.1016/j.matdes.2017.05.082

183. Rosa MF, Medeiros ES, Malmonge JA, Gregorski KS, Wood DF, Mattoso LHC, … Imam SH (2010) Cellulose nanowhiskers from coconut husk fibers: effect of preparation conditions on their thermal and morphological behavior. Carbohyd Polym 81(1):83–92. https://doi.org/10.1016/j.carbpol.2010.01.059
184. Berber MR (2020) Current advances of polymer composites for water treatment and desalination. J Chem. https://doi.org/10.1155/2020/7608423
185. Huang Y, Wang Z (2018) Preparation of composite aerogels based on sodium alginate, and its application in removal of Pb^{2+} and Cu^{2+} from water. Int J Biol Macromolecules 107(Part A):741–747. https://doi.org/10.1016/j.ijbiomac.2017.09.057
186. Sharma S, Kaur J, Sharma G, Thakur KK, Chauhan GS, Chauhan K (2013) Preparation and characterization of pH-responsive guar gum microspheres. Int J Biol Macromol 62:636–641. https://doi.org/10.1016/j.ijbiomac.2013.09.045
187. Wurzburg OB (1986) Modified starches: properties and uses. CRC Press, Boca Raton, FL
188. Kweon DK, Choi JK, Kim EK, Lim ST (2001) Adsorption of divalent metal ions by succinylated and oxidized corn starches. Carbohyd Polym 46(2):171–177. https://doi.org/10.1016/S0144-8617(00)00300-3
189. Xu SM, Feng S, Peng G, Wang JD, Yushan A (2005) Removal of Pb (II) by crosslinked amphoteric starch containing the carboxymethyl group. Carbohyd Polym 60(3):301–305. https://doi.org/10.1016/j.carbpol.2005.01.018
190. Crini G (2005) Recent developments in polysaccharide-based materials used as adsorbents in wastewater treatment. Progr Polym Sci (Oxford) 30(1):38–70. https://doi.org/10.1016/j.progpolymsci.2004.11.002
191. Wesslén KB, Wesslén B (2002) Synthesis of amphiphilic amylose and starch derivatives. Carbohyd Polym 47(4):303–311. https://doi.org/10.1016/S0144-8617(01)00196-5
192. Naushad M, Ahamad T, Sharma G, Al-Muhtaseb AH, Albadarin AB, Alam MM, … Ghfar AA (2016) Synthesis and characterization of a new starch/SnO_2 nanocomposite for efficient adsorption of toxic Hg^{2+} metal ion. Chem Eng J 300:306–316. https://doi.org/10.1016/j.cej.2016.04.084
193. Karaki N, Aljawish A, Humeau C, Muniglia L, Jasniewski J (2016) Enzymatic modification of polysaccharides: mechanisms, properties, and potential applications: a review. Enzyme Microb Technol 90:1–18. https://doi.org/10.1016/j.enzmictec.2016.04.004

Algal Biomass Valorization for the Removal of Heavy Metal Ions

Latifa Boukarma, Rachid Aziam, Mhamed Abali, Gabriela Carja, Amina Soudani, Mohamed Zerbet, Fouad Sinan, and Mohamed Chiban

Abstract Pollution generated by wastewater containing inorganic pollutants, such as heavy metals, has always been considered a real problem for our planet. Therefore, the removal of these micropollutants from polluted water is a valuable intervention to preserve human health and the environment. Many conventional methods are used today to treat wastewater, such as membrane filtration, chemical precipitation, ion exchange, and adsorption by activated carbon, but the operating cost they generate has restricted their use. To overcome this limitation, scientists have focused for always on the application of marine resources to clean up the environment. The adsorption of heavy metals by biosorbents obtained from algae has been widely studied for wastewater treatment, as the exploitation of this biomass has the advantage of being a low cost, renewable and abundant biological raw material, and its use as a biosorbent is also a great alternative to activated carbon. The sorption capacity of the vegetable adsorbent is depending on chemical constitution of their cell wall and the presence of macromolecules with various functional groups that interact with metal ions. We review in this chapter, (1) the challenges associated with heavy metals, such as water pollution, hazardous effects, and their removal techniques including biosorption based on algae biopolymers, such as alginate and carrageenan, (2) the main chemical and structural compounds of macroalgae responsible for the metal ions removal, (3) current knowledge on the potential of macroalgae regarding their pharmacological applications and possible biosorbents prepared from them for the removal of metal ions from aqueous solutions.

L. Boukarma · R. Aziam · M. Abali · M. Zerbet · F. Sinan · M. Chiban (✉)
Faculty of Science, Department of Chemistry, Ibn Zohr University, Agadir, Morocco

G. Carja
Faculty of Chemical Engineering and Environmental Protection, Gheorghe Asachi, Technical University, Iași, Romania

A. Soudani
Faculty of Applied Sciences, University Campus of Ibn Zohr University, Ait Melloul, Morocco

E. Lichtfouse et al. (eds.), *Inorganic-Organic Composites for Water and Wastewater Treatment*, Environmental Footprints and Eco-design of Products and Processes,
https://doi.org/10.1007/978-981-16-5928-7_8

Keywords Algal biomass · Classification of macro algae · Pharmacological potential · Biopolymers · Low-cost adsorbents · Heavy metal · Biosorption · Industrial wastewater

1 Introduction

Heavy metal pollution is among the phenomena that received throughout the world an increasing attention, it has become a serious global environmental problem and a major concern. The interest in these pollutants is mainly linked to the harmful effects they have on human health, as well as on other organisms living in the aquatic environment, because of their bioaccumulation, toxicity, and non-biodegradability properties, these pollution problems are due to the environmental disorder caused by the fast growth of industrial activities [102].

The adverse effects of these molecules on humans depend on some factors such as dosage, emission rate, and exposure period. Moreover, the toxicity level of some heavy metals for humans was found to follow the order Co < Al < Cr < Pb < Ni < Zn < Cu < Cd < Hg [81]. Among these toxic micropollutants, Hg, Cd, and Pb have received more attention for the last decades, because of their toxicity and their effects on the environment and the living organisms [157]. Few of these inorganic pollutants such as zinc and iron are needed for human defense, plants, and other living beings, but although the importance occupied by them, they can pose health hazards if their concentrations exceed allowable limits. As an example, the excess of zinc has main symptoms for human, such as nausea, dizziness, electrolyte imbalance, and muscle stiffness [5, 53].

If some metals have demonstrated their main function in human body, there are some that are toxic even at low concentration as the case of cadmium, which is a non-biodegradable element in nature, and up to day it has not been proven to have any physiological function in human body [54]. Another point is that a chronic exposure to Pb(II) even to low concentration decreases the intelligence capacity of children [81]. As a result of the critical effects of heavy metals, concerned environmental agencies, such as the World Health Organization (WHO) and the United States Environmental Protection Agency (USEPA) set safe limits for heavy metals in recycled, in drinking water, and in wastewater [108]. The safe limits of heavy metals in wastewater samples as well as in drinking water according to the WHO guidelines are listed in Table 8.1.

For many decades, various conventional technologies were employed to remove metals from aqueous effluents, such as chemical precipitation (hydroxide and sulfide precipitation), ion exchange, membrane filtration, coagulation/flocculation, flotation, and electrochemical treatment [9]. Each of these processes has its disadvantages [48]. These limitations have led researchers to find other alternative methods such as adsorption, especially which focus on the use of abundant and less expensive biomaterials. The widely used adsorbent around the world is activated carbons, but the increased cost of these materials and some complexities involved in their synthesis limited their use. Therefore, low-cost adsorbents must be developed to

Table 8.1 World Health Organization guidelines for safe limits of heavy metals

Sample	Metal ions	Safe limits (ppm)	References
Wastewater	Pb(II)	0.01	Ayeni [14]
	Ni(II)	0.02	Ayeni [14]
	Cr(III)	0.05	Ayeni [14]
	Hg(II)	0.001	Onuegbu et al. [113]
	Cd(II)	0.003	Ayeni [14]
Drinking water	As(V)	0.01	Sayato [137]
	Cd(II)	0.003	Sayato [137]
	Pb(II)	0.01	Sayato [137]
	Hg(II)	0.006	Sayato [137]
	Ni(II)	0.07	Sayato [137]
	Cr(III)	0.05	Sayato [137]
	Zn(II)	5	King et al. [80]

replace the current expensive adsorption method of removing heavy metals from solution. Hence, searching for an alternative of activated carbon from abundant and inexpensive sources is of concern [48].

A number of researches have been carried out on biosorption using both the micro and macroalgae biomass [19, 68, 69, 121]. The marine macroalgae have demonstrated their ability toward the removal of inorganic pollutants including heavy metal ions [93]. The good removal capacity of these resources is due to the surface structure, which contains active functional groups involved in the biosorption process [69]. Biosorption by dried seaweeds has been intensively studied in recent years as an economical treatment for the removal or recovery of metals from industrial effluents [59, 70, 140]. Applying dried seaweeds in the adsorption of these ions present several advantages such as wide availability, low cost, high metal sorption capacity, and reasonably regular quality [12]. This process implies the use of dried macroalgae or their derivatives to adsorb the metal ions with the ligands or functional groups located on the external surface of them. The passive elimination of some of these toxic pollutants as the case of Cd^{2+}, Cu^{2+}, Zn^{2+}, Pb^{2+}, Cr^{3+}, and Hg^{2+} by inexpensive biomaterials requires that the adsorbent has certain adsorbent properties such as high selectivity and high metal uptake [33].

2 Metal Pollution and Health Effects

Heavy metals are the most toxic pollutants, and the resulting pollution is considered a worldwide environmental problem. The discharge of industrial wastewater without prior treatment or with insufficient treatment leads to dangerous effects on the one hand, on marine organisms living in aquatic environment and on human health as the main consumer in the food chain of marine resources, on the other hand, on the state of the wastewater treatment plant and that of the sanitation network.

Heavy metals are defined as metallic elements that have a relatively high density compared to water, they are characterized by a high atomic weight and a density approximately five times higher than that of water [151]. They present in many industrial effluents generated by numerous anthropogenic activities, such as plating facilities, mining operations, and tanneries, as well as, they are continuously released into the biosphere by volcanoes due to natural weathering of rocks [17, 157]. Heavy metals, such as cadmium, arsenic, lead, and copper, are among the common inorganic pollutants of serious concern in wastewater treatment, they are particularly characterized by their high toxicity, and even in trace amounts, they can cause serious disturbance to aquatic organisms and health problems for humans [142]. Several studies have focused on the exploitation of marine resources as a source of alternative applications more suited to current environmental challenges including heavy metal pollution [86, 95, 96]. The following paragraphs will highlight some of the effects of heavy metals, which have taken a large part in the recently published studies regarding the biosorption of inorganic pollutants by marine macroalgae.

Cadmium, which is known under the symbol Cd, it is a chemical element from group IIB of the transition metals of Mendeleev's periodic table, with atomic number 48 and atomic mass of about 112.4 g/mol [112]. This metal is used in various domains such as accumulators or alkaline batteries, pigments for paints or plastics and in electrolytic process by deposit or by cadmium plating on metals or to reduce melting points. Cadmium is a known cumulative toxic substance whose disposal half-life is about 20–40 years; it is stored primarily in the liver and kidneys after entering the body [11]. This metal poses serious risks to human health, and it has not been shown to have any physiological function in human body [54].

Arsenic is another heavy metal, which occupies a considerable place in the published data concerning the field of biosorption due to its toxicity and danger. Arsenic is a chemical element with atomic number 33 designated by the symbol As, and it is generally considered a metalloid. Arsenic occurs naturally in its two forms of oxyanions, namely, arsenite As(III) and arsenate As(V) [138], as well as it occurs in the $-3, 0, +3$, and $+5$ oxidation states [143]. Numerous studies have been conducted as an attempt to solve the problem of arsenic pollution and to find the most efficient technology to remove its species from drinking water and industrial wastewater [28, 61, 138]. The main source of pollution caused by arsenic in the environment is the smelting of ores, such as those of gold, silver, and copper. As a result, arsenic from these sources is distributed in the air, water, soil and finds its way into human body through direct inhalation or contamination of food and consumer products.

Lead, known as Pb, it is a chemical element with atomic number 82. It is a natural constituent of the earth's crust, and it is commonly found in soils, plants, and water in trace amounts. This metal usually exists in ores, which also contain other metals such as copper and zinc, which are extracted as a co-product of these metals. Lead is a toxic heavy metal that can reach human body by inhalation and ingestion from various sources, such as contaminated air, water, soil, and food. Lead has become the most common toxic metal in the world due to human actions, and today, this metal is widely used in different sectors including, building construction, fusible alloys, bullets, and lead-acid batteries [26].

Copper is another kind of inorganic pollutants that have been extensively studied by researchers regarding the field of bio-absorption [29, 97]. Copper is a chemical element with atomic number 29 known as Cu. It is a member of the family of metals included in the periodic table of the elements. Copper is an essential nutrient for humans, animals, and plants, but it can be hazardous to human health at high exposures. This metal can exist in the form of free cationic Cu^{2+} under acidic conditions, and trace amounts of $[Cu(OH)]^{+}$ and water-soluble $Cu(OH)_2$ under neutral and basic conditions [72].

3 Technologies of Wastewater Treatment

Currently, various technologies are applied for the removal of heavy metals from industrial wastewater in large conventional treatment plants. The following section highlights the most commonly used methods for this purpose including chemical precipitation, ion exchange, membrane filtration, electrochemical, flotation, and adsorption. For a brief comparison, the advantages and disadvantages of each treatment method are listed in Table 8.2.

3.1 Chemical Precipitation

Due to its simplicity, chemical precipitation is considered among the most widely used conventional processes to remove heavy metals from inorganic effluents [84]. This process is based on a mechanism in which an insoluble metal is produced by the reaction of the metals dissolved in solution and the precipitant [56]. In this process, the positively charged molecules and those negatively charged (anions) are combined. As well as, the dissolved metal precipitation is compelled by increasing the soluble anion concentration [61]. Very fine particles are then generated at the end of this process. Therefore, chemical precipitants, coagulants, and flocculation processes are used to increase their size and in order to remove them as sludge [56].

3.2 Ion Exchange

Ion exchange is another example of the technologies used to treat polluted effluents. This technique is based on the attraction of soluble ions from the liquid to the solid phase. In addition, the resins used in this method, which are hydrophobic solid

Table 8.2 Comparison of applied technologies to heavy metal ions removal from wastewaters

Process	Advantages	Disadvantages	References
Chemical Precipitation	• Low capital cost • Simple operation	• High operating cost • High cost of disposal of produced sludge	Wang et al. [159]
Ion exchange	• Metal-selective, • eLimited-pH tolerance • High regeneration	• High-initial-capital • High-maintenance cost	Gao et al. [51]
Membrane filtration	• High efficiency • Appropriate for a variety of wastewater compositions	• Membrane fouling • Secondary pollution • Short lifespan of membranes • High operation cost	Hube et al. [67]
Electrochemical treatment	• Cost effective • Effective at ambient temperature • Electrode materials with long operating life	• Blockage of electrodeposition due to the metal hydroxide precipitation • Wasted energy and unstable process	Chaplin [25] Liu et al. [95, 96] Tran et al. [153]
Adsorption	• High efficiency • Capacity to remove metals • Less chemical consumption	• High operating cost, • The efficiency decrease with the presence of other metals	Da'na [31]

substances, can retain positively or negatively charged ions from an electrolyte solution and at the same time release other ions of a similar charge in the solution in an equivalent quantity [56]. The process of ion exchange should not be confused with solvent extraction, the difference being that the first indicates the separation of solid from liquid, and the second means the separation of liquid from liquid. Furthermore, this technique is considered a valuable technique for recovering minerals from the mining process and mining tailings using chelating resins [71].

3.3 Membrane Filtration

This technology makes it possible to obtain purified water by passing it through special filter membranes, which physically retain the impurities present in water. Various types of membrane filtration such as ultrafiltration, nanofiltration, and reverse osmosis can be used to treat wastewater containing heavy metals. In general, these methods of separation have been of growing interest both for the treatment of drinking water and wastewater. Ultrafiltration and nanofiltration are effective for the removal of all classes of pollutants. Whereas, microfiltration is not much use for the treatment of these effluents because of its large pore size [124]. Reverse osmosis is another membrane separation technique that uses pressure to allow the solution to pass through a membrane that retains the solute on one side and allows the solvent to move to the other side [56].

3.4 Electrochemical Treatment

The electrochemical method is a technology that allows to eliminate heavy metals from wastewater. It consists in making these micropollutants precipitate in a weakly acidic or neutralized catholyte in the form of hydroxides [56]. The main reagent used during this treatment is the electron, which is a clean reagent. Thus, there is no need to add any additional reagent [125]. The electrochemical wastewater treatment involves electroplating, electrocoagulation, electroflotation, and electrooxidation [56]. This technology is chosen to treat these effluents because it offers ideal tools for addressing environmental problems.

3.5 Adsorption

Adsorption process is defined as a surface phenomenon by which one or more than one adsorbate are fixed to the adsorbent surface and form binding through physical or chemical bonds. This process is recognized as the most promising and efficient fundamental approach in the wastewater treatment processes [44]. One of the main adsorbents employed to treat waste effluents is the activated carbon, because it contains a developed surface area with large porous. For commercial carbons, usually, their surface area is ranging from 500 to 1500 m^2/g and sometimes even up to 3000 m^2/g. Coal, lignite, bone charcoal, and wood are among the main raw materials extensively employed in the preparation of these adsorbents [134].

Despite the advantages of the previously listed physicochemical methods in the treatment of polluted effluents, they can sometimes load the natural environment with organic pollutants more toxic than the original ones. As well as, these methods are known to be very expensive and require sophisticated equipment. Another point is that activated carbon is recognized as the most powerful adsorbent, its industrial application is prevented by the high cost associated with its production [10]. These limitations, led researchers and industrialists to focus on finding other more efficient and less expensive methods, including biosorption based on the use of biomaterials prepared from cheap and abundant resources.

Biosorption is defined as a process in which substances of aqueous phase are removed by passive bonds created between the substrate and the dead biomass or derived materials [2]. Many data in the literature have been used to describe the biosorption process the following terms; sorption, bio-adsorption, and removal regarding organic and inorganic pollutants [15, 27, 45, 49, 139]. This interfacial phenomenon, should not be confused with the accumulation or bioaccumulation that indicates the absorption. As explained above, the biosorption involves a surface phenomenon where metal ions are attached to the dried algae, while bioaccumulation requires the metabolic activity of a living organism to sequester these pollutants [2].

Recently, research has intensified on the mechanisms of biosorption by which biomass is used to remove or to recover precious metals from processing solutions [2]. Adsorption is still considered as a phenomenon in which a complex and poorly understood mechanism is involved. It depends on the type of organism whether is alive or not, the type of microorganism and the elemental species [106]. This mechanism explains how the metal ion is binding to the biomass. Commonly, these mechanisms are classified as either physical, chemical, or electrostatic adsorption [118]. Ion exchange has been shown to be an important concept in biosorption, as it can explain many observations made in heavy metal uptake experiments [33]. Ion exchange, complexation, and coordination are of the main mechanisms behind the uptake of heavy metals by macroalgae [136]. In addition to the previously mentioned mechanisms, adsorption, precipitation, and covalent binding may also be involved, but the most likely is ion exchange [90].

The functional groups of algae cell walls play vital roles in biosorption. Carboxylic and sulfate groups are known to be the main metal bending functional ionic groups in the cell wall of algae [90]. Carboxylate groups of alginate have been identified as the main binding site for metals [98]. Thus, there is a scientific and practical interest in identifying the active sites of the bio polymeric structures involved in the sorption. For this purpose, the determination of different functional groups that may be involved in biosorption process is mandatory. The Fourier transform infrared spectroscopy analysis (FTIR) provides information desired on these functional groups and the molecular bonds of biosorbent are investigated [8]. This spectroscopy over the last decades has proven and accepted to be a powerful tool for studying biological samples [37]. FTIR analysis coupled with potentiometric titration was used to identify and quantify surface functional groups of algae. In this regard, some functional groups were identified to be the predominant in the surface of the three classes of marine algae, namely, carboxyl, hydroxyl, amino, and sulfate groups [136]. These functional groups work as binding sites for metal and are located at the surface of the cell wall as polysaccharides, proteins and lipids.

Among the previously discussed groups, the hydroxyl groups especially present in all polysaccharides can be negatively charged, which contributes to the adsorption of metals at a high level [46]. The possible functional groups that participate in heavy metals biosorption are, hydroxyl, carboxyl, amino, ester sulfhydryl, carbonyl, and phosphate groups, they are found in alcohols and carbohydrates, fatty acids, proteins and organic acids, proteins and nucleic acids, lipids, cysteine, aldehydes and polysaccharides, deoxyribonucleic acid and tissue plasminogen activator, respectively [9]. Phaeophyceae cell wall has been characterized by its composition rich in chemical functional groups, such as hydroxyl, carboxylic phenolic acid, and amine involved in the biosorption process that is induced by the selective binding and interaction created between the metals or any other pollutants and the biosorbent [108]. To study various aspects involved in the biosorption of metal ions, the potential of biosorption and some other parameters must be calculated based on the equations.

$$q_t = \frac{V(C_i - C_t)}{m} \tag{8.1}$$

$$\%Removal = 100\frac{C_i - C_f}{C_i} \tag{8.2}$$

where q_t is the biosorption capacity usually expressed in mg of metal per g of biosorbent, subscript t indicates adsorption capacity at a given time (t). C_i, C_t, and C_f indicate the initial concentration, the concentration at given time, and at equilibrium time (ppm), respectively. V is the volume of metal ion solution (L), and m shows the biosorbent mass (g).

4 Marine Algae

Macro and microalgae are a diverse group of photosynthetic and aquatic living organisms that are lacking advanced structures in their cell alignment and morphology [147]. They range from unicellular to multicellular. Algae can grow both in fresh and marine water. As well as, and compare to agricultural plants, they require only less space [147]. In addition, they can be harvested throughout the year compared with other crops that are usually harvested once or twice a year [107].

Algae are able to produce oxygen, consume carbon dioxide, act as the base of the aquatic food chain, remove nutrients and pollutants from water, and to stabilize sediments [89]. Moreover, these photoautotrophic prokaryotic and eukaryotic organisms are capable to assimilate nitrogen and phosphorus from the medium in the biomass during their growth, and the biomass generated can then be converted into various bioproducts following the apposite process [95, 96].

Algae are classified into two main types depending on their size: (i) microalgae, which are a diverse group of photosynthetic microscopic organisms. Chemical energy is produced by microalgae by converting the solar energy based on photosynthesis like terrestrial plants. These microorganisms are suitable candidates to produce oil. There are several methods to retrieve this oil, but the choice of which one will be used is madding according to the properties of the algae [21], (ii) macroalgae, which concern us much in this chapter, also known as seaweeds, as their name indicates macroalgae are a group of aquatic organisms that are visible to the naked eye. It should be noted that there is another type of classification, which suggests that algae can be classified into three categories according to their growth habits, namely, microscopic algae, filamentous mat-forming algae, and the Chara/Nitella group, and each of the previously mentioned groups causes its own unique problems to water system [89].

5 Macroalgae

5.1 Classification and Use of Macroalgae

Based on their pigmentation, botanists divided marine macroalgae into three main categories, which include (i) red algae (rhodophyceae), which include 390 genera with 1810 intraspecific species or taxa, (ii) brown algae (phaeophyceae), that include 96 genera with 596 infraspecific species or taxa, and finally, (iii) green algae (chlorophyceae), which comprises 77 genera with 585 infraspecific species or taxa [141]. The pigment responsible for the red color of rhodophyta is phycobilins, and the brown color of phaeophyta comes from fucoxanthin, while; many pigments (e.g., chlorophyll a and b, carotenes and xanthophylls) are behind the green color of chlorophyta, the main characteristics of marine macroalgae are summarized in Table 8.3 [114]. Moreover, the chemical structures of some chlorophyll behind the color of these species are shown in Fig. 8.1. Seaweeds are often used as a source of food and as a promising sustainable alternative to conventional terrestrial animal feed resources, for their mineral content or the functional properties of their polysaccharides. The high content of some components in the macroalgae, such as alginate and carrageenan, hinders the use of these seaweeds in the feed of monogastric animals, because the passage of these polysaccharides in their digestive tract is often without any digestion [114].

Marine macroalgae are well recognized in Asia, in Japan for example, seaweed is an important part of many daily meals, they represent up to 10% of the total nutritional intake of some cities as Kombu, Nori, Wakame, and Hijiki [115]. Moreover, the current human consumption in terms of green, brown and red algae is 5%, 66.5%, and 33%, respectively, this is high in Asia mainly in Japan, China, and Korea [85].

According to FAO estimates of the year 2014, 38% of the 23.8 million algae recorded in the 2012 world harvest were consumed by people in recognized forms as seaweeds, such as kelp, nori, or lava not counting the consumption of agar, alginates, and carrageenans extracted from these algae and used as thickening food additives [161]. This marine biomass is not limited to human consumption and feed production,

8.3 Main characteristics of macroalgae

Common name	Divisions	Pigments	Type of cell wall	Main components of cell wall
Brown algae	Phaephycophyta	Chl a, Chl c, Phycoxanthin	Double	Cellulose, alginic acid
Red algae	Rhodopheacophyta	Chl a, Chl d, Beta-carotene, Zeaxanthin	Double	Cellulose
Green algae	Chloropheacophyta	Chl a, Chl b, Beta-carotene	Single	Cellulose

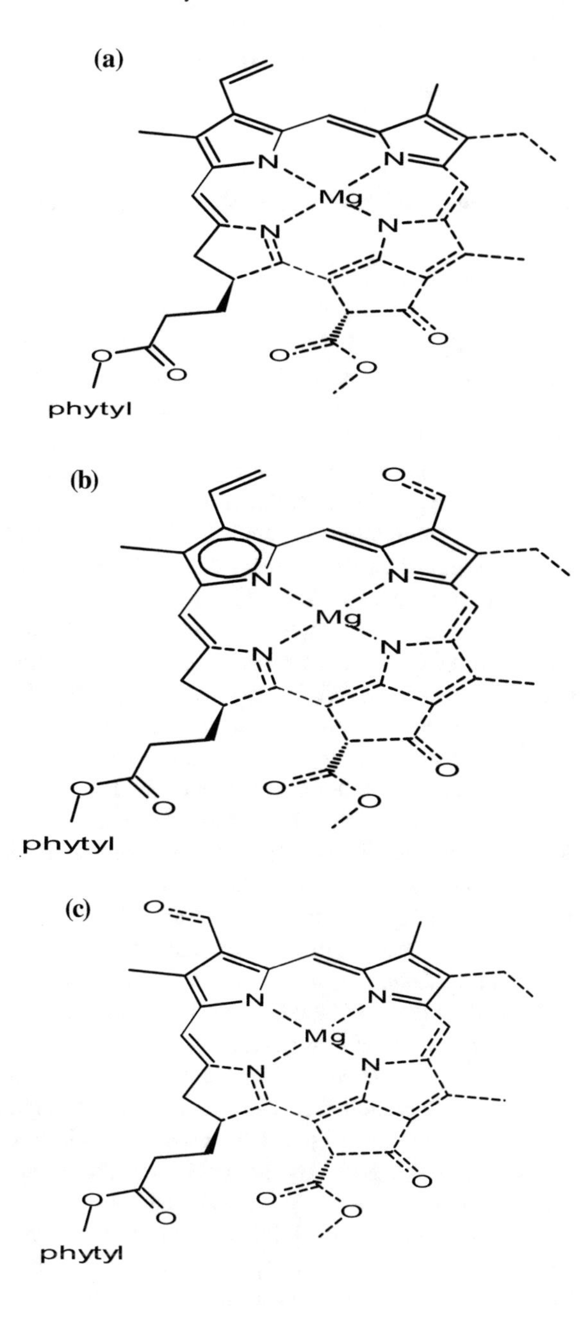

Fig. 8.1 Structures of chlorophylls: **a** Chlorophyll a, **b** Chlorophyll b, **c** Chlorophyll d

but it can also be exploited as water purifier because it recycles polluted water from fish waste in aquaculture [85].

5.2 Chemical and Structural Composition of Macroalgae

Algae are considered a promising material to be used as a biosorbent to clean wastewater from inorganic and organic pollutants, such as heavy metals and industrial dyes, this is mainly due to their rich biochemical composition [108]. Besides the pigmentation, they present various chemical and structural compositions [39]. This composition is generally constituted of three main components that are proteins, lipids, and carbohydrates [148]. Some of these constituents especially carbohydrates, proteins, and polysaccharides are commonly used to identify the type of macroalgae.

Lipids, the composition of macroalgae is characterized by different classes of fats, such as polyunsaturated fatty acids (PUFAs), triglycerides (TAGs), unsaturated acids and sterols, and the content of each class is depending on the season in which the algae were harvested. Nelson et al. [111] investigated three macroalgae species, namely, *Egregia menziesii* (phaeophyta), *Chondracanthus canaliculatus* (rhodophyta), and *Ulva lobata* (chlorophyta) to examine their total lipids content as well as their lipid classes in terms of diacylglyceryl ethers, wax esters, free fatty acid, sterols and polar lipid, this content was monitored for 4 months, namely December, March, July, and October 1997–98 that represent winter, spring, summer, and autumn, respectively. It was found that the dominant class of lipids in all algal samples was polar lipids, this result normally means that most lipids are structurally bound in membranes. Regarding the total lipid content, *U. lobata* showed the highest lipid content recorded in the spring (29 mg/g), while *C. canaliculatus* has been characterized with a low lipid content ranging from 1, 7 to 3.1 dry mass (Table 8.4).

In the same context, three macroalgae were investigated, namely, the chlorophyta *Ulva lactuca Linnaeus*, *Jania rubens*, and the rhodophyta *Pterocladia capillacea* to evaluate their major components content including lipid, all samples were harvested in April, August, and October 2010, corresponding to spring, summer, and autumn, respectively [75]. *Ulva lactuca* has shown to contain more lipids (4.09 $\pm$ 0.2%) than *Jania rubens* and *Pterocladia capillacea.* The green algae *Ulva lactuca* seems to be among the most lipid-rich macroalgae, with a content of 3–4% of the algal dry weight (Table 8.4). The lipid class that makes seaweeds more appropriate for biodiesel applications is TAGs [149]. The red alga *Gracilaria verrucosa* was studied in order to evaluate its total lipid content and the proportion of lipid fractions during the different stages of development. Male gametophytes were shown to have the highest amount of TAGs that are contained in the lowest total lipid content measured compared with all stages of development studied [77].

Proteins, regarding the protein content of marine algae, vary depending on the species type. In general, algae show such a marked variation in their constituents due to the effect of certain factors, such as the season of the year, the habitat, and the depth in which they grow [22]. Protein-rich algae are used in industry as food

Table 8.4 Chemical composition of macroalgae (% dry weight)

Macroalgae	Lipids	Proteins	Carbohydrates	References
U. lactuca	3–4	17–20	42–46	Khairy and El-Shafay [75]
J. rubens	1–2	10–13	34–42	Khairy and El-Shafay [75]
P. capillacea	2–3	17–24	48–51	Khairy and El-Shafay [75]
E. intestinalis	0.22	9.67	44.71	Pramanick et al. [122]
U. lactuca	0.38	8.77	37.87	Pramanick et al. [122]
C. repens	0.16	7.1	31.45	Pramanick et al. [122]
G. cervicornis	0.43	22.96	63.12	Marinho-Soriano et al. [101]
S. vulgare	0.45	15.76	67.80	Marinho-Soriano et al. [101]
U. lactuca	1.64	7.06	14.60	Wong and Cheung [162]
H. japonica	1.42	19.00	4.28	Wong and Cheung [162]
H. charoides	1.48	18.4	7.02	Wong and Cheung [162]
U. armoricana	2.6	–	–	Kendel et al. [74]
S. chordalis	3.0	–	–	Kendel et al. [74]

additives, as well as a high protein content has been reported in some red algae [43]. The biochemical composition of some macroalgae showed that the red alga *Gracilaria Cervicornis* has the highest protein content 22.96% compared with other species harvested during the same seasons of the year (Table 8.4). Even if, some published data gave a value of 32% for the protein content of the green alga *Ulva lactuca*, but this was judged to be a high seasonal value [43]. It was reported that the red macroalga *Porphyra yezoensis* can have up to 47% protein expressed on a dry weight basis [50]. Moreover, this content was found to be higher even than that of legumes recognized for their high-protein content such as soybeans [43].

The extraction of proteins from seaweeds is difficult due to the complex polysaccharide of the cell wall and extracellular matrix, which is somewhat species dependent [114]. The protein fraction containing in the macroalgae may be calculated based on the determination of elemental nitrogen using the nitrogen–protein conversion factor of 6.25 according to AOAC method [91].

Carbohydrates and Polysaccharides, each algal division has its typical carbohydrates. For example, the carbohydrates in brown algae species consist mainly of alginates, laminaran (β-1.3-glucan), cellulose, fucoidan, and mannitol [82]. The cell wall of marine algae and other components of the cell matrix consist mainly of structural polysaccharides existing in the form of heteropolysaccharide complex [116]. A large amount of these polysaccharides is in their sulfated form, which includes the three main phycocolloids, namely brown algae alginates, red algae carrageenan, and agar, and these biopolymers are in high demand in the hydrocolloid industry [116]. These polysaccharides have a significant importance both technologically and economically. Furthermore, Agar, alginate, and carrageenan are the three high-value algal hydrocolloids, they are used as gelling and thickening agents in different food, pharmaceutical, and biotechnology applications [129].

Pramanick et al. [122] investigated three macroalgae, namely, *Enteromorpha intestinalis*, *Ulva Lactuca,* and *Catenella Repens* to compare their biochemical components. Their proximate composition was found to follow the order carbohydrate > protein > fat > astaxanthin (Table 8.4). The estimation of carbohydrate fraction is achieved by the phenol–sulfuric acid method developed by Dubois et al. [35].

Earlier studies have indicated the effect of some environmental factors, namely, light, temperature, salinity, and nitrogen availability on the growth and biochemical compounds of marine macroalgae [36, 63, 76, 117, 128, 152]. Similarly, some researches have displayed that the growth conditions, such as phosphate limitation, nitrogen deprivation, and high salinity would affect more specifically lipid content [126, 127, 132]. For example, the protein and carbohydrate contents of *Ulva lactuca* are strongly affected by temperature and incubation time with an optimum at 30 °C for 24 h of incubation [120]. The nutritional content of marine algae is also strongly dependent on geographic locations and seasonal variations. Table 8.4 presents the average seasonal composition of some macroalgae species reported in the literature.

6 Macroalgae Potential and Applications

6.1 Pharmacological Potential

The search for metabolites with pharmacological potential in different divisions of macroalgae has developed considerably and has become one of the researcher's concerns worldwide. In addition, a continuous and great effort is being made by academic and corporate institutions to identify biological activities in extracts of this biomass with the desired potential [7, 16, 47] (Al-Malki 2020). The interest in seaweeds may be explained by different advantages that they present compared with other biomasses. The present section brings into light the macroalagae extracts as a source of bioproducts exhibiting biological activities, such as antibacterial, antitumoral, and antileishmanial activity for possible pharmacological applications.

Ainane [8] investigated two marine algae namely, *Cystoseira tamariscifolia* and *Bifurcaria bifurcata,* to isolate extracts in order to evaluate their pharmacological potential according to a range of biological activities, namely, antibacterial, antileishmanian, antitumoral, and cytotoxicity activity. The technique used during the extraction process is the Soxhlet, it allows a continuous solid–liquid extraction using cycles of vaporization—condensation of the solvent. Four organic extracts were obtained from successive Soxhlet extraction with solvents of increased polarity, namely, hexane, ether, chloroform, and methanol. Each obtained extract was tested with the following activities; antibacterial activity tested using four bacterial strains, *Escherichia coli, Enterobacter cloacae, Klebsiella pneumoni,* and *Staphylococcus aureus,* while the antitumor activity tested based on the interaction with the DNA of the calf thymus, the antileishmanial activity tested toward *Leishmaniain fantum,* and finally the cytotoxicity activity tested toward the larvae of a saltwater shrimp: Artemia Salina. All the extracts from *C. tamariscifolia* and *B. bifurcata* showed an interesting inhibition zone against the three strains *E. cloacae*, *K. pneumoni*, and *E. colis*, except the hexanic extracts, which do not show remarkable activity against the latter strain.

Salari et al. [133] examined the antibacterial activity of nanoparticles synthesized through the bio-reduction of silver ions into the desired silver nanoparticles (SNPs) using the green alga Spirogyra varians. The bacterial strains used to evaluate the SNPs activity are *Bacillus cereus, Staphylococcus aureus, Escherichia coli, Listeria monocytogenes, Pseudomonas aeruginosa,* and *Klebsiella*. The reported results have shown a remarkable antibacterial effect against *Klebsilla, P. aeruginosa,* and *B. cereus*, and this activity was greater even than that of standard antibiotics.

The inhibition concentrations relative to the different extracts of *Bifurcaria bifurcata* where 50% of cells are inhibited (IC50) have a significant activity compared with those of *Cystoseira tamariscifolia*, as well as IC50 of the hexanic extracts of *Bifurcaria bifurcata* are much better than other extracts of the same species with a remarkable value of 46.83 ppm [8].

Freile-Pelegrin et al. [47] used 27 species of macroalgae that belong to the three algal categories, namely, red, brown, and green algae to test their antileishmanial activity in vitro against Leishmania mexicana (L. Mexicana). The extracts from *Dictyota caribaea, Turbinariat urbinata,* and *Lobophora variegata* (phaeophyta) and from *Laurencia microcladia* (rhodoyta) showed a promising activity against L. Mexicana with a lethal concentration of 50% (LC50) values ranging from 10.9 to 49.9 ppm.

Whereas, some researchers were unable to demonstrate antitumoral activity in some of the macroalgae extracts [8], a group of researchers have proved that this biomass is a source of various bioproducts with biomedical properties that are capable to treat various types of diseases, such as cancer and inflammatory bowel diseases [135]. The extracts of *Sphaerococcus corono pifolius*, namely n hexane, dichloromethane, and acetone/methanol, exhibit the inhibition effect of diseases, namely, the proliferation of cancer cells of the cervix, breast cancer cells, and pancreatic cancer cells by inducing apoptosis [135]. Moreover, ethyl acetate fraction of *Tubunaria conoides* was tested and showed high anticancer activity through exhibiting synergistic effects over the respective standard compounds [119].

Abourriche et al. [4] proved that methylene chloride (CH_2Cl_2) fraction of *Cystoseira tamariscifolia* has a cytotoxicity activity against *Artemia salina*, with a lethal dose 50 (LD50) of 41.7 ppm. Furthermore, the extracts of *Bifurcaria bifurcata* were found to be toxic more than those of *Cystoseira tamariscifolia* toward the larvae of Artemiasalina [8]. To know the cytotoxic potential of macroalgae, the results of their LD50 must be compared with those of the active products used as a reference, such as digitalin 151 ppm, podophyllotoxin 2.4 ppm, and bebeerine chloride 22.5 ppm [103].

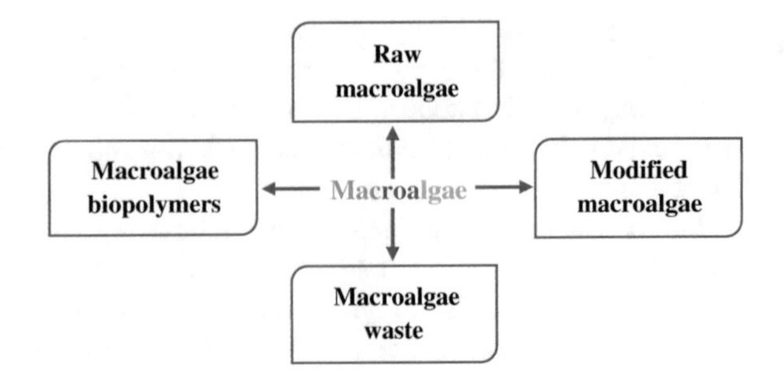

Fig. 8.2 Possible forms of dried macroalgae for biosorbent preparation

6.2 Macroalgae Biosorbents for Heavy Metal Removal

Numerous studies have been conducted looking at the use of marine macroalgae as low-cost biosorbent materials for the removal of heavy metals from wastewater [8, 30, 33, 99, 133, 164]. While, some researchers have focused on the use of another type of biomasses such as plants to remove these pollutants from aqueous solutions. In this context, Chiban et al. [28] investigated the Moroccan plant *Withania frutescens* to remove arsenic (V) from aqueous solution. The authors have shown the effectiveness of using this plant as biosorbent with a removal rate of arsenic reaching 73%, obtained by controlling some physicochemical parameters. In the present section, the possible biosorbents obtained from marine macroalgae are reviewed considering their application in the removal of heavy metal ions. According to our humble summary of the data reviewed in this study, the various potential sources of sorbents prepared from dried biomass can be grouped into four main categories as shown in Fig. 8.2.

6.2.1 Raw Macroalgae

Apart from marine algae, there are many other forms of raw biomass, which are currently used either to enhance the removal rate of adsorbents or as a source to prepare adsorbent materials for further use in removing organic and inorganic pollutants from aqueous solution, such as clays, plants, fungi, bacteria, and yeasts [15, 23, 110, 123, 165]. The complexity of the raw biomass structure indicates that there are many ways involved in the removal of pollutants, but these remain poorly understood [2].

Biosorbents can be derived from dried raw macroalgae without undergoing any significant pretreatment that is normally applied to increase their adsorption capacity. In this case, macroalgae are exposed only to a preliminary preparation during which undesirable objects are removed from the recovered samples after their collection. This preparation step remained similar in all data covered by this study with slight differences. The use of dried algae in their raw form seems to be advantageous as they are a natural and renewable material, and if they are directly applied in the biosorption process as a biosorbent, which indicates no costs are involved other than transport and simple approaches for their preparation [136]. It should also be noted that even after a simple preparation, such as washing and drying, these adsorbents have important chemical properties that contribute to the removal of many pollutants from the aqueous solution. Usually, during the preparation step, samples go through four main steps: washing with distilled water in order to remove any adhering debris, drying in the sun or in an oven, grinding/crushing, sieving to select particles of the desired size, and finally the obtained samples are kept for later use. The exploitation of algal biomass in its raw form for biosorbent preparation has proven its effectiveness in the removal of heavy metals, and this can be seen through the various publications regarding the potential of this non-activated biomass. Kumar et al. [86] studied the adsorption potential of adsorbent prepared from *Ulva fasciata* without prior pretreatment for the retention of Zn ions. This alga has demonstrated a good zinc biosorption capacity with a maximum adsorption of 13.5 mg/g, which highlights the potential of this alga in the treatment of wastewater.

6.2.2 Modified Macroalgae

Numerous attempts have been carried out regarding the use of efficient techniques to give new functionalities to bioserbent's surface, as well as to understand the fundamental aspects of the biosorption mechanism [5, 34, 98]. Research and development of new biosorption materials involve the use of seaweeds, especially those popularly named brown algae, because of their high sorption capacity, which is similar to that of commercial ion exchange resins, and their availability in almost unlimited quantities in the ocean [98]. This biomass can be exploited in its original form with a sample preliminary preparation (washing and drying) as mentioned above, as it can be used after passing through important chemical and sometimes physical treatments to increase its adsorption capacity toward metal ions dissolved in an aqueous solution. On the whole, surface modification methods can be categorized into two main types: chemical and physical modification.

(a) Chemical Modifications

Marine algae are chemically treated to enhance their sorption ability and reinforce their applications. The biosorbent prepared from this marine biomass can be modified by various functional groups depending on the nature of the metal ion to adsorb. These modifications can be made by a cross-linking reaction using epichlorohydrin to harden the cell wall structure [79], or by NaOH treatment to increase the negative charge on the cell surface [58], or by acid to make the actives sites open for adsorption [163], or by Ca^{2+} solution treatment in order to enhance the ion exchange [83]. One of the examples of algal biomass chemically modified used in the biosorption process, is the one obtained after the crosslinking with epichlorohydrin and oxidizing by potassium permanganate. In this approach, *Laminaria japonica* has undergone a chemical modification by crosslinking treatment (Fig. 8.3), and before processing to this step, the biomass selected for lead sorption uptake experiments was first: dried in an oven at 40 °C until constant weight, ground and then sieved into a fraction 0.30–0.4 mm. 10 g of the prepared biomass was made in contact with dimethyl sulphoxide

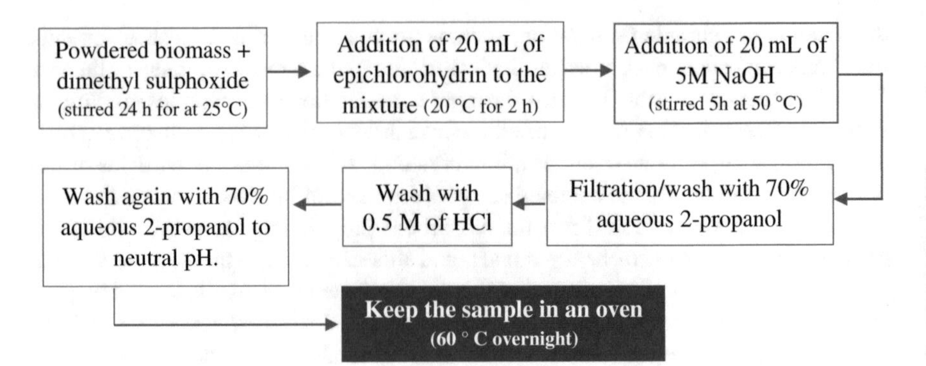

Fig. 8.3 Crosslinking with epichlorydrin of *L. japonica* biomass

(DMSO) to expose the hidden metal-binding groups before crosslinking. Then, the diluted 2-propanol was added to remove DMSO and epichlorohydrin excess [98].

Macroalgae can be crosslinked using other reagents as well. Crosslinking of the raw brown alga *Ascophyllum nodosum* with bis(etheny1) sulfone reagent proved to increase the sorption capacity of the biosorbent material due to the sulfone group's incorporation. In addition, this treatment has been shown to improve algae's physical properties, such as strength, hardness, and swelling characteristics without adversely affecting the sorption [64].

Deniz and Karabulut [34] used various macroalgae, *Polysiphonia sp* (red algae), *Cystoseira sp* (brown algae), and *Chaetomorpha sp*, *Ulva sp* (green algae) and *Cystoseira sp* (brown algae) to prepare a unique adsorbent, the biomass after having undergone a preparation step (washing, drying, sieving), it was treated with a NaOH solution before testing its behavior regarding the elimination of zinc. The adsorbent prepared from various types of macroalgae was judged to have a great potential for the uptake of zinc ions from aqueous solution. The maximum adsorption capacity of zinc was found to be 115.198 mg/g and obtained under some optimal physicochemical parameters (Table 8.5).

Among the interest of researchers is to compare the biosorption behavior of macroalgae and understand the main causes behind their differences in regard to the elimination of heavy metal ions. In this approach, Hamdy [60] used three brown macroalgae *Turbinaria decurrens*, *Cystoseira trinode,* and *Sargassum asperifolium*, and one red alga *Laurencia obtusa* to test their adsorption ability towards Cr^{3+}, Co^{2+}, Ni^{2+}, Cu^{2+}, and Cd^{2+}. The prepared algal samples were then treated with chloridric acid until evolution of CO_2 to remove calcium carbonates present in the algal cell matrix before being used in the adsorption study. Compared with the good adsorption capacities of the algae tested without HCl treatment for the elimination of different metal ions, the treated algae showed a low capacity, especially for Cr^{3+} adsorption with *Laurencia obtusa*, where there was practically no adsorption and that was explained to the loss of $CaCo_3$ and the pH of the solution.

8.5 Maximum adsorption capacities (mg/g) for the biosorption of metal ions on macroalgae

Macroalga	Biosorbent	Ion	pH	T (°C)	q_{max} (mg/g)	References
Callithamnion corymbosum sp	Raw alga	Cu(II)	4.4	25	47.6	Lucaci et al. [97]
	Iron nanoparticles /Alginate	Cu(II)			52.6	
	Waste alga	Cu(II)			83.3	
	Alginate	Cu(II)			166	
Undaria pinnatifida	Alga physically modified	Cu(II)	4.5–5.5	20	126	Cho et al. [29]
		Cu(II)			14	
Cystfoseira barbata	Raw alga	Cu(II)	–	25	279	Trica et al. [154]
	Raw alga	Pb(II)			69.3	
	Alginate	Cu(II)			454	
	Alginate	Pb(II)			107	
Hydroclathrus clathratus	Waste	Cd(II)	6.2	25	96.5	Soliman et al. [144]
		Cu(II)			43.4	
Jania rubens	Raw alga	Co(II)	5	25	32.6	Ibrahim [68]
		Pb(II)			30.6	
		Cr(III)			28.5	
		Cd(II)			30.5	
Pterocladia capillacea	Raw alga	Co(II)	–	–	52.6	Ibrahim [68]
		Pb(II)			34.1	
Corallina mediterranea	Raw alga	Co(II)	–	–	76.2	Ibrahim [68]
		Pb(II)			70.3	
		Cr(III)			64.3	
		Cd(II)			64.1	
Galaxaura oblongata	Raw alga	Co(II)	–	–	74.2	Ibrahim [68]
		Pb(II)			105	
		Cr(III)			88.6	
		Cd(II)			85.5	
Schizomeris leibleinii	Raw alga	Cu(II)	6	25	55	Tavana et al. [150]
Undaria pinnatifida	Alga chemically modified	Pb(II)	5.5	25	980	Kim et al. [79]
Syzygium cumini L	Raw alga	Zn(II)	6	-	35.8	King et al. [80]
Ulva fasciata sp	Raw alga	Zn(II)	30	5	13.5	Kumar et al. [86]

(b) Physical Modifications

Chemical modification is not the only option to improve the properties of biosorbents, physical modification has also been used to some extent. One of the main examples of this type of modification is physical activation. This treatment is based on the elimination of a large amount of internal carbon mass, which is a necessary step to obtain a well-developed carbon structure. While, in chemical activation, all the chemicals employed are dehydrating agents that affect pyrolytic decomposition and prevent tar formation, thus, enhancing the yield of carbon [13]. The production of an activated carbon by the physical activation includes a high temperature around 1100–1250 K, this processing can be carried out on the basis of steam, carbon dioxide, and air, or a combination of these [131]. The physical activation takes place in two main steps: the first is a pyrolysis step of the carbon product with a temperature ranges from 300 °C to 500 °C. At this first stage, the resulting char has a porosity not well developed. The second step is to attack the char with steam or carbon dioxide with a temperature of 900 °C–1000 °C. This step allows the porosity to develop completely [6]. This kind of modification has been the subject of some published studies and this is perhaps due to its remarkable influence on the adsorption capacity of bioadsorbents prepared from macroalgae and other biomasses regarding inorganic and even organic pollutants [13, 29, 65, 78].

Cho et al. [29] evaluated the adsorption characteristics of charcoal derived from the brown macroalga *Undaria pinnatifida* in the removal of copper from aqueous solution. The charcoal was physically and chemically modified to compare the effect of the chosen modification method on the capacity of the prepared biosorbent to remove copper ions. This study demonstrated the effectiveness of using the physical activated char prepared from seaweeds (Fig. 8.4) in the adsorption of Cu^{2+}, which is higher even than that activated as a result of the chemical process (activation by KOH). The physically activated char showed a great adsorption capacity toward Cu^{2+}. As well as, the authors have explained the low adsorption capacity of char prepared through the chemical activation by the low amount of exchangeable cations, and this was due to the use of an alkaline solution, which decreased the adsorption capacity. The activation by KOH contributes to the removal of the exchangeable alkali or alkali earth metals. The authors noted that the ion exchange might be the main mechanism

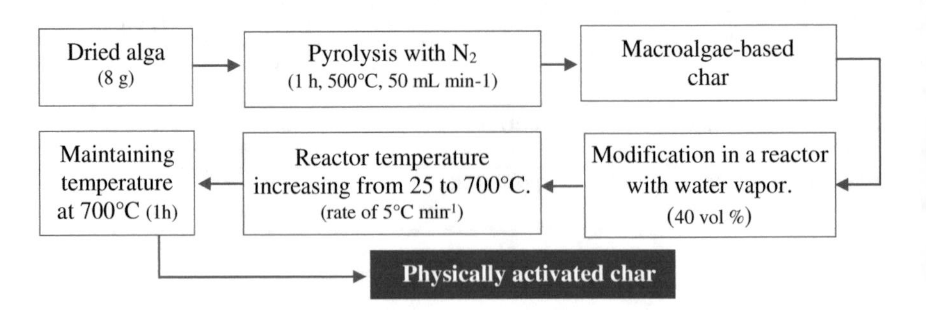

Fig. 8.4 Physical activation of char is derived from *U. pinnatifida* biomass

involved in the removal of copper ions. Also, it was suggested that the heavy metal ion removal procedure using biochar takes place in two steps, the first being the adsorption that takes place in the porous structure of this biosorbent. While, in the second step the ion exchange occurs [142].

6.2.3 Macroalgae Biopolymers

Agar, carrageenan, and alginates are seaweeds hydrocolloids and the most recognized constituents of the algae cell wall. The term "hydrocolloid" refers to any substance allowing the formation of colloidal systems in the form of a gel or sol system of solubilized particles when it is in contact with water [129]. This section focuses mainly on the alginates and carrageenans, which are biopolymers extracted from seaweeds very suitable for the removal of heavy metal pollution. Carrageenans are isolated from the cell wall of red algae, while alginates are obtained from those of brown algae.

(a) **Alginic acid and Alginates**

Alginic acids are ionic polysaccharides and are abundant in the brown algae cell walls [41]. These polysaccharides belong to a family of copolymers of b-D-mannuronic acid (M) and a-L-guluronic acid (G) [87]. The content of these polysaccharides in the biomass can reach 40% of the dry weight; this strongly depends on two main factors, namely the species and growth conditions.

Alginate is a salt of alginic acid that refers to the common name given to the family of linear polysaccharides containing 1,4-linked b-d-mannuronic (M) and a-l- guluronic (G) [32, 82]. Apart from seaweeds, some microorganisms are able to produce alginates as well, such as *Azotobacter vinelandii* and some strains of *Pseudomonas* [57]. The extracts of brown seaweeds containing alginates were not sold as thinking and gelling agents until 1930s [82]. The total world production of this polymer was estimated to be around 30 000 million/year [57]. The difference between alginic acid and alginate can be observed in their chemical structures (Fig. 8.5).

The functional groups that are most abundant in brown algae are carboxylic groups present in alginate polysaccharides, they are responsible for the ion exchange capacity and that of adsorption [136]. These hydrocolloids that are currently used in various domains, have a high degree of physicochemical heterogeneity, which affects their quality and determines their possible applicability.

Sodium and calcium alginate are the main known forms of alginic salts extracted from macroalgae, and the difference between these salts depends mainly on the nature of the reagent used during the extraction step. The adapted method to extract sodium alginate by several researchers is that developed by Calumpong et al. [24] with some slight modifications [42, 158]. In accordance with this procedure, 25 g of the dried algae is soaked in 800 mL of 2% (v/v) formaldehyde under steering at room temperature for 1 day to remove phenolic compounds and pigments. Later, the sample is thoroughly rinsed with water, then added to 0.2 M hydrochloric acid

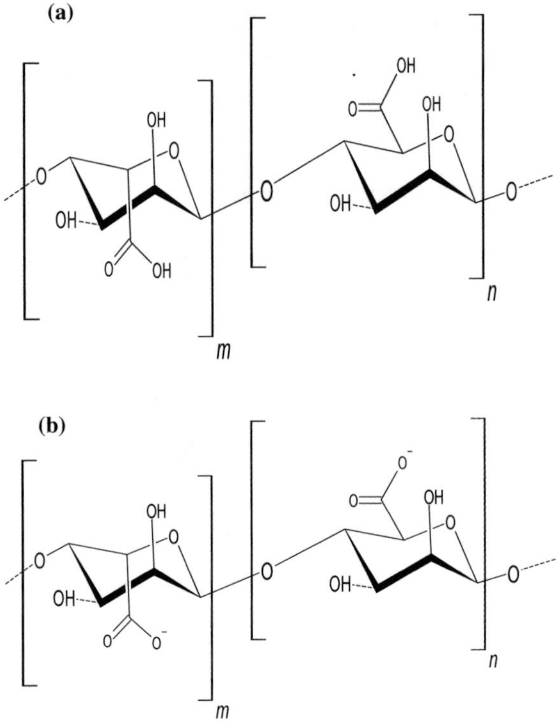

Fig. 8.5 **a** Chemical structure of alginic acid, **b** Chemical structure of sodium alginate

(800 mL) and soaked overnight in HCL. Then, washed with pure water and start the extraction step with the addition of 2% Na_2Co_3 during 3 h at 100 °C. After this time, the mixture is centrifuged (10,000 × g, 30 min) to separate the soluble fraction from the obtained mixture, while the polysaccharides are precipitated by 3 volumes of ethanol 95% (v/3v). Then, the collected sodium alginate is washed twice by 100 mL of acetone, dried at 65 °C, dissolved again in 100 mL of pure water, and before drying the alginate at 65 °C, it was first reprecipitated with ethanol (v/3v). The different steps of sodium alginates extraction are listed in Fig. 1.6.

The difference between the alginates obtained from different sources is in their guluronic (G) and mannuronic (M) residues and the length of each block [88]. These extracts can exist in different forms, which do not have the same properties. Sodium alginate is soluble in water, but calcium alginate can only be soluble if it is dissolved in an alkaline medium because of their physico-chemical and rheological properties. This polymer has a selective affinity toward the cations according to their nature. This affinity depends on some factors including the composition of alginate and electrostatics forces [3]. Physicochemical and rheological properties of the polymer are influenced by the composition in terms of G and M. In addition, the M/G ratio, which ranges between 0.25 and 2.25, allows to appreciate the quality of the alginate. This ratio varies depending on two important factors: the season and the type of the brown alga specie [104]. For instance, the alginate M/G of *Laminaria digitata*,

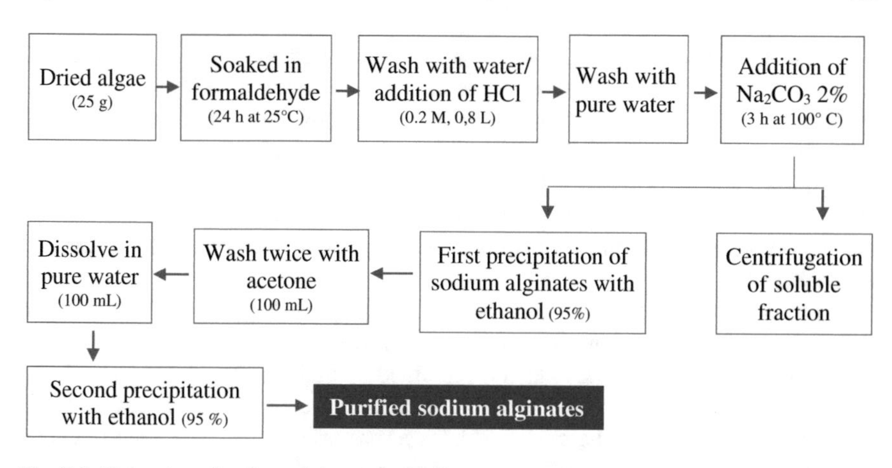

Fig. 8.6 Extraction of sodium alginates from a brown macroalga

Laminaria longicruris, Ascophyllum Nodosum, Macrocystis Pyrifera, Fucus Serratus is 1,44; 2,03; 1,77 1,56; and 1,06, respectively [130]. M/G is calculated using the below equation [55]. Alginates are widely used as additives and ingredients in the food industry [52]. Furthermore, the appreciated extract alginate in food applications is that rich in G acid thanks to its ability to form a gel.

$$\frac{M}{G} = \frac{1 - F_G}{F_G}. \tag{8.3}$$

where M: mannuronic acid, G: guluronic, and F_G represent the mole fraction of G.

After the extraction, the characterization step of the isolated alginate is essential. In sum, the techniques employed for this purpose are: (i) FTIR spectroscopic analysis that is used to show characteristic peaks corresponding to the various groups present in the alginates extracted from seaweeds. It was recommended to quasi-quantitatively determine the M/G value in alginates by measuring the ratio of the absorption band intensities at 808 (M) and 787 cm^{-1} (G) in the infrared spectra [156]. (ii) NMR spectroscopy; this technique of characterization is considered as the most reliable method for the determination of the structure and composition of alginates [20]. The information acquired by this analysis in terms of the uronic acid composition of the alginate allow the calculation of the ratio Mannuronic acid/Guluronic acid (M/G) [55] as well as the distribution of M and G units throughout the polymer chain [146]. (iii) Reological analysis: this technique allows to measure the viscosity of alginate extract using a rheometer with a Peltier temperature control system at 25 °C [40].

Classical methods of alginate extraction seem to present certain disadvantages as they can alter the chemical structure of the biopolymer [164]. Therefore, this situation has pushed some researchers to look for other effective techniques, which do not affect the physicochemical properties, especially if these alginates extracts are made for future pharmaceutical or medical application. Ultrasound-assisted extraction is a process that does not affect the chemical structure of the biopolymer and its molar mass distribution, and it also allows to reduce the extraction time [164].

Apart from its use in the removal of heavy metals, sodium alginate obtained from brown seaweeds exhibits various biological properties, such as antitumour and anti-inflammatory activity [119]. As well, alginate was described as a smart polysaccharide. More information on alginate can be found in a good discussion on the smarter behavior of this polymer and its applications; carried out by Gupta and Raghava [57].

There is a wealth of literature on the use of alginates derived from brown algae to remove heavy metal ions from aqueous solution [32, 38, 97, 118]. The huge exploitation of this algae class may be explained by their performance toward these cations, which is generally better than it is of other seaweeds divisions because of the strong presence of carboxyl groups [136]. It was proved that these linking groups can be increased by an oxidation of marine macroalgae with potassium permanganate. Moreover, this treatment is considered to be one of the main means of increasing the number of carboxylic groups of alginic acids. Consequently, the adsorption behavior of this biomass increases [98]. The evaluation of this behavior in the raw brown alga *Cystoseira barbata* and in its alginate extract regarding the removal of Cd(II) and Pb(II) was the aim of a study conducted by [154]. According to this research, *C. barbata alginate showed* a good performance, and the ratio M/G was found to be 0.64, this normally indicates the dominance of block G over M. Regarding the adsorption capacity, the alginate beads prepared from *C. barbata* showed a maximum adsorption capacity of 454 mg/g for Pb^{2+} and 107 mg/g for Cu^{2+}. These capacities were found to be higher even than those obtained when *C. barbata* was used as adsorbent at its raw form. The obtained adsorption capacities of raw alga toward Cu^{2+} and Pb^{2+} are 279.2 mg/g and 69.3 mg/g, respectively.

Sodium alginates as adsorbents are usually undergoing an essential chemical modification during their preparation, due to the high solubility and weak chemical resistance. This treatment consists of surface grafting and crosslinking [51]. As well as, it was reported that the alginate extracted from *Laminaria digitata* in the form of calcium alginate beads has a high copper and cadmium uptake capacity compared with other bioadsorbents. This is usually due to its high M/G ratio [118]. One of the solutions to overcome the difficulties of adsorbent materials based on sodium alginates, such as their hard structure, limited solubility and high viscosity of their solution, is to compositing them with those with suitable low viscosity and rotation of chains such as poly(vinyl alcohol). Ebrahimi et al. [38] prepared poly(vinyl alcohol) and sodium alginate composite nanofibers (PVA/SA) through the electro spinning method to remove cadmium ions from aqueuse solution. The maximum adsorption capacity related to the use of this adsorbent is 93.163 mg/g, which was obtained with Langmuir model and under optimal experimental conditions.

Brown algae are not the only class from which alginates can be extracted, in 1982, alginates were unexpectedly detected in one of the calcareous red algae species belonging to the *Corallinaceae* family [156]. Therefore, this discovery has encouraged the researchers to focus on the red algae species as a source of extraction of these polysaccharides. In this approach, and for a better valorization of marine resources, a recent study conducted by Lucaci et al. [97] has been carried out on the extraction of alginate from red alga *C. corymbosum sp*, in order to test its potential as biosorbent toward the removal of Cu(II) ions. The maximum adsorption capacity found based on the Langmuir model is 166.66 mg/g reported with the use of *C. corymbosum sp* alginate as adsorbent. Adsorption process of Cu^{2+} was 10 times higher than when using raw algae as the adsorbent. The scanning electron microscope (SEM) characterization of the alginate showed a porous structure and opening shapes. Furthermore, FTIR spectroscopy indicated that the alginate extracted from the raw alga has a large number of functional groups on its surface compared with the all adsorbent materials investigated.

A pre-extraction step has been recommended in order to obtain a high viscosity of alginate, in which the alga is treated with 0.1% formaldehyde overnight and then washed once using hydrochloric acid at pH 4 in a batch system with continuous stirring lasting 15 min [62]. It was found that the alginate presents in algae biomass cannot be all precipitated, but there is a significant amount still after the separation and the precipitation steps due to its high solubility in water [42]. In this context, and in order to not lose soluble alginate, Lucaci et al. [97] proposed the iron nanoparticles functionalized with this polymer as adsorbent to test its behavior with regard to the elimination of copper ions contained in the aqueous solution..

(b) **Carageenans**

Carrageenans are a group of linear sulfated polysaccharides obtained by extraction from certain species of red algae [155]. Carrageenan is composed of β-1,3 D-galactose and α-1,4 D-(anhydro) galactose and contains about 24% ester sulfate. There are different types of carrageenan, which depends on the number and the position of sulfate group. Carrageenan is divided into variety of types, such as lambda, kappa, iota, theta, and mu carrageenan, and all containing 22–35% of sulfate groups. This classification has been developed on the basis of the solubility of this polymer in potassium chloride [109]. The chemical structure of some of these carrageenans, as well as that of agarose, is shown in Fig. 8.7 [18, 109].

The world production of carrageenan is estimated at 16,500 ton/year, and only commercially available forms are kappa, iota, and lambda carrageenan [130]. Kappa-carrageenan is distinguished from agarose, which is a sulfated polysaccharide of brown algae by the configuration of the a-linked galactose residues and by the presence of one sulfate substituent at C_4 on the b-linked D galactose residues (Fig. 8.7). Related to their rheological properties, agarose gives rigid and turbid gels, while κ- and ι-carrageenans form clear gels [130]. The rigidity of the gels is directly related to the molecular structure and decreases when the sulfate content increases [130]. Carrageenan is widely used as texturing and moisturizing agents in various industries and its use depends mainly on its rheological properties [18, 109].

Carrageenan can be isolated from red algae by the following two main steps; the first is the extraction step using chemicals, such as strong bases KOH and NaOH, the second is alcohol precipitation with ethanol, and then the carrageenan yield is separated from the ethanol–water mixture by a filtration membrane [155]. The precipitation can be also conducted using isopropyl alcohol [130]. Despite the emphasis on the

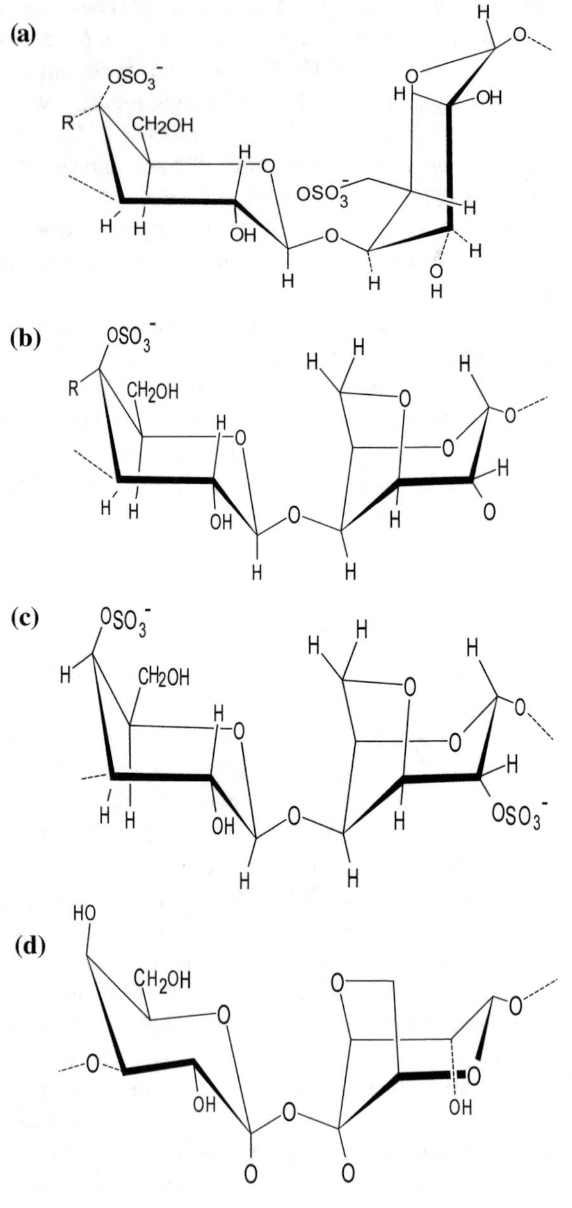

Fig. 8.7 (**a**) Chemical structure of μ-carageenan, (**b**) Chemical structure of k-carrageenan, (**c**) Chemical structure of i-carrageenan, (**d**) Chemical structure of agarose

extraction of carrageenan from algal biomass, its direct use in wastewater treatment whether laden with organic or inorganic pollution, remains poorly documented. Thus, more research is needed on the use of carrageenan prepared from red macroalgae as a natural sorbent. Contrariwise, commercial grade carrageenan is well used as materials to prepare effective adsorbents for organic and inorganic pollutants removal [1,

92, 94, 100, 105]. Biopolymers such as carrageenans have often been considered as adsorbents for removing metals from solution, which is mainly explained by their intrinsic properties in which several functional groups have a significant role [145]. Specifically, hydroxide and sulfate groups have been reported as dominant binding groups in this polymer that are involved in the adsorption process.

6.2.4 Macroalgae Waste

Several studies have focused on the removal of heavy metals using agricultural and industrial wastes, such as coal ash, rice husk, activated carbon [66, 73, 160]. In this regards, algal marine macroalgae as adsorbents have also been widely studied and appreciated. As well as, it was recommended that these biomaterials undergo modifications and be treated with specific chemicals to increase their adsorption capacity and make their surfaces more reactive [136].

The macroalgal waste from alginate extraction is a valuable biomass that can be used as an adsorbent to remove Cu^{2+} from aqueous solution. A recent study conducted by Lucaci et al. [97] proved that waste resulting from the alginate extraction process contained in the marine red alga *C. corymbosum sp* can be successfully used in the removal of these ions with a considerable maximum adsorption capacity of 83.33 mg/g. In fact, this capacity was found to be higher even than it's of the raw alga and the composite material (iron nanoparticles functionalized with alginate). In the same approach, Soliman et al. [144] studied the residue of *Hydroclathrus clathratus* after extracting most of its phytochemicals. The adsorption capacities of the prepared adsorbent from this alga were 96.46 mg/g and 43.4 mg/g for Cd^{2+} and Cu^{2+}, respectively.

7 Concluding Remarks

This chapter has highlighted the possibility of using marine macroalgae as inexpensive bio-adsorbents to reduce and remove heavy metal ions from industrial wastewater, and through our humble analysis of the data reviewed in this topic, the following remarks can be made:

- Seaweeds showed a great capacity to eliminate these inorganic pollutants from water, thus, they can be used as adsorbents.
- Alginate and carrageenan have turned out to be the main polymers responsible for the selectivity of macroalgae for these ions removal, due to their active binding sites, such as carboxylic, sulfonic, and hydroxyl groups.
- Therefore, a particular attention should be paid to the determination of the chemical and structural composition of algae, because it is a crucial step to which any

researcher has to proceed before considering applying these algae in the biosorption process or in any other field, and such information makes it possible to find out which category of algae is more suited to the removal of these ions.

- The great diversity and rich composition of this biomass may lead to the discovery of many new algal bioproducts and processes in the future with potential removal rate of metal ions.
- Also, the use of this resource in the field of biosorption has been seen as a promising solution to the high cost associated with the traditional technologies currently used to treat polluted effluents.
- Even if the number of attempts concerning the removal of these micropollutants by dead marine algae, but still poorly treated their application in the treatment of industrial wastewater.
- Despite the satisfactory results of algal biosorbents in removing heavy metals, there is still a need to develop from this biomass effective and suitable sorbents for the treatment of industrial wastewater containing various types of inorganic impurities.

References

1. Abdellatif FHH, Abdellatif MM (2020) Bio-based i-carrageenan aerogels as efficient adsorbents for heavy metal ions and acid dye from aqueous solution. Cellulose 27(1):441–453. https://doi.org/10.1007/s10570-019-02818-x
2. Abdi O, Kazemi M (2015) A review study of biosorption of heavy metals and comparison between different biosorbents. J Mater Environ Sci 6(5):1386–1399
3. Abi Nassif L (2019) Elaboration et caractérisation de biomatériaux antimicrobiens à base d'alginate pour des applications dans les domaines médicaux et marins. Doctoral dissertation, Brest, France
4. Abourriche A, Charrouf M, Berrada M, Bennamara A, Chaib N, Francisco C (1999) Antimicrobial activities and cytotoxicity of the brown alga Cystoseira tamariscifolia. Fitoterapia 70(6):611–614. https://doi.org/10.1016/s0367-326x(99)00088-x
5. Afroze S, Sen TK, Ang HM (2016) Adsorption removal of zinc (II) from aqueous phase by raw and base modified Eucalyptus sheathiana bark: kinetics, mechanism and equilibrium study. Proc Saf Environ Prot 102:336–352. https://doi.org/10.1016/j.psep.2016.04.009
6. Ahmed Hared I (2007) Optimisation d'un procédé de pyrolyse en four tournant: application à la production de charbons actifs. Doctoral dissertation
7. Ainane T, Abourriche A, Kabbaj M, Elkouali M, Bennamara A, Charrouf M, Talbi M, Lemrani M (2014) Biological activities of extracts from seaweed Cystoseira tamariscifolia: Antibacterial activity, antileishmanial activity and cytotoxicity. J Chem Pharm Res 6(4):607–611
8. Ainane T (2011) Valorisation de la biomasse algale du Maroc: Potentialités pharmacologiques et Applications environnementales, cas des algues brunes Cystoseira tamariscifolia et Bifurcaria bifurcata. Doctoral dissertation, Faculté des Sciences Ben M'sik Université Hassan II Casablanca
9. Ali Redha A (2020) Removal of heavy metals from aqueous media by biosorption. Arab J Basic Appl Sci 27(1):183–193. https://doi.org/10.1080/25765299.2020.1756177
10. Ameri A, Tamjidi S, Dehghankhalili F, Farhadi A, Saati MA (2020) Application of algae as low cost and effective bio-adsorbent for removal of heavy metals from wastewater: a review study. Environ Technol Rev 9(1):85–110

11. Andujar P, Bensefa-Colas L, Descatha A (2010) Intoxication aiguë et chronique au cadmium. Rev Med Interne 31(2):107–115. https://doi.org/10.1016/j.revmed.2009.02.029
12. Apiratikul R, Pavasant P (2008) Batch and column studies of biosorption of heavy metals by Caulerpa lentillifera. Biores Technol 99(8):2766–2777. https://doi.org/10.1016/j.biortech.2007.06.036
13. Aravindhan R, Rao JR, Nair BU (2009) Preparation and characterization of activated carbon from marine macro-algal biomass. J Hazard Mater 162(2–3):688–694. https://doi.org/10.1016/j.jhazmat.2008.05.083
14. Ayeni O (2014) Assessment of heavy metals in wastewater obtained from an industrial area in Ibadan, Nigeria. RMZ–M&G [Internet] 61:19–24
15. Aziam R, Chiban M, Eddaoudi H, Soudani A, Zerbet M, Sinan F (2017) Kinetic modeling, equilibrium isotherm and thermodynamic studies on a batch adsorption of anionic dye onto eco-friendly dried Carpobrotus edulis plant. Eur Phys J Spec Topics 226(5):977–992. https://doi.org/10.1140/epjst/e2016-60256-x
16. Badal S, Gallimore W, Huang G, Tzeng T-R, Delgoda R (2012) Cytotoxic and potent CYP1 inhibitors from the marine algae Cymopolia barbata. Org Med Chem Lett 2(1):21. https://doi.org/10.1186/2191-2858-2-21
17. Bailey SE, Olin TJ, Bricka RM, Adrian DD (1999) A review of potentially low-cost sorbents for heavy metals. Water Res 33(11):2469–2479. https://doi.org/10.1016/S0043-1354(98)00475-8
18. Barbeyron T, Michel G, Potin P, Henrissat B, Kloareg B (2000) ι-Carrageenases constitute a novel family of glycoside hydrolases, unrelated to that of κ-carrageenases. J Biol Chem 275(45):35499–35505. https://doi.org/10.1074/jbc.M003404200
19. Bayramoğlu G, Tuzun I, Celik G, Yilmaz M, Arica MY (2006) Biosorption of mercury (II), cadmium (II) and lead (II) ions from aqueous system by microalgae Chlamydomonas reinhardtii immobilized in alginate beads. Int J Miner Process 81(1):35–43. https://doi.org/10.1016/j.minpro.2006.06.002
20. Belattmania Z, Engelen AH, Pereira H, Serrão EA, Barakate M, Elatouani S, Zrid R, Bentiss F, Chahboun N, Reani A (2016) Potential uses of the brown seaweed Cystoseira humilis biomass: 2-Fatty acid composition, antioxidant and antibacterial activities. J Mater Environ Sci 7(6):2074–2081
21. Bergthorson JM, Kunst L, Levin DB, McVetty PBE, Smith DL, Vessey JK (2011) Biodiesel–an integrated approach for a highly efficient biofuel.https://doi.org/10.1016/B978-0-08-088504-9.00257-9
22. Black W (1950) The seasonal variation in weight and chemical composition of the common British Laminariaceae. J Mar Biol Assoc UK 29(1):45–72. https://doi.org/10.1017/S0025315400056186
23. Boukhemkhem A, Rida K (2017) Improvement adsorption capacity of methylene blue onto modified Tamazert kaolin. Adsorpt Sci Technol 35(9–10):753–773. https://doi.org/10.1177/0263617416684835
24. Calumpong HP, Maypa AP, Magbanua M (1999) Population and alginate yield and quality assessment of four Sargassum species in Negros Island, central Philippines. In: Sixteenth International Seaweed Symposium, vols. 398–399. Springer Netherlands, pp 211–215. https://doi.org/10.1023/A:1017015824822
25. Chaplin BP (2018) Advantages, disadvantages, and future challenges of the use of electrochemical technologies for water and wastewater treatment. In: Electrochemical water and wastewater treatment. Elsevier, pp 451–494. https://doi.org/10.1016/B978-0-12-813160-2.00017-1
26. Cheng H, Hu Y (2010) Lead (Pb) isotopic fingerprinting and its applications in lead pollution studies in China: a review. Environ Pollut 158(5):1134–1146
27. Chiban M, Carja G, Lehutu G, Sinan F (2016) Equilibrium and thermodynamic studies for the removal of As (V) ions from aqueous solution using dried plants as adsorbents. Arab J Chem 9:S988–S999. https://doi.org/10.1016/j.arabjc.2011.10.002

28. Chiban M, Lehutu G, Sinan F, Carja G (2009) Arsenate removal by Withania frutescens plant from the south–western Morocco. Environ Eng Manag J 8(6):1377–1383
29. Cho HJ, Baek K, Jeon JK, Park SH, Suh DJ, Park YK (2013) Removal characteristics of copper by marine macro-algae-derived chars. Chem Eng J 217:205–211. https://doi.org/10.1016/j.cej.2012.11.123
30. Crini G, Torri G, Lichtfouse E, Kyzas GZ, Wilson LD, Morin-Crini N (2019) Dye removal by biosorption using cross-linked chitosan-based hydrogels. Environ Chem Lett 17(4):1645–1666
31. Da'na EAM (2012) Amine-modified SBA-15 (prepared by co-condensation) for adsorption of copper from aqueous solutions. Doctoral dissertation, University of Ottawa, Canada
32. Davis TA, Ramirez M, Mucci A, Larsen B (2004) Extraction, isolation and cadmium binding of alginate from Sargassum spp. J Appl Phycol 16(4):275–284. https://doi.org/10.1023/b:japh.0000047779.31105.ec
33. Davis TA, Volesky B, Mucci A (2003) A review of the biochemistry of heavy metal biosorption by brown algae. Water Res 37(18):4311–4330. https://doi.org/10.1016/S0043-1354(03)00293-8
34. Deniz F, Karabulut A (2017) Biosorption of heavy metal ions by chemically modified biomass of coastal seaweed community: studies on phycoremediation system modeling and design. Ecol Eng 106:101–108
35. Dubois M, Gilles KA, Hamilton JK, Rebers PA, Smith F (1956) Colorimetric method for determination of sugars and related substances. Anal Chem 28(3):350–356
36. Duke ClS, Lapointe BE, Ramus J (1986) Effects of light on growth, rubpcase activity and chemical composition of ulva species (Chlorophyta) 1. J Phycol 22(3):362–370. https://doi.org/10.1111/j.1529-8817.1986.tb00037.x
37. Dumas P, Miller L (2003) The use of synchrotron infrared microspectroscopy in biological and biomedical investigations. Vib Spectrosc 32(1):3–21
38. Ebrahimi F, Sadeghizadeh A, Neysan F, Heydari M (2019) Fabrication of nanofibers using sodium alginate and Poly (Vinyl alcohol) for the removal of Cd2+ ions from aqueous solutions: adsorption mechanism, kinetics and thermodynamics. Heliyon 5(11):e02941. https://doi.org/10.1016/j.heliyon.2019.e02941
39. Fabre E, Dias M, Costa M, Henriques B, Vale C, Lopes CB, Pinheiro-Torres J, Silva CM, Pereira E (2020) Negligible effect of potentially toxic elements and rare earth elements on mercury removal from contaminated waters by green, brown and red living marine macroalgae. Sci Total Environ 724:138133. https://doi.org/10.1016/j.scitotenv.2020.138133
40. Fenoradosoa TA, Ali G, Delattre C, Laroche C, Petit E, Wadouachi A, Michaud P (2010) Extraction and characterization of an alginate from the brown seaweed Sargassum turbinarioides Grunow. J Appl Phycol 22(2):131–137
41. Fernández-Martín F, López-López I, Cofrades S, Colmenero FJ (2009) Influence of adding Sea Spaghetti seaweed and replacing the animal fat with olive oil or a konjac gel on pork meat batter gelation. Potent Protein Alginate Assoc Meat Sci 83(2):209–217
42. Fertah M, Belfkira A, Dahmane E, montassir, Taourirte, M., & Brouillette, F. (2017) Extraction and characterization of sodium alginate from Moroccan Laminaria digitata brown seaweed. Arab J Chem 10:S3707–S3714. https://doi.org/10.1016/j.arabjc.2014.05.003
43. Fleurence J, Morançais M, Dumay J (2018) Seaweed proteins. In: Proteins in food processing. Second Edition (Second Edi). Elsevier Ltd. https://doi.org/10.1016/B978-0-08-100722-8.00010-3
44. Foo KY, Hameed BH (2009) An overview of landfill leachate treatment via activated carbon adsorption process. J Hazard Mater 171(1–3):54–60
45. Foroutan R, Esmaeili H, Abbasi M, Rezakazemi M, Mesbah M (2018) Adsorption behavior of Cu(II) and Co(II) using chemically modified marine algae. Environ Technol 39(21):2792–2800. https://doi.org/10.1080/09593330.2017.1365946
46. Forsberg Å, Söderlund S, Frank A, Petersson LR, Pedersen M (1988) Studies on metal content in the brown seaweed, Fucus vesiculosus, from the Archipelago of Stockholm. Environ Pollut 49(4):245–263

47. Freile-Pelegrin Y, Robledo D, Chan-Bacab MJ, Ortega-Morales BO (2008) Antileishmanial properties of tropical marine algae extracts. Fitoterapia 79(5):374–377
48. Fu F, Wang Q (2011) Removal of heavy metal ions from wastewaters: a review. J Environ Manage 92(3):407–418. https://doi.org/10.1016/j.jenvman.2010.11.011
49. Fuat G, Cumali Y (2020) Synthesis, characterization, and lead (II) sorption performance of a new magnetic separable composite: MnFe2O4@ wild plants-derived biochar. J Environ Chem Eng 104567
50. Fujiwara-Arasaki T, Mino N, Kuroda M (1984) The protein value in human nutrition of edible marine algae in Japan. Eleventh International Seaweed Symposium, 513–516
51. Gao X, Guo C, Hao J, Zhao Z, Long H, Li M (2020) Adsorption of heavy metal ions by sodium alginate based adsorbent-a review and new perspectives. Int J Biol Macromol
52. Gheorghita Puscaselu R, Lobiuc A, Dimian M, Covasa M (2020) Alginate: from food industry to biomedical applications and management of metabolic disorders. Polymers 12(10):2417
53. Girardi F, Hackbarth FV, De Souza SMAGU, De Souza AAU, Boaventura RAR, Vilar VJP (2014) Marine macroalgae Pelvetia canaliculata (Linnaeus) as natural cation exchanger for metal ions separation: a case study on copper and zinc ions removal. Chem Eng J 247:320–329. https://doi.org/10.1016/j.cej.2014.03.007
54. Godt J, Scheidig F, Grosse-Siestrup C, Esche V, Brandenburg P, Reich A, Groneberg DA (2006) The toxicity of cadmium and resulting hazards for human health. J Occupational Med Toxicol 1(1):1–6. https://doi.org/10.1186/1745-6673-1-22
55. Grasdalen H, Larsen B, Smidsrød O (1979) A pmr study of the composition and sequence of uronate residues in alginates. Carbohyd Res 68(1):23–31
56. Gunatilake SK (2015) Methods of removing heavy metals from industrial wastewater. J Multidiscip Eng Sci Studies 1(1):14
57. Gupta MN, Raghava S (2008) Smart systems based on polysaccharides. In: Natural-based polymers for biomedical applications. Elsevier, pp 129–161
58. Gürişik E, Arica MY, Bektaş S, Genç Ö (2004) Comparison of the heavy metal biosorption capacity of active, heat-inactivated and NaOH-treated phanerochaete chrysosporium biosorbents. Eng Life Sci 4(1):86–89
59. Gutiérrez C, Hansen HK, Hernández P, Pinilla C (2015) Biosorption of cadmium with brown macroalgae. Chemosphere 138:164–169. https://doi.org/10.1016/j.chemosphere.2015.06.002
60. Hamdy AA (2000) Biosorption of heavy metals by marine algae. Curr Microbiol 41(4):232–238. https://doi.org/10.1007/s002840010126
61. Harper TR, Kingham NW (1992) Removal of arsenic from wastewater using chemical precipitation methods. Water Environ Res 64(3):200–203
62. Hernández-Carmona G, McHugh DJ, Arvizu-Higuera DL, Rodríguez-Montesinos YE (1998) Pilot plant scale extraction of alginate from Macrocystis pyrifera. 1. Effect of pre-extraction treatments on yield and quality of alginate. J Appl Phycol 10(6):507–513
63. Ho S-H, Chen C-Y, Chang J-S (2012) Effect of light intensity and nitrogen starvation on CO2 fixation and lipid/carbohydrate production of an indigenous microalga Scenedesmus obliquus CNW-N. Biores Technol 113:244–252
64. Holan ZR, Volesky B (1994) Biosorption of lead and nickel by biomass of marine algae. Biotechnol Bioeng 43(11):1001–1009
65. Hoslett J, Ghazal H, Ahmad D, Jouhara H (2019) Removal of copper ions from aqueous solution using low temperature biochar derived from the pyrolysis of municipal solid waste. Sci Total Environ 673:777–789
66. Hubadillah SK, Othman MHD, Harun Z, Ismail AF, Rahman MA, Jaafar J (2017) A novel green ceramic hollow fiber membrane (CHFM) derived from rice husk ash as combined adsorbent-separator for efficient heavy metals removal. Ceram Int 43(5):4716–4720
67. Hube S, Eskafi M, Hrafnkelsdóttir KF, Bjarnadóttir B, Bjarnadóttir MÁ, Axelsdóttir S, Wu B (2020) Direct membrane filtration for wastewater treatment and resource recovery: a review. Sci Total Environ 710:136375
68. Ibrahim WM (2011) Biosorption of heavy metal ions from aqueous solution by red macroalgae. J Hazard Mater 192(3):1827–1835

69. Ibrahim WM, Hassan AF, Azab YA (2016) Biosorption of toxic heavy metals from aqueous solution by Ulva lactuca activated carbon. Egyptian J Basic Appl Sci 3(3):241–249
70. Jayakumar V, Govindaradjane S, Rajasimman M (2020) Efficient adsorptive removal of Zinc by green marine macro alga Caulerpa scalpelliformis–characterization, optimization, modeling, isotherm, kinetic, thermodynamic, desorption and regeneration studies. Surf Interf 100798
71. Junior ABB, Dreisinger DB, Espinosa DCR (2019) A review of nickel, copper, and cobalt recovery by chelating ion exchange resins from mining processes and mining tailings. Mining Metall Explor 36(1):199–213
72. Kadirvelu K, Faur-Brasquet C, Cloirec PL (2000) Removal of Cu (II), Pb (II), and Ni (II) by adsorption onto activated carbon cloths. Langmuir 16(22):8404–8409
73. Karnib M, Kabbani A, Holail H, Olama Z (2014) Heavy metals removal using activated carbon, silica and silica activated carbon composite. Energy Proc 50:113–120
74. Kendel M, Wielgosz-Collin G, Bertrand S, Roussakis C, Bourgougnon N, Bedoux G (2015) Lipid composition, fatty acids and sterols in the seaweeds Ulva armoricana, and Solieria chordalis from Brittany (France): an analysis from nutritional, chemotaxonomic, and antiproliferative activity perspectives. Mar Drugs 13(9):5606–5628
75. Khairy HM, El-Shafay SM (2013) Seasonal variations in the biochemical composition of some common seaweed species from the coast of Abu Qir Bay, Alexandria. Egypt Oceanol 55(2):435–452
76. Khoeyi ZA, Seyfabadi J, Ramezanpour Z (2012) Effect of light intensity and photoperiod on biomass and fatty acid composition of the microalgae. Chlorella vulgaris Aquacul Int 20(1):41–49
77. Khotimchenko SV (2006) Variations in lipid composition among different developmental stages of Gracilaria verrucosa (Rhodophyta). Bot Mar 49(1):34–38
78. Kim B-S, Lee HW, Park SH, Baek K, Jeon J-K, Cho HJ, Jung S-C, Kim SC, Park Y-K (2016) Removal of Cu 2+ by biochars derived from green macroalgae. Environ Sci Pollut Res 23(2):985–994
79. Kim YH, Park JY, Yoo YJ, Kwak JW (1999) Removal of lead using xanthated marine brown alga Undaria pinnatifida. Proc Biochem 34(6–7):647–652
80. King P, Rakesh N, Lahari SB, Kumar YP, Prasad V (2008) Biosorption of zinc onto Syzygium cumini L.: equilibrium and kinetic studies. Chem Eng J 144(2):181–187
81. Kinuthia GK, Ngure V, Beti D, Lugalia R, Wangila A, Kamau L (2020) Levels of heavy metals in wastewater and soil samples from open drainage channels in nairobi, Kenya: community health implication. Sci Rep 10(1):1–13. https://doi.org/10.1038/s41598-020-65359-5
82. Kılınç B, Cirik S, Turan G, Tekogul H, Koru E (2013) Seaweeds for food and industrial applications. In: Food industry. IntechOpen.
83. Kratochvil D, Pimentel P, Volesky B (1998) Removal of trivalent and hexavalent chromium by seaweed biosorbent. Environ Sci Technol 32(18):2693–2698
84. Ku Y, Jung I-L (2001) Photocatalytic reduction of Cr (VI) in aqueous solutions by UV irradiation with the presence of titanium dioxide. Water Res 35(1):135–142
85. Kumar CS, Ganesan P, Suresh PV, Bhaskar N (2008) Seaweeds as a source of nutritionally beneficial compounds-a review. J Food Sci Technol 45(1):1
86. Kumar YP, King P, Prasad V (2007) Adsorption of zinc from aqueous solution using marine green algae—Ulva fasciata sp. Chem Eng J 129(1–3):161–166
87. Kuo CK, Ma PX (2001) Ionically crosslinked alginate hydrogels as scaffolds for tissue engineering: part 1. Structure, gelation rate and mechanical properties. Biomaterials. https://doi.org/10.1016/s0142-9612(00)00201-5
88. Lee KY, Mooney DJ (2012) Alginate: properties and biomedical applications. Prog Polym Sci 37(1):106–126
89. Lembi CA (2003) Control of nuisance algae. In: Freshwater algae of North America. Elsevier, pp 805–834
90. Leusch A, Holan ZR, Volesky B (1995) Biosorption of heavy metals (Cd, Cu, Ni, Pb, Zn) by chemically-reinforced biomass of marine algae. J Chem Technol Biotechnol 62(3):279–288. https://doi.org/10.1002/jctb.280620311

91. Li G-Y, Luo Z-C, Yuan F, Yu X (2017) Combined process of high-pressure homogenization and hydrothermal extraction for the extraction of fucoidan with good antioxidant properties from Nemacystus decipients. Food Bioprod Process 106:35–42. https://doi.org/10.1016/j.fbp.2017.08.002
92. Liang X, Duan J, Xu Q, Wei X, Lu A, Zhang L (2017) Ampholytic microspheres constructed from chitosan and carrageenan in alkali/urea aqueous solution for purification of various wastewater. Chem Eng J 317:766–776
93. Lin Z, Li J, Luan Y, Dai W (2020) Application of algae for heavy metal adsorption: a 20-year meta-analysis. Ecotoxicol Environ Saf 190:110089. https://doi.org/10.1016/j.ecoenv.2019.110089
94. Liu C, Omer AM, Ouyang X (2018) Adsorptive removal of cationic methylene blue dye using carboxymethyl cellulose/k-carrageenan/activated montmorillonite composite beads: Isotherm and kinetic studies. Int J Biol Macromol 106:823–833. https://doi.org/10.1016/j.ijbiomac.2017.08.084
95. Liu C, Wu T, Hsu P-C, Xie J, Zhao J, Liu K, Sun J, Xu J, Tang J, Ye Z (2019) Direct/alternating current electrochemical method for removing and recovering heavy metal from water using graphene oxide electrode. ACS Nano 13(6):6431–6437
96. Liu J, Pemberton B, Lewis J, Scales PJ, Martin GJO (2019) Wastewater treatment using filamentous algae—a review. Biores Technol 122556. https://doi.org/10.1016/j.biortech.2019.122556
97. Lucaci AR, Bulgariu D, Popescu MC, Bulgariu L (2020) Adsorption of Cu(II) ions on adsorbent materials obtained from marine red algae callithamnion corymbosum sp. Water 12(2):372. https://doi.org/10.3390/w12020372
98. Luo F, Liu Y, Li X, Xuan Z, Ma J (2006) Biosorption of lead ion by chemically-modified biomass of marine brown algae Laminaria japonica. Chemosphere 64(7):1122–1127. https://doi.org/10.1016/j.chemosphere.2005.11.076
99. Mahamadi C, Mawere E (2013) High adsorption of dyes by water hyacinth fixed on alginate. Environ Chem Lett 12(2):313–320. https://doi.org/10.1007/s10311-013-0445-z
100. Mahdavinia GR, Mosallanezhad A (2016) Facile and green rout to prepare magnetic and chitosan-crosslinked κ-carrageenan bionanocomposites for removal of methylene blue. J Water Proc Eng 10:143–155. https://doi.org/10.1016/j.jwpe.2016.02.010
101. Marinho-Soriano E, Fonseca PC, Carneiro MAA, Moreira WSC (2006) Seasonal variation in the chemical composition of two tropical seaweeds. Biores Technol 97(18):2402–2406
102. Mashkoor F, Nasar A (2020) Carbon nanotube-based adsorbents for the removal of dyes from waters: a review. Environ Chem Lett 18(3):605–629. https://doi.org/10.1007/s10311-020-00970-6
103. Meyer BN, Ferrigni NR, Putnam JE, Jacobsen LB, Nichols DE, McLaughlin JL (1982) Brine shrimp: a convenient general bioassay for active plant constituents. Planta Med 45(05):31–34
104. Minghou J, Yujun W, Zuhong X, Yucai G (1984) Studies on the M:G ratios in alginate. In: Bird CJ, Ragan MA (eds) Eleventh International Seaweed Symposium. Developments in Hydrobiology, vol 22. Springer, Dordrecht. https://doi.org/10.1007/978-94-009-6560-7_114
105. Mola Ali Abasiyan S, Dashbolaghi F, Mahdavinia GR (2019) Chitosan cross-linked with κ-carrageenan to remove cadmium from water and soil systems. Environ Sci Pollut Res 26(25):26254–26264. https://doi.org/10.1007/s11356-019-05488-1
106. Montazer-Rahmati MM, Rabbani P, Abdolali A, Keshtkar AR (2011) Kinetics and equilibrium studies on biosorption of cadmium, lead, and nickel ions from aqueous solutions by intact and chemically modified brown algae. J Hazard Mater 185(1):401–407
107. Nautiyal P, Subramanian KA, Dastidar MG (2014) Recent advancements in the production of biodiesel from algae: a review. Ref Module Earth Syst Environ Sci. https://doi.org/10.1016/b978-0-12-409548-9.09380-5
108. Nazal MK (2019) Marine algae bioadsorbents for adsorptive removal of heavy metals. In: Advanced sorption process applications. IntechOpen. https://doi.org/10.5772/intechopen.80850

109. Necas J, Bartosikova L (2013) Carrageenan: a review. Veterinarni Medicina 58(4):187–205. https://doi.org/10.17221/6758-VETMED
110. Negm NA, Abd El Wahed MG, Hassan ARA, Abou Kana MTH (2018) Feasibility of metal adsorption using brown algae and fungi: effect of biosorbents structure on adsorption isotherm and kinetics. J Mol Liq 264(2017):292–305. https://doi.org/10.1016/j.molliq.2018.05.027
111. Nelson MM, Phleger CF, Nichols PD (2002) Seasonal lipid composition in macroalgae of the northeastern Pacific Ocean. Bot Mar 45(1):58–65
112. Nriagu JO (1980) Production, uses, and properties of cadmium. New York, Wiley, pp 35–70
113. Onuegbu TU, Umoh ET, Onwuekwe IT (2013) Physico—chemical analysis of effluents from Jacbon chemical industries limited, makers of bonalux emulsion and gloss paints. Int J Sci Technol 2(2):169–173
114. Øverland M, Mydland LT, Skrede A (2019) Marine macroalgae as sources of protein and bioactive compounds in feed for monogastric animals. J Sci Food Agric 99(1):13–24
115. Packer MA, Harris GC, Adams SL (2016) Food and feed applications of algae. In: Algae biotechnology. Springer, pp 217–247
116. Padam BS, Chye FY (2020) Seaweed components, properties, and applications. Elsevier Inc., In Sustainable Seaweed Technologies. https://doi.org/10.1016/b978-0-12-817943-7.00002-0
117. Pal D, Khozin-Goldberg I, Cohen Z, Boussiba S (2011) The effect of light, salinity, and nitrogen availability on lipid production by Nannochloropsis sp. Appl Microbiol Biotechnol 90(4):1429–1441. https://doi.org/10.1007/s00253-011-3170-1
118. Papageorgiou SK, Kouvelos EP, Katsaros FK (2008) Calcium alginate beads from Laminaria digitata for the removal of Cu+2 and Cd+2 from dilute aqueous metal solutions. Desalination 224(1–3):293–306. https://doi.org/10.1016/j.desal.2007.06.011
119. Ponnan A, Ramu K, Marudhamuthu M, Marimuthu R, Siva K, Kadarkarai M (2017) Antibacterial, antioxidant and anticancer properties of Turbinaria conoides (J Agardh) Kuetz. Clinic Phytosci 3(1). https://doi.org/10.1186/s40816-017-0042-y
120. Postma PR, Cerezo-Chinarro O, Akkerman RJ, Olivieri G, Wijffels RH, Brandenburg WA, Eppink MHM (2017) Biorefinery of the macroalgae Ulva lactuca: extraction of proteins and carbohydrates by mild disintegration. J Appl Phycol 30(2):1281–1293. https://doi.org/10.1007/s10811-017-1319-8
121. Pradhan D, Sukla LB, Mishra BB, Devi N (2018) Biosorption for removal of hexavalent chromium using microalgae Scenedesmus sp. J Clean Prod. https://doi.org/10.1016/j.jclepro.2018.10.288
122. Pramanick P, Bera D, Banerjee K, Zaman S, Mitra A (2016) Seasonal variation of proximate composition of common seaweeds in Indian Sundarbans. Int J Life Sci Scienti. Res 2(5)
123. Qu C, Ma M, Chen W, Cai P, Yu X-Y, Feng X, Huang Q (2018) Modeling of Cd adsorption to goethite-bacteria composites. Chemosphere 193:943–950
124. Rahman IMM, Begum ZA, Yahya S, Lisar S, Motafakkerazad R, Cell AS-P (2016) Complimentary contributor copy (Issue July)
125. Rajkumar D, Palanivelu K (2004) Electrochemical treatment of industrial wastewater. J Hazard Mater 113(1–3):123–129
126. Rao AR, Dayananda C, Sarada R, Shamala TR, Ravishankar GA (2007) Effect of salinity on growth of green alga Botryococcus braunii and its constituents. Biores Technol 98(3):560–564
127. Reitan KI, Rainuzzo JR, Olsen Y (1994) Effect of nutrient limitation on fatty acid and lipid content of marine microalgae 1. J Phycol 30(6):972–979
128. Renaud SM, Thinh L-V, Lambrinidis G, Parry DL (2002) Effect of temperature on growth, chemical composition and fatty acid composition of tropical Australian microalgae grown in batch cultures. Aquaculture 211(1–4):195–214
129. Rhein-Knudsen N, Ale MT, Meyer AS (2015) Seaweed hydrocolloid production: an update on enzyme assisted extraction and modification technologies. Mar Drugs 13(6):3340–3359
130. Rinaudo M (2002) Les alginates et les carraghénanes. Actual Chim 11(12):35–38
131. Rodriguez-Reinoso F, Molina-Sabio M (1992) Activated carbons from lignocellulosic materials by chemical and/or physical activation: an overview. Carbon 30(7):1111–1118

132. Ruangsomboon S (2012) Effect of light, nutrient, cultivation time and salinity on lipid production of newly isolated strain of the green microalga, Botryococcus braunii KMITL 2. Biores Technol 109:261–265
133. Salari Z, Danafar F, Dabaghi S, Ataei SA (2016) Sustainable synthesis of silver nanoparticles using macroalgae Spirogyra varians and analysis of their antibacterial activity. J Saudi Chem Soc 20(4):459–464
134. Saleem J, Shahid UB, Hijab M, Mackey H, McKay G (2019) Production and applications of activated carbons as adsorbents from olive stones. Biomass Convers Biorefin 9:775–802
135. Salhi G, Zbakh H, Moussa H, Hassoun M, Bochkov V, Ciudad CJ, Noé V, Riadi H (2018) Antitumoral and anti-inflammatory activities of the red alga Sphaerococcus coronopifolius. Eur J Integrat Med 18:66–74
136. Santos SCR, Ungureanu G, Volf I, Boaventura RAR, Botelho CMS (2018) Macroalgae biomass as sorbent for metal ions. In: Biomass as renewable raw material to obtain bioproducts of high-tech value. Elsevier, pp 69–112
137. Sayato Y (1989) WHO guidelines for drinking-water quality. Eisei Kagaku 35(5):307–312. https://doi.org/10.1248/jhs1956.35.307
138. Shakya AK, Ghosh PK (2018) Simultaneous removal of arsenic and nitrate in absence of iron in an attached growth bioreactor to meet drinking water standards: importance of sulphate and empty bed contact time. J Clean Prod 186:304–312
139. Shilina AS, Bakhtin VD, Burukhin SB, Askhadullin SR (2017) Sorption of cations of heavy metals and radionuclides from the aqueous media by new synthetic zeolite-like sorbent. Nucl Energy Technol 3(4):249–254
140. Shobier AH, El-Sadaawy MM, El-Said GF (2020) Removal of hexavalent chromium by ecofriendly raw marine green alga Ulva fasciata: kinetic, thermodynamic and isotherm studies. Egypt J Aqua Res
141. Silva PC, Basson PW, Moe RL (1996) Catalogue of the benthic marine algae of the Indian Ocean, vol 79. Univ of California Press
142. Singh S, Wasewar KL, Kansal SK (2020) Low-cost adsorbents for removal of inorganic impurities from wastewater. In: Inorganic pollutants in water. INC. https://doi.org/10.1016/b978-0-12-818965-8.00010-x
143. Smedley PL, Nicolli HB, Macdonald DMJ, Barros AJ, Tullio JO (2002) Hydrogeochemistry of arsenic and other inorganic constituents in groundwaters from La Pampa Argentina. Appl Geochem 17(3):259–284
144. Soliman NK, Mohamed HS, Ahmed SA, Sayed FH, Elghandour AH, Ahmed SA (2019) Cd2+ and Cu2+ removal by the waste of the marine brown macroalga Hydroclathrus clathratus. Environ Technol Innov 15:100365
145. Son BC, Park K, Song SH, Yoo YJ (2004) Selective biosorption of mixed heavy metal ions using polysaccharides. Korean J Chem Eng 21(6):1168–1172
146. Stokke BT, Smidsroed O, Bruheim P, Skjaak-Braek G (1991) Distribution of uronate residues in alginate chains in relation to alginate gelling properties. Macromolecules 24(16):4637–4645
147. Sudhakar MP, Viswanaathan S (2019) Algae as a sustainable and renewable bioresource for bio-fuel production. In: New and future developments in microbial biotechnology and bioengineering. Elsevier, pp 77–84
148. Suganya T, Varman M, Masjuki HH, Renganathan S (2016) Macroalgae and microalgae as a potential source for commercial applications along with biofuels production: a biorefinery approach. Renew Sustain Energy Rev 55:909–941
149. Suutari M, Leskinen E, Fagerstedt K, Kuparinen J, Kuuppo P, Blomster J (2015) Macroalgae in biofuel production. Phycol Res 63(1):1–18. https://doi.org/10.1111/pre.12078
150. Tavana M, Pahlavanzadeh H, Zarei MJ (2020) The novel usage of dead biomass of green algae of Schizomeris leibleinii for biosorption of copper (II) from aqueous solutions: equilibrium, kinetics and thermodynamics. J Environ Chem Eng 104272
151. Tchounwou PB, Yedjou CG, Patlolla AK, Sutton DJ (2012) Heavy metal toxicity and the environment. In: Molecular, clinical and environmental toxicology. Springer, pp 133–164

152. Teoh M-L, Chu W-L, Marchant H, Phang S-M (2004) Influence of culture temperature on the growth, biochemical composition and fatty acid profiles of six Antarctic microalgae. J Appl Phycol 16(6):421–430
153. Tran T-K, Chiu K-F, Lin C-Y, Leu H-J (2017) Electrochemical treatment of wastewater: selectivity of the heavy metals removal process. Int J Hydrogen Energy 42(45):27741–27748
154. Trica B, Delattre C, Gros F, Ursu AV, Dobre T, Djelveh G, Michaud P, Oancea F (2019) Extraction and characterization of alginate from an edible brown seaweed (Cystoseira barbata) harvested in the Romanian Black Sea. Mar Drugs 17(7):405
155. Tuvikene R, Truus K, Vaher M, Kailas T, Martin G, Kersen P (2006) Extraction and quantification of hybrid carrageenans from the biomass of the red algae Furcellaria lumbricalis and Coccotylus truncatus. In: Proceedings of the Estonian Academy of Sciences, Chemistry, 55(1)
156. Usov AI (1999) Alginic acids and alginates: analytical methods used for their estimation and characterisation of composition and primary structure. Russ Chem Rev 68(11):957–966. https://doi.org/10.1070/rc1999v068n11abeh000532
157. Valavanidis A, Vlachogianni T (2010) Metal pollution in ecosystems. Ecotoxicology studies and risk assessment in the marine environment. University of Athens University Campus Zografou, p 15784
158. Vijayaraghavan G, Shanthakumar S (2015) Removal of sulphur black dye from its aqueous solution using alginate from Sargassum sp.(Brown algae) as a coagulant. Environ Progress Sustain Energy 34(5):427–1434
159. Wang LK, Vaccari DA, Li Y, Shammas NK (2005) Chemical precipitation. In: Wang LK., Hung YT, Shammas NK (eds) Physicochemical treatment processes. handbook of environmental engineering, vol 3. Humana Press. https://doi.org/10.1385/1-59259-820-x:141
160. Wang S, Soudi M, Li L, Zhu Z (2006) Coal ash conversion into effective adsorbents for removal of heavy metals and dyes from wastewater. J Hazard Mater 133(1–3):243–251. https://doi.org/10.1016/j.jhazmat.2005.10.034
161. Wells ML, Potin P, Craigie JS, Raven JA, Merchant SS, Helliwell KE, Smith AG, Camire ME, Brawley SH (2017) Algae as nutritional and functional food sources: revisiting our understanding. J Appl Phycol 29(2):949–982
162. Wong KH, Cheung PCK (2000) Nutritional evaluation of some subtropical red and green seaweeds: part I—proximate composition, amino acid profiles and some physico-chemical properties. Food Chem 71(4):475–482
163. Yang J, Volesky B (1999) Modeling uranium-proton ion exchange in biosorption. Environ Sci Technol 33(22):4079–4085
164. Youssouf L, Lallemand L, Giraud P, Soulé F, Bhaw-Luximon A, Meilhac O, Couprie J (2017) Ultrasound-assisted extraction and structural characterization by NMR of alginates and carrageenans from seaweeds. Carbohydrate Polymers 166:55–63. https://doi.org/10.1016/j.carbpol.2017.01.041
165. Zeraatkar AK, Ahmadzadeh H, Talebi AF, Moheimani NR, McHenry MP (2016) Potential use of algae for heavy metal bioremediation, a critical review. J Environ Manage 181:817–831. https://doi.org/10.1016/j.jenvman.2016.06.059
166. Zhang W, Zhang Y (2020) Development of ZnFe2O4 nanoparticle functionalized baker's yeast composite for effective removal of methylene blue via adsorption and photodegradation. J Water Proc Eng 37:101234. https://doi.org/10.1016/j.jwpe.2020.101234

Electrochemical Synthesis of Titania-Containing Composites with a Metallic Matrix for Photochemical Degradation of Organic Pollutants in Wastewater

V. S. Protsenko and F. I. Danilov

Abstract Immobilized TiO_2 photocatalyst is considered as a promising approach to fabricate high-performance and available materials for wastewater treatment. The immobilization of TiO_2 particles can be performed by various manufacturing techniques, the electrochemical deposition being the most convenient and tunable method. This chapter surveyed the literature data on electrochemical synthesis of TiO_2-containing composites with photocatalytic performance. The electrochemical deposition and characterization of iron/titania composite coatings were considered in detail. The content of TiO_2 dispersed phase in the fabricated composites was up to about 5 wt.%. The electrodeposited Fe–TiO_2 composites exhibited photocatalytic properties regarding the photochemical degradation of methylene blue and methyl orange organic molecules in water solution. The apparent rate constants of photochemical degradation of the examined organic dyes on the electrodeposited Fe–TiO_2 photocatalyst were in the range of $(1.8\text{–}12.8) \times 10^{-3}$ min^{-1}.

Keywords TiO_2 heterogeneous photocatalyst · Coatings · Electrodeposition · Composite · Fe–TiO_2 · Methanesulfonate electrolyte · Protective ceria layer · Photochemical degradation · Organic dye · Wastewater treatment

1 Introduction

It is well known that titania has wide applications in photocatalytic degradation of organic pollutants, this technique is associated with promising and efficient advanced oxidation processes [1, 9, 37, 39]. Advanced oxidation processes imply that different strongly oxidizing agents (for instance, hydroxyl radicals •OH) are formed in the reaction zone, which ensures almost the total mineralization of various organic contaminants. A number of treatment processes have been designed, which yield •OH radicals and other oxidizing agents: O_3 and H_2O_2 photolysis, Fenton and photo-Fenton techniques, peroxonation, and heterogeneous photochemical catalysis [1, 32, 37,

V. S. Protsenko (✉) · F. I. Danilov
Ukrainian State University of Chemical Technology, Gagarin Avenue, 8, Dnipro 49005, Ukraine

E. Lichtfouse et al. (eds.), *Inorganic-Organic Composites for Water and Wastewater Treatment*, Environmental Footprints and Eco-design of Products and Processes,
https://doi.org/10.1007/978-981-16-5928-7_9

39]. Although all these kinds of advanced oxidation processes are of great value and importance in the light of their practical application, the heterogeneous photocatalysis, involving the use of photo-excited semiconductor TiO_2, attracts special attention. This is due to the fact that titanium dioxide is considered actually as an ideal photocatalyst for the treatment of industrial sewage [8, 13, 22, 27, 39, 41, 52].

TiO_2 particles in heterogeneous photocatalysis can be used both in the form of fine dispersed powders suspended in water solutions and in the form of layers in which TiO_2 particles are fixed on an appropriate substrate (immobilized TiO_2 films) [1, 39]. TiO_2 photocatalytic particles in a suspended form demonstrate quite a few advantages: they can be easily and reliably prepared and utilized, and their specific surface area is commonly very high. In addition, slurry TiO_2-containing systems can be readily aerated, thereby reinforcing the photocatalytic behavior via impeding the recombination of electron–hole pairs [39].

Unfortunately, the TiO_2-based slurry systems are not stable enough; the processes of aggregation and sedimentation can quickly occur in these systems, which reduces the photocatalytic activity. In addition, there is a need to carry out an expensive and time-consuming post-treatment of titania particles in order to maintain slurry systems. At the same time, immobilized titania films are stable enough and exhibit reliable photocatalytic activity during long-run tests [53]. In addition, immobilized TiO_2 thin layers do not require the filtration and separation of photocatalytic particles. However, a problem arises concerning the search for a suitable and trouble-free support for titania catalytic films.

Various kinds of materials have been developed to support the titania photocatalyst particles: activated carbon, silica, glasses, polymeric materials, zeolites, alumina, different steels, synthetic fibers, etc. [2, 36, 50, 53]. Immobilization of TiO_2 particles can be performed by using both transparent substrates (such as glass-based materials and fused silica) and lightproof substrates (such as fibers, metals, and activated carbon). Numerous methods were developed to prepare supported titania: chemical vapor deposition, thermal treatment, sol–gel technique, electrochemical and electrophoretic deposition, sol-spray, and hydrothermal procedure [53]. These production techniques commonly yield composites, which are regarded as high-efficient materials to remove different pollutants from wastewater [14].

Generally speaking, the electrodeposition of composites coatings can be conducted both on cathode and on anode. Cathodic electrochemical deposition yields composite films with a metallic matrix [28], whereas anodic electrodeposition allows producing oxide/hydroxide layers, which can also be used in wastewater treatment [4, 23, 40].

We confine this chapter to the processes of cathodic electrodeposition of composite films, in which finely dispersed TiO_2 particles are immobilized by a growing electrodeposited metallic matrix. This technique seems to be very promising to develop efficient supports for TiO_2 heterogeneous photocatalysts.

2 Electrodeposition of Titania-Containing Composites with Metallic Matrices

The electrochemical deposition of composite coatings often provides better physicochemical and performance properties than the electrodeposition of individual metals and alloys [28, 60]. Electrochemical synthesis of composites includes incorporation of finely dispersed phase into a growing metallic matrix by using electrolytes containing proper metal ions together with colloidal particles. This technique can provide controlled, flexible, and reliable production of composite materials with targeted chemical and phase compositions and physicochemical and service characteristics. In addition, electrodeposition is considered a simple, available, and not expensive technique [59]. Obviously, entrapping TiO_2 nano- or micro-particles by an electrodeposited metal matrix will ensure photocatalytic behavior of the obtained composite layers.

There are a number of papers, which report the electrochemical synthesis of photocatalytic titania-containing composites with a metallic matrix. Thus, nanostructured composites Zn–TiO_2 coatings were electrodeposited on a steel substrate [21]. The obtained composite films, Zn–TiO_2, were shown to be extremely active in the photocatalytic oxidation of CH_3CHO.

Reference [55] considered immobilization of TiO_2 by co-electrodeposition with a nickel matrix from a common Watt's nickel plating electrolyte. The electrodeposition was performed using a rotating disc electrode (brass). It was shown that the changes in Ni surface morphology and microstructure have an influence upon the photocatalytic behavior of the composite Ni–TiO_2 electrodeposits under ultraviolet illumination. The photocatalytic performance was reinforced by an increase in the TiO_2 loading in deposited layers and a decrease in the value of grain size in the Ni matrix.

Ni–TiO_2/TiO_2 multilayer electrodeposits were prepared via the sol-modified pulse plating deposition method [34]. The activity of the multilayer coatings was estimated by measuring the rate of the photochemical destruction of methyl orange dye. The multilayers annealed at 450 °C showed optimum photocatalytic efficiency: 53.64% methyl orange degradation after 5 h of ultraviolet illumination was observed.

It was stated that Ni–TiO_2 composite films containing up to 2.35 wt.% of titania can be prepared from an electrochemical system based on a deep eutectic solvent (ethaline) [10]. These coatings were successfully used to perform the destruction of methylene blue organic dye induced by ultraviolet light.

Reference [33] described the electrodeposition of Al–TiO_2 composite layers from a dimethyl sulfone–aluminum chloride plating bath with suspended titania particulates. The electrodeposited Al–TiO_2 composite layers were subsequently anodized in an oxalic aqueous solution. It was stated that anodization step yielded the conversion of Al matrix into an alumina film with a great number of nanopores. Thus, a porous alumina layer with the embedded titania dispersed phase was prepared. The anodized composite deposits exhibited better photocatalytic efficiency than the as-deposited layers.

Ag–TiO_2 nanocomposites were electrochemically deposited using an alkaline cyanide-free electrolyte based on the solution of $AgNO_3$ with colloidal TiO_2 particles [31]. Silver–TiO_2 coatings showed photocatalytic and antimicrobial properties; such behavior was comparable with the one obtained by other fabrication techniques. The electrodeposition of Ag–SiO_2–TiO_2 and Ag–TiO_2 composite layers on a graphite substrate yielded materials with an improved photocatalytic performance under the action of visible or ultraviolet radiation [49].

Reference [26] reported an interesting approach to the preparation of composite films containing Ag, TiO_2, and bamboo charcoal. The stepwise method included sol–gel synthesis, wet impregnation, and electrodeposition. The as-fabricated coatings exhibited photocatalytic and antibacterial activity. Bamboo-type TiO_2 nanotubes embedded in electrodeposited nanostructured silver matrix showed photocatalytic activity toward the photochemical destruction of methylene blue organic molecules induced by ultraviolet illumination [29].

A gold–titania heterojunction nanotube composite having a tube-in-tube nanostructure was synthesized using a pulsed electrolysis [30]. The electrodeposited composite demonstrated an enhanced photocatalytic behavior with respect to destruction of acid orange 7 organic dye; this phenomenon being resulted in the synergetic effect in the system Cr(VI)—acid orange 7 organic dye.

Reference [51] showed that doped N,S–TiO_2 particles can be immobilized into Sn–Ni alloy matrix by direct current electrodeposition technique. The highest content of embedded N,S–TiO_2 particles in composite coatings was about 3.25 wt.%. The prepared composite films manifested an improved photocatalytic activity toward the destruction of Rhodamine B and methylene blue dyes.

Thus, various metals can be used as a matrix in TiO_2-containing electrodeposited composites with photocatalytic performance. A number of them suffer from high costs (Ag, Au) and some are toxic and hazardous to people's health and the environment (Ni, Zn). From our point of view, the iron matrix is an excellent choice to produce this kind of composite coatings, since iron and its soluble compound are easily available, nontoxic, and inexpensive.

Iron-based coatings can be produced from plating electrolytes of various kinds and compositions. Among them, a methanesulfonate plating bath proved to be very successful [45, 57]. These plating electrolytes imply the application of methanesulfonic acid (CH_3SO_2OH) and its salts. Methanesulfonic acid is an eco-friendly product and is nowadays regarded as a "green acid" (Walsh and Ponce de Leon [61]; [3, 15]. Most methanesulfonates well dissolve in water, the electroconductivity of the electrolytes based on methanesulfonic acid and its salts is relatively high, and the treatment of the resulting wastewater does not pose any difficulties and potential danger to the environment [15]. Therefore, aqueous solutions of methanesulfonic acid seem to be practically ideal electrolytes for many electrochemical processes.

Furthermore, we will overview the main results of our studies of the electrochemical preparation of iron–titania coatings by using a methanesulfonate plating bath and their use for photochemical catalysis in wastewater treatment [11, 43, 44], Protsenko et al. [47, 48]; [12, 46, 58].

3 Electrodeposition and Characterization of Titania-Containing Composites with an Iron Matrix

To electrodeposit iron–TiO_2 composite films, a methanesulfonate plating bath was used [12, 46, 58]. Plating iron-containing electrolytes were prepared by dissolving reagent grade chemicals in double-distilled water. The electrolyte contained 1.25 M $Fe(CH_3SO_3)_2$ and 1–12 g dm^{-3} TiO_2 nanopowder. The synthesis of aqueous solution of Fe (II) methanesulfonate was performed according to the procedure defined elsewhere [15]. A specified volume of Na_2CO_3 solution was added to the plating bath to adjust its pH to a designated value (pH 1.3). This electroplating bath allowed obtaining nanocrystalline iron-based electrodeposits, the current efficiency and electrodeposition rate being relatively high (about 95% and ca. 2–4 μm min^{-1}, respectively) [45].

It is known that TiO_2 may form different crystal structures determining its physicochemical properties [8]. In this work, we used Degussa P 25 nanopowder (Evonik) as a TiO_2 source for the plating bath preparation. The average diameter of particulates in Degussa P 25 is about 25 nm. This kind of TiO_2 nanopowder includes mixed anatase and rutile crystalline modifications (at the weight ratio of about 80:20, respectively), microquantity of an amorphous TiO_2 phase being also detectable [38]. After adding a required portion of TiO_2 powder immediately to the iron plating electrolyte, the colloidal methanesulfonate electrolyte was intensively stirred by a magnetic stirrer for 1 h and then sonicated (an UZDN ultrasonic generator, 22.4 kHz, 1 h). This treatment allowed ensuring the homogeneous distribution of dispersed TiO_2 particulates throughout the system. In addition, the colloidal plating bath was uninterruptedly agitated by a magnetic stirrer (~60 revs per minute) during the electrochemical deposition of iron–titania composite films, which prevents the sedimentation of TiO_2 agglomerated particles.

Fe–TiO_2 composite coatings were electrodeposited at steady values of cathodic current density (5–20 A dm^{-2}). A usual thermostated glass cell was used. The temperature was 298 K. Plates made of mild steel (1 cm^2) were used as substrates to electrodeposit iron-based composite coatings. Prior to electrodepositing the coating, the surface of the sample was degreased, etched for 1–2 min in aqueous HCl solution, and washed with double distilled water. The anodes were made of mild steel.

Nanopowders of metal oxides in aqueous electrolyte are known to tend toward aggregation. In compliance with DLVO (Derjaguin, Landau, Verwey, Overbeek) theory, the double electric layer forming at the interface of colloidal particulates will be compressed in the electrolytes having a relatively high ionic strength. This will lead to aggregation occurring in any lyophobic colloidal system. Therefore, we observed the particle coagulation in iron-based plating baths, and the average diameter of the colloidal particles was evaluated by the sedimentation in the field of gravitation (digital analytical balance Vibra HT-120 (Shinko Denshi, Japan)). It was stated that the most probable diameter of the TiO_2 dispersed particles in methanesulfonate iron plating electrolytes is about 2 μm.

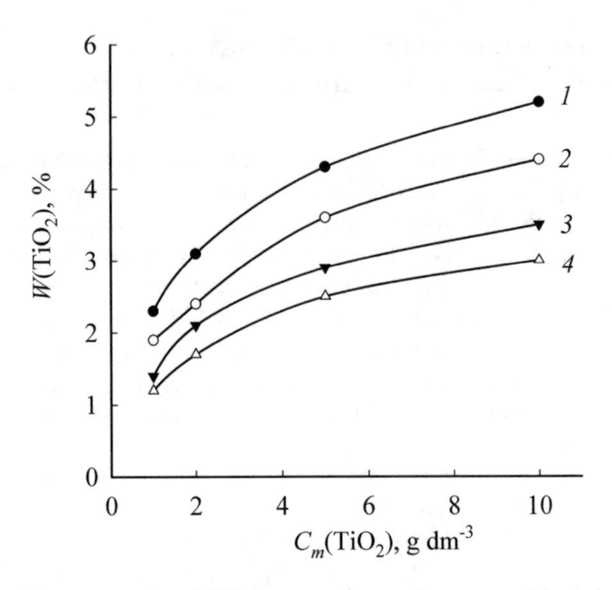

Fig. 1 Effect of the concentration of TiO_2 in suspensions on the content of titania in the composite films electrodeposited at various current densities (A m^{-2}): (1) 0.05, (2) 0.10, (3) 0.15, and (4) 0.20. The electrolytes contained 1.25 M $Fe(CH_3SO_3)_2$, pH 1.3. The electrodeposition was conducted at the temperature of 298 K for 20 min. (Reprinted from [58] by permission of Springer Nature)

An increase in the content of TiO_2 dispersed phase in the composite electrodeposits is observed with both increasing the titania concentration in the colloidal iron plating bath and decreasing the cathodic current density (Fig. 1). Similar dependences of the content of a dispersed phase are often detected in the case of electrochemical deposition of different composite coatings [28].

The inclusion of colloidal particles in an electrochemically deposited matrix is ascribed to their adsorption on the cathode surface, this process is described by a kinetic model developed by ref.[16]. This concept implies that the introduction of dispersed particulates into an electrodeposited matrix occurs via the two-step adsorption mechanism. A Langmuir-like mathematical expression is involved to interpret the adsorption of colloidal particles. At the first stage of adsorption, colloidal particles become attached to the cathode surface, essentially saving their adsorption-solvate shells. Therefore, the physical adsorption at the first stage is referred to as a “weak adsorption”. The second stage implies that the adsorption-solvate shells at the interface of colloidal particles disappear, and the particles become strongly fixed on the electrode. Thus, the adsorption at the second stage is referred to as a “strong adsorption”. Furthermore, the colloidal particulates that are firmly attached to the metallic substrate are gradually incorporated into the growing composite film.

The kinetics of electrochemical codeposition of Fe–TiO_2 composite electrodeposits in methanesulfonate plating baths complies with the advanced model suggested by ref.[5]. The main kinetic equation of this model can be written as follows:

$$\frac{C(1-\alpha)^{\left(2-\frac{B}{A}\right)}}{\alpha} = \frac{Mi_0^{\frac{B}{A}}}{nF\upsilon_0\rho_m}\left(\frac{1}{k}+C\right)i^{\left(1-\frac{B}{A}\right)} \tag{1}$$

where α is the volume concentration of the dispersed particles in the electrodeposit; C is the volume concentration of dispersed particles in the electrolyte; A is the coefficient taken from the exponential-type kinetic equation for the electrode process of metal deposition $i = i_0 e^{A\eta}$ (here η is the cathodic overpotential and i_0 is the exchange current density);n is the number of electrons participating in electrochemical deposition process; M is the relative atomic mass of deposited metal; F is the Faraday constant; ρ_m is the metal density; k is the equilibrium adsorption constant; and B and υ_0 are some constants relating to the process of embedding the dispersed particulates in the co-deposited metal (these constants are analogous to the coefficients A and i_0, respectively).

If we take the logarithm of Eq. (1), then the following expression can be obtained:

$$\lg\frac{C(1-\alpha)}{\alpha} = \lg\left(\frac{Mi_0^{B/A}}{nF\rho_m\upsilon_0}\right) + \left(1-\frac{B}{A}\right)\lg\left(\frac{i}{1-\alpha}\right) + \lg\left(\frac{1}{k}+C\right) \tag{2}$$

As follows from Eq. (2), the graph in the $\lg\frac{C(1-\alpha)}{\alpha}$ vs. $\lg\frac{i}{1-\alpha}$ coordinates should give straight lines (Fig. 2), the slopes of these lines allow determining the values of $\frac{B}{A}$.

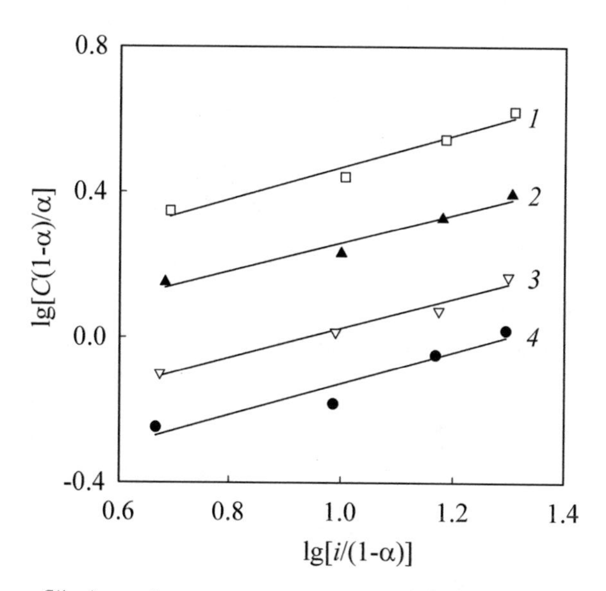

Fig. 2 Plots of $\lg\frac{C(1-\alpha)}{\alpha}$, $\lg\frac{i}{1-\alpha}$ dependences for iron–titania composite layers electrodeposited at different TiO_2 loadings in the electrolyte (g dm^{-3}): (1) 10, (2) 5, (3) 2, and (4) 1. The electrolytes contained 1.25 M of iron (II) methanesulfonate, pH 1.3. The electrodeposition was performed at the temperature of 298 K. (Reprinted from [58] by permission of Springer Nature)

These straight lines are parallel, which indicates that the $\frac{B}{A}$ ratio does not depend on both the cathode current density and the content of TiO_2 in colloidal plating bath. The calculated $\frac{B}{A}$ value was stated to be about 0.583; hence, B is less than A. Considering that $B < A$, it can be concluded that iron (II) cations that are on the surface of titania particles in an adsorbed state are reduced with a decreased rate as compared with the iron (II) cations in the bulk of plating bath [6].

On the basis of the calculated $\frac{B}{A}$ values, $C(1-\alpha)^{(2-\frac{B}{A})}/\alpha$ versus dependences were plotted at varied cathodic current densities (Fig. 3). In conformity with Eq. (1), the extrapolation of the series of these lines to the horizontal axis yields the value $\left(-\frac{1}{k}\right)$. Calculation showed that the adsorption coefficient is $k = 2.8$ [58].

It is important to emphasize that the established value of k is much lower than that in case of the use of a methanesulfonate plating bath with pure rutile modification of titania at the same ionic composition ($k = 49.8$) [46]. This difference can be attributed to some dissimilarity of chemical and colloidal properties of anatase and rutile.

It is reasonable to compare the content of TiO_2 in composite electrodeposits fabricated using colloidal electrolytes with Degussa P 25 titania particles on the one hand, and pure rutile powder on the other hand. The experimental results revealed that the mass concentration of TiO_2 dispersed phase in the electrodeposited composites from the electrolyte with the addition of Degussa P 25 powder is much lower than that when the powder of solely rutile phase was used (all other conditions being

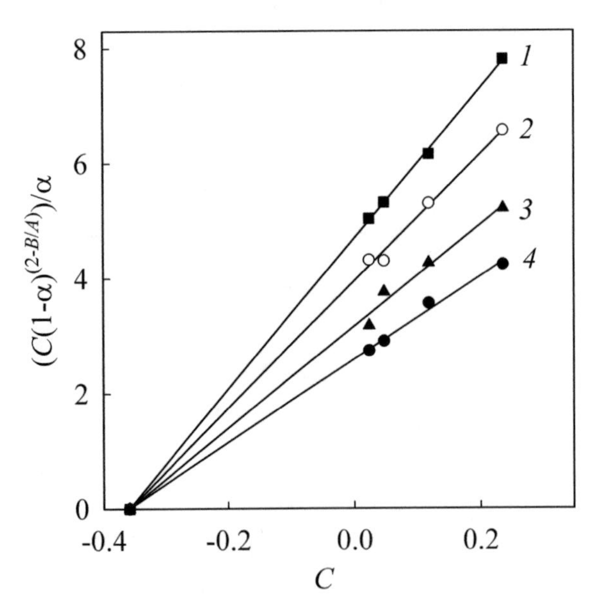

Fig. 3 Plots of $\left(C(1-\alpha)^{\left(2-\frac{B}{A}\right)}\right)/\alpha$ vs. dependences for iron–titania composite layers electrodeposited at varied current densities (A m^{-2}): (1) 0.20, (2) 0.15, (3) 0.10, and (4) 0.05. The electrolytes contained 1.25 M of iron (II) methanesulfonate, pH 1.3. The electrodeposition was performed at the temperature of 298 K. (Reprinted from [58] by permission of Springer Nature)

equal). For instance, at the TiO_2 loading of 5 g dm^{-3} and the current density of 0.20 A m^{-2}, the content of TiO_2 particles embedded in the electrodeposits was about 12.1 and 2.5% for colloidal electrolytes containing rutile and Degussa P 25, respectively [46, 58]. This may be due to the difference in the relevant adsorption coefficients.

To examine the morphological features of the surface of iron and iron–TiO_2 composite electrodeposits, scanning electron microscopy technique was used (Fig. 4). The surface images of "pure" iron coatings exhibited the occurrence of solitary non-uniform grains growing out of the surface layer (Fig. 4A). The presence of iron (about 97%) was demonstrated by the electron probe X-ray spectrum

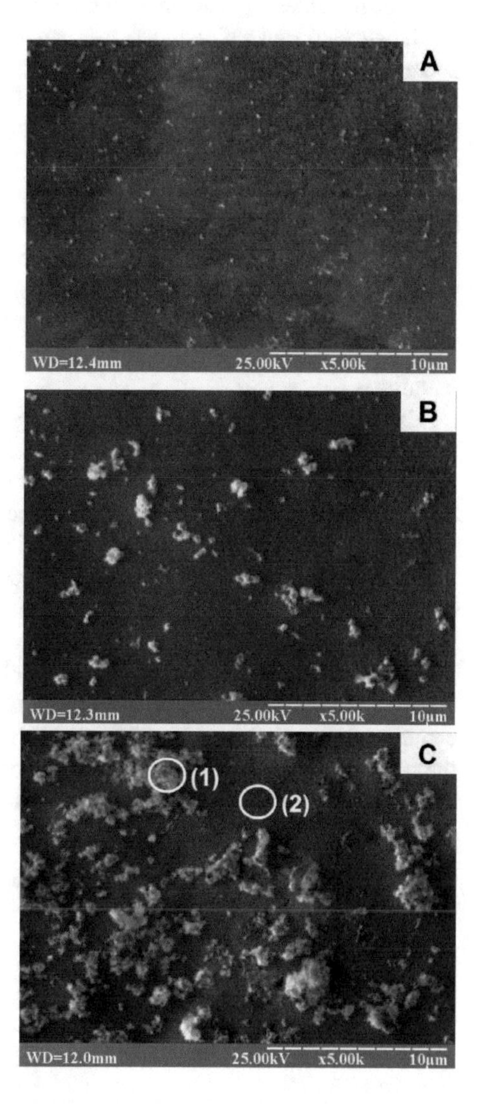

Fig. 4 Scanning electron microscopy micrographs of pure iron (A), iron–titania (2 wt.%) (B), and iron–titania (5 wt.%) (C) layers electrodeposited at the current density of 0.10 A m^{-2}. The coatings were deposited from the plating bath containing 1.25 M of iron (II) methanesulfonate (pH 1.3) at the temperature of 298 K for 20 min. (Reprinted from [12], Copyright 2016, with permission from Elsevier)

Table 1 Content of different chemical elements (wt.%) in the places corresponding to points (1) and (2) in Fig. 4C. (Reprinted from [12], Copyright 2016, with permission from Elsevier)

Point	Iron	Titanium	Oxygen	Other elements
(1)	1.21	58.49	40.14	0.16
(2)	97.27	1.41	1.17	0.15

microanalysis, small traces of oxygen and some other elements on the surface being observed too.

Some flakelike agglomerations consisting of TiO_2 dispersed particles are observed on the surface images obtained by scanning electron microscopy (Fig. 4b,c); the average size of these agglomerations is about 1–5 μm. The electron probe X-ray spectrum microanalysis indicates that the flake-like agglomerated particles contain mainly Ti and O (Table 1), and the ratio of titanium to oxygen practically coincides with their stoichiometric ratio in titanium dioxide. The growth of the concentration of TiO_2 particulates in the colloidal plating bath promotes the expansion of these agglomerations on the electrodeposits surface.

The results of electron probe X-ray spectrum microanalysis indicate that a limited amount of titanium (about 1–1.5%) is detectable on the surface areas where the entrapped agglomerations of TiO_2 are absent. It means that some nanosized particles of titania are not conglomerated and can be embedded directly in the electrodeposited metallic matrix.

The surface hardness of electrodeposits grows when titania dispersed particles are entrapped by electrochemically deposited Fe matrix (Table 2). This may be ascribed to the effect of dispersion strengthening [20]. This phenomenon is due to the incorporation of fine particles of a dispersed phase, which impedes the movement of dislocations in crystal structure, ensuring a hardening of composite materials. When compared with "pure" iron electrodeposits, the microhardness of Fe–TiO_2 composites is relatively high, which is extremely favorable in respect of their possible practical use.

The presence of TiO_2 dispersed phase in electrodeposited layers imparts the photocatalytic activity to the surface of an electrodeposited composite. A process of photochemical decomposition of methyl orange dye in wastewater [54] was used to evaluate the photocatalytic activity of Fe–TiO_2 composite electrodeposits fabricated from a plating bath based on methanesulfonate salts.

Table 2 Effect of introduction of TiO_2 dispersed phase into electrodeposited iron matrix on surface microhardness (the current density of 0.10 A m^{-2}. (Reprinted from [12], Copyright 2016, with permission from Elsevier)

Electrodeposit	Surface hardness (HV)
Iron	470
Iron/titania (2 wt.%)	535
Iron/titania (5 wt.%)	665

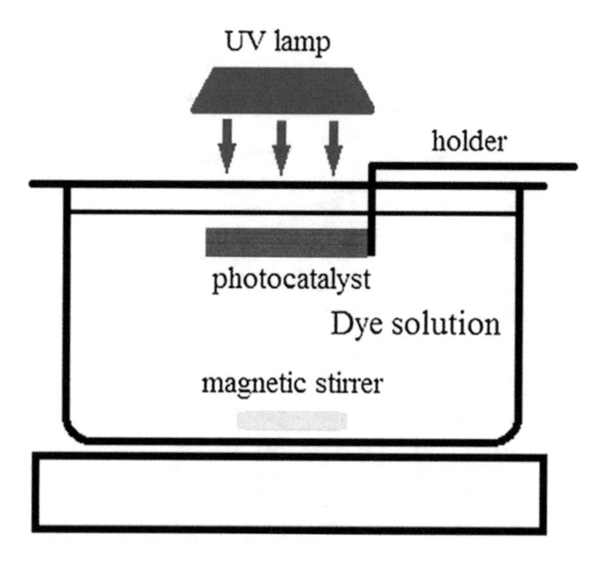

Fig. 5 Installation used to study photochemical efficiency of Fe–TiO_2 composite films. UV lamp was used as a [48] source of ultraviolet radiation. The sample of a photocatalyst (Fe–TiO_2 electrodeposited composite film) was immersed in the aqueous solution of an organic dye by a plastic holder. The solution was continuously stirred by a magnetic stirrer (~60 revs per minute). (Reprinted from Protsenko et al., Copyright 2017, with permission from Elsevier)

The kinetics of the destruction of above-mentioned organic dye in 0.1 M NaOH was examined upon exposure to ultraviolet irradiation (λ = 180–275 nm) emitted by a DKB-9 ultraviolet lamp [58]. The lamp was fixed above the solution of methyl orange dye. Plates made of steel with deposited Fe–TiO_2 films as photocatalysts were immersed in the solution at a depth of 0.2 cm (Fig. 5). The volume of methyl orange aqueous solution was 20 cm^3. In all experiments, methyl orange solution was uninterruptedly agitated. The concentration of methyl orange was determined via photometric analysis (the wavelength was λ = 490 nm).

The molecules of methyl orange dye do not decompose in a 1 M NaOH solution if they are not subjected to ultraviolet radiation (Fig. 6, curve 1). However, the influence of ultraviolet light causes the photochemical degradation (i.e., photolysis) of methyl orange molecules (Fig. 6, curve 2). The rate of photochemical decomposition considerably increases when the Fe–TiO_2 electrodeposited film was used (Fig. 6, curve 3). Hence, the electrochemically synthesized Fe–TiO_2 composite electrodeposits demonstrate photocatalytic activity. It should be observed that the iron coatings without embedded titania particles do not exhibit any photocatalytic behavior.

Kinetic curves plotted as "logarithm of the methyl orange concentration versus time" dependences show a linear relation. This behavior indicates that the photochemically induced destruction of methyl orange obeys the pseudo-first reaction order. The linearized kinetic curves allow calculating the apparent rate constants: 5.4 $\times$ 10^{-3} min^{-1} and 13.5 $\times$ 10^{-3} min^{-1} for organic dye decay in the absence and in the

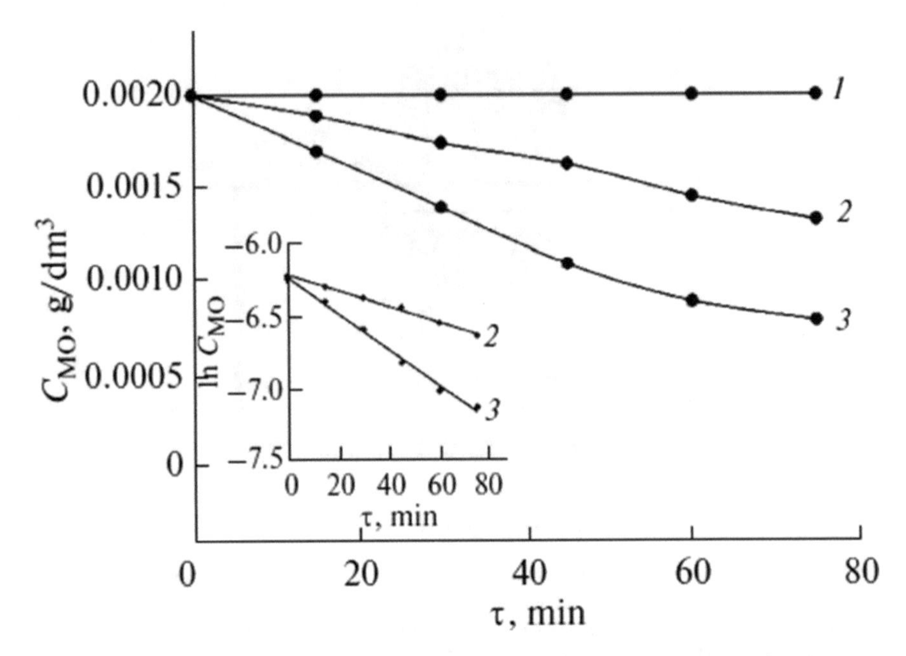

Fig. 6 Kinetic curves characterizing the reaction of methyl orange dye decolorization in a 1 M NaOH solution: (1) in the dark, (2) under the action of ultraviolet light without any catalyst, and (3) under the action of ultraviolet light on the electrodeposited iron–TiO_2 photocatalyst (10 wt.% TiO_2). The insert shows the first-order linear transforms of the kinetic curves in coordinates lnC vs. τ. (Reprinted from [58] by permission of Springer Nature)

presence of electrodeposited composite iron–titania (Degussa P 25) photocatalytic layer, respectively [58].

It is worth comparing the value of apparent rate constant measured for the photocatalyst involving the use of TiO_2 in the form of anatase (Degussa P 25) with that type of the deposited Fe–TiO_2 photocatalyst that includes titania in the form of "pure" rutile: 9.8×10^{-3} min^{-1} [46]. Composite iron–titania films containing predominantly anatase exhibit considerably better photocatalytic efficiency than those based solely on rutile phase, this observation coincides with the results given elsewhere [7, 42].

As far as the mechanism of photodestruction of methyl orange organic dye is concerned, we believe that this process occurs on the surface of semiconductor titania particles incorporated into deposited films in accordance with a basic reaction scheme of heterogeneous photocatalysis. This mechanism includes the adsorption of a photon followed by the step of a generation of an electron and positively charged hole. Then the water molecules are oxidized yielding •OH radicals; in addition, molecular oxygen originating from air is reduced forming hydrogen peroxide (H_2O_2), hydroperoxyl radicals ($\bullet OOH$), and superoxide radical anions ($O_2^{\bullet -}$) [8, 41]. These oxygen-containing species are extremely reactive, they take part in the subsequent oxidative destruction of various organic compounds. Generally, heterogeneous photochemical catalysis involving titania particles ensures virtually total mineralization of organic molecules and provides their destruction with the formation of CO_2, H_2O, sulfate, and nitrate anions [25, 54].

The corrosion stability of iron-based electrochemically deposited composite layers is not satisfactory enough, and iron matrix will corrode quickly in aqueous solutions and humid atmosphere. Obviously, the use of composite iron–titania films deposited in a methanesulfonate electrolyte as photocatalysts will be associated with their rapid corrosion destruction and failure. Therefore, the search for ways to increase the corrosion stability of the iron matrix is an important task, the solution of which is a necessary condition for the successful application of iron–titania composites as a photocatalytic material to purify the sewage from organic contaminants.

Various treatment methods can be used to produce a protective film on the iron-based coatings in order to enhance its corrosion resistance; cerium oxide layers appear to be very promising from this standpoint. Deposited ceria layers were shown to protect effectively iron and steel from corrosion damage [18, 19, 56]. They are regarded as a proper replacement for chromate coatings that are very hazardous to people's health and environment (the issue of hexavalent chromium). In this context, it should be mentioned that cerium oxide has been widely used in different water treatment processes [24].

An environmentally friendly methanesulfonate electrolyte was developed to prepare a protective thin ceria layer for the iron–titania composite films, Protsenko et al. [48]. The plating bath for ceria electrodeposition contained 0.5 mol dm^{-3} $Ce(CH_3SO_3)_3$. The cathodic electrosynthesis was conducted at the electrolyte pH of 1.3. The temperature was 298 K, and the cathodic current density was 5–25 A m^{-2}.

When electrical current passes through the acid aqueous methanesulfonate solution, several reactions proceed on the cathode resulting in evolution of hydrogen and electrochemical reduction of dissolved oxygen [18]:

$$H_2O + e^- = 1/2H_2 + OH^- \quad (3)$$

$$H_3O^+ + e^- = 1/2H_2 + H_2O \quad (4)$$

$$O_2 + 2H_2O + 2e^- = H_2O_2 + 2OH^- \quad (5)$$

$$O_2 + 2H_2O + 4e^- = 4OH^- \quad (6)$$

Since methanesulfonic acid is a very strong electrolyte [15] and its water solutions do not exhibit any noticeable buffering action, an increase in the value of pH occurs near the cathode surface due to the reactions [3–6] . This causes the formation of cerium (III) hydroxides and hydroxo-complexes in the near electrode solution layer [18]:

$$Ce^{3+} + 3OH^- = Ce(OH)_3 \quad (7)$$

$$2Ce^{3+} + 1/2O_2 + H_2O + 2OH^- = 2[Ce(OH)]_2^{2+} \quad (8)$$

Cerium (III) compounds can be easily oxidized forming cerium (IV) complexes and cerium (IV) oxide near the cathode surface:

$$Ce^{3+} + 1/2H_2O_2 + OH^- = [Ce(OH)]_2^{2+} \quad (9)$$

$$Ce(OH)_3 + O_2 = 4CeO_2 + 6H_2O \quad (10)$$

$$[Ce(OH)]_2^{2+} = CeO_2 + 2H_2O \quad (11)$$

It is shown that the surface appearance of iron–TiO_2 composite films changes after their cathodic processing in a methanesulfonate solution. As-deposited Fe–TiO_2 coatings are light grey, whereas their surface becomes greyish-blue after the cathodic treatment. Electron probe X-ray spectrum microanalysis indicated the formation of CeO_2 layers. Calculations revealed that the average thickness of the obtained ceria films is about 60–70 nm Protsenko et al. [48]. The synthesized ceria films demonstrate a good adhesion to the steel substrate: they cannot be separated from the surface of iron–TiO_2 electrodeposited composite by intensive rinsing with water.

According to the data obtained by the voltammetry method and electrochemical impedance spectroscopy, Protsenko et al. [48], the formation of a thin ceria layer causes an appreciable improvement in the corrosion resistance of iron–TiO_2 electrodeposited composite in an aqueous medium. Thus, the Fe–TiO_2 composite coatings with a cathodically deposited protective CeO_2 layer may serve as corrosion-resistant photochemical catalysts to decompose organic dyes in wastewater having natural pH. These composites modified with a ceria thin film will not corrode quickly in the treated wastewater.

The photocatalytic behavior of the composite iron–TiO_2 coatings with a thin protective CeO_2 film was characterized in the reactions of photodestruction of two organic dyes, methyl orange and methylene blue, in water solutions, Protsenko et al. [48]. Spontaneous decomposition of methyl orange dye is not detected without external irradiation, although photolysis is observed when the system is exposed to ultraviolet light. However, the photochemically induced decay of methyl orange molecules considerably accelerates by composite Fe–TiO_2 catalyst. Thus, the rate constant of photochemical decolorization was stated to increase from 1.1×10^{-3} min^{-1} to 1.8×10^{-3} min^{-1} when the composite electrodeposited iron–titania film modified by ceria layer was introduced into the system.

Iron matrix without protective ceria layer corrodes rapidly in an aqueous medium at neutral pH. Corrosion products of iron are accumulated in a solution as intensively colored fine-dispersed iron hydroxide; therefore, the kinetics of decolorization cannot be estimated reliably by means of photometric measurements. Thus, the influence of electrochemical deposition of CeO_2 film on the catalytic properties of electrodeposited iron–titania photocatalyst was investigated in a water solution containing 0.1 M NaOH, in which Fe is highly resistant to corrosion and colored corrosion products are not formed.

Kinetic study of the decolorization process in an alkaline solution containing methyl orange, which was subjected to ultraviolet irradiation, showed that the deposition of a thin ceria layer on the surface of iron–titania composite has no noticeable effect on its photochemical performance (Fig. 7). Respective kinetic curves plotted in linearized coordinates lnC *vs.* time practically coincide with each other, the calculated rate constant being about 12.8×10^{-3} min^{-1}. This value may be compared with that typical of the photolysis (5.4×10^{-3} min^{-1}, curve 1 in Fig. 7).

Several important conclusions can be drawn from the results described above. An increase in solution pH accelerates both photolysis and photocatalytic destruction of the dye, other conditions being equal. It is believed that improved catalytic activity of titania at higher pH is caused by the growth of the surface concentration of active adsorption centers rather than by an increment in •OH radicals content that is generated from OH^- ions due to interaction with positively charged holes [17].

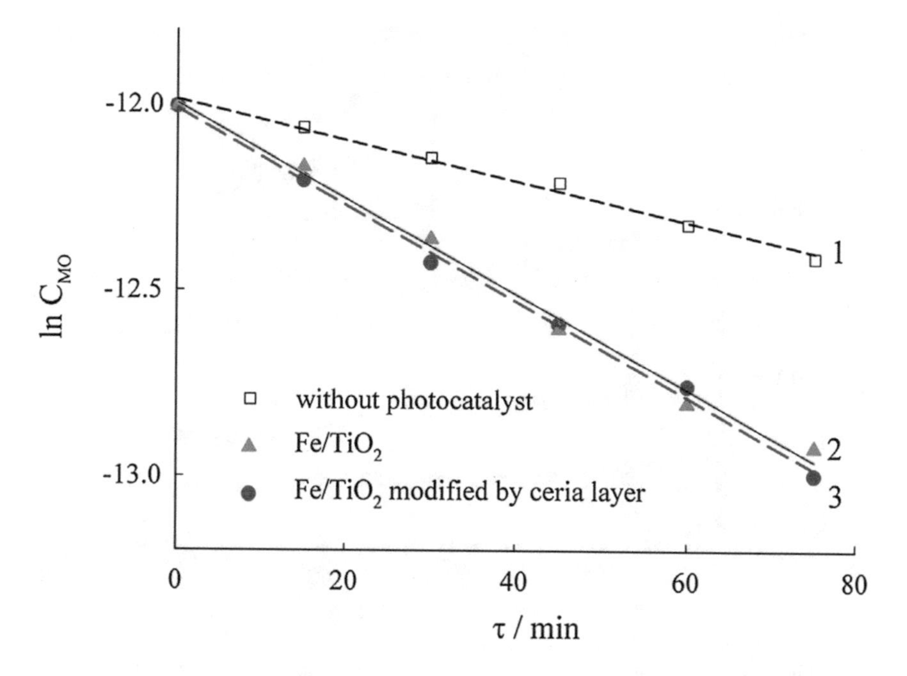

Fig. 7 Plot in linearized coordinates lnC *vs.* time characterizing the kinetics of decolorization induced by ultraviolet light: (1) in the absence of catalyst, (2) under the action of iron–titania catalyst (without modification), and (3) under the action of iron–titania catalyst [48] modified by CeO_2. Methyl orange dye was dissolved in 0.1 M NaOH (Reprinted from Protsenko et al., Copyright 2017, with permission from Elsevier)

The obtained data also indicate that the photocatalytic efficiency of composite $Fe–TiO_2$ deposits is not affected by the electrochemical modification of its surface via formation of a thin protective CeO_2 film (the rate constant of photochemically induced decomposition of methyl orange remains practically unchanged). However, it is well known that doping of semiconducting titania with some dopants (including cerium) commonly enhances its photocatalytic activity [8, 35]. It should be remembered in this context that the changes in chemistry, electronic structure, and crystal lattice is a decisive factor when doping titanium dioxide [8]. The electrodeposition of ceria films in accordance with the utilized procedure is conducted after the fabrication of titania particles; therefore, there is no influence on both chemistry and electronic and crystal structures of as-deposited $Fe–TiO_2$ photocatalyst. Therefore, this procedure cannot provide an efficient doping.

The photocatalytic performance of iron–TiO_2 composite catalysts was also investigated in the process where the molecules of methylene blue dye were subjected to destruction under the action of ultraviolet light (Fig. 8). The absorbance of aqueous solutions was measured at the wavelengths of 670 nm in these experiments.

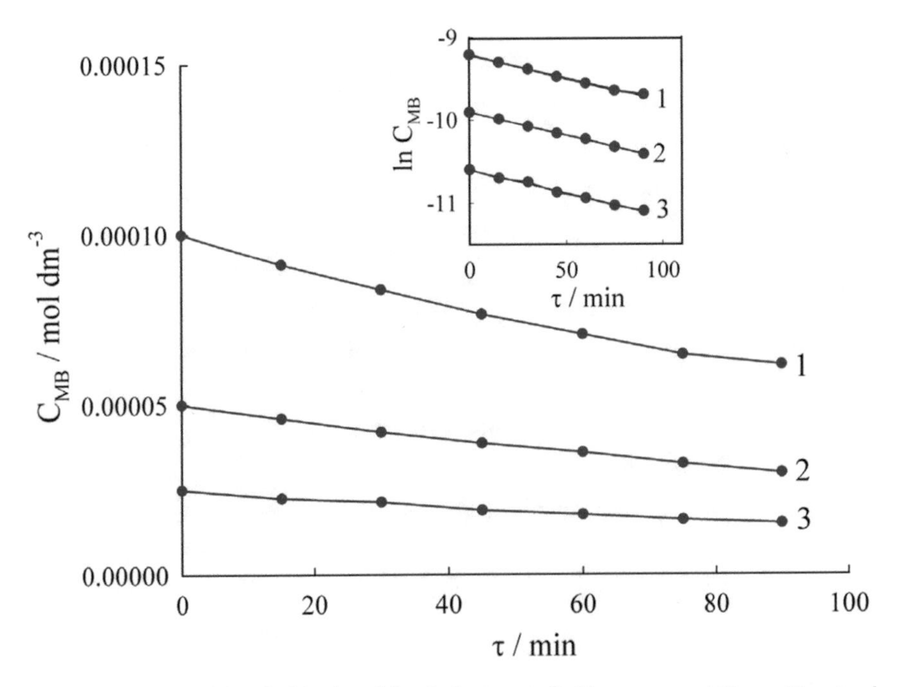

Fig. 8 Plot characterizing the kinetics of decolorization studied in aqueous solutions with natural pH at different initial contents of organic dye (methylene blue): 1—1.00×10^{-4}, 2—0.50×10^{-4}, and 3—0.25×10^{-4} M. The insert shows the first-order linear transforms of the kinetic curves in coordinates lnC vs. τ. (Reprinted from Protsenko et al. [48], Copyright 2017, with permission from Elsevier)

Table 3 Kinetic parameters characterizing the photochemical decay of methylene blue and methyl orange molecules on iron–titania catalyst with electrodeposited thin CeO_2 layer. (Reprinted from Protsenko et al. [48], Copyright 2017, with permission from Elsevier)

System	k (min^{-1})	$\tau_{1/2}$ (min)
Methylene blue (at natural pH)	5.6×10^{-3}	123.8
Methyl orange (at natural pH)	1.8×10^{-3}	385.1
Methyl orange (0.1 M NaOH)	12.8×10^{-3}	54.2

It should be noted that the rate of the photolysis of methylene blue dye is negligibly small as compared with the rate of its photochemical destruction on the electrodeposited Fe–TiO_2 composite photocatalysts Protsenko et al. [48]. The rate constant of the degradation of methylene blue was stated to be 5.6×10^{-3} min^{-1}, irrespective of the initial concentration of the dye.

Table 3 summarizes the data on the rate constant and half-life of photochemical destruction of the examined organic dyes on the Fe–TiO_2 photocatalyst. As can be seen, the electrodeposited iron–titania composites with a CeO_2 protective film are promising photocatalyst with respect to the photochemical decay of methyl orange and methylene blue dyes in aqueous media at neutral pH values.

4 Conclusion

High-performance heterogeneous photocatalysts based on semiconducting titania can be fabricated by the electrochemical deposition technique via co-deposition with an iron matrix. The developed procedure involves the use of eco-friendly aqueous methanesulfonate plating baths. The concentration of titania dispersed phase in electrodeposited composite films may reach 5 wt.%. The mechanism of the co-deposition complies with the improved Guglielmi's mechanism.

The synthesized films with immobilized TiO_2 particles exhibited pronounced catalytic behavior in the photochemically induced decomposition of some organic contaminants (organic dyes) under the action of ultraviolet radiation. Thus, these heterogeneous composite photocatalysts can be successfully used in wastewater treatment.

In order to protect the iron matrix from rapid corrosion damage, the modification of the surface of as-deposited iron–titania composite was suggested via cathodic electrochemical treatment in methanesulfonate solutions containing Ce(III) ions. A deposited thin film of CeO_2 protective layer appreciably improves corrosion stability of the catalytic iron–TiO_2 composite and does not affect its catalytic performance toward the photochemical decomposition of methylene blue and methyl orange in wastewater.

References

1. Ahmad R, Ahmad Z, Khan AU, Mastoi NR, Aslam M, Kim J (2016) Photocatalytic systems as an advanced environmental remediation: Recent developments, limitations and new avenues for applications. J Environ Chem Eng 4:4143–4164. https://doi.org/10.1016/j.jece.2016.09.009
2. Alhaji MH, Sanaullah K, Khan A, Hamza A, Muhammad A, Ishola MS, Rigit ARH, Bhawani SA (2017) Recent developments in immobilizing titanium dioxide on supports for degradation of organic pollutants in wastewater—a review. Int J Environ Sci Technol 14:2039–2052. https://doi.org/10.1007/s13762-017-1349-4
3. Allen HC, Raymond EA, Richmond GL (2001) Surface structural studies of methanesulfonic acid at air /aqueous solution interfaces using vibrational sum frequency spectroscopy. J Phys Chem A 105:1649–1655. https://doi.org/10.1021/jp0032964
4. Amadelli R, Samiolo L, Velichenko AB, Khysh VA, Luk'yanenko TV, Danilov FI (2009) Composite PbO_2–TiO_2 materials deposited from colloidal electrolyte: electrosynthesis, and physicochemical properties. Electrochim Acta 54:5239–5245. https://doi.org/10.1016/j.electacta.2009.04.024
5. Bahadormanesh B, Dolati A (2010) The kinetics of Ni-Co/SiC composite coatings electrodeposition. J Alloys Compd 504:514–518. https://doi.org/10.1016/j.jallcom.2010.05.154
6. Berçot P, Peña-Muñoz E, Pagetti J (2002) Electrolytic composite Ni–PTFE coatings: an adaptation of Guglielmi's model for the phenomena of incorporation. Surf Coat Technol 157:282–289. https://doi.org/10.1016/S0257-8972(02)00180-9
7. Bickley RI, Gonzalez-Carreno T, Lees JS, Palmisano L, Tilley RJD (1991) A structural investigation of titanium dioxide photocatalysts. J Solid State Chem 92:178–190. https://doi.org/10.1016/0022-4596(91)90255-G
8. Chen X, Mao SS (2007) Titanium dioxide nanomaterials: synthesis, properties, modifications, and applications. Chem Rev 107:2891–2959. https://doi.org/10.1021/cr0500535
9. Crini G, Lichtfouse E (2019) Advantages and disadvantages of techniques used for wastewater treatment. Environ Chem Lett 17:145–155. https://doi.org/10.1007/s10311-018-0785-9
10. Danilov FI, Kityk AA, Shaiderov DA, Bogdanov DA, Korniy SA, Protsenko VS (2019) Electrodeposition of Ni–TiO_2 composite coatings using electrolyte based on a deep eutectic solvent. Surf Eng Appl Electrochem 55(2):138–149. https://doi.org/10.3103/S106837551902008X
11. Danilov FI, Tsurkan AV, Vasil'eva EA, Korniy SA, Cheipesh TA, Protsenko VS (2017) Electrochemical synthesis and properties of iron–titanium dioxide composite coatings. Russ J Appl Chem 90:1148–1153. https://doi.org/10.1134/S1070427217070199
12. Danilov FI, Tsurkan AV, Vasil'eva EA, Protsenko VS (2016) Electrocatalytic activity of composite Fe/TiO_2 electrodeposits for hydrogen evolution reaction in alkaline solutions. Int J Hydrogen Energy 41:7363–7372. https://doi.org/10.1016/j.ijhydene.2016.02.112
13. Diebold U (2003) The surface science of titanium dioxide. Surf Sci Rep 48:53–229. https://doi.org/10.1016/S0167-5729(02)00100-0
14. Dutt MA, Hanif MA, Nadeem F, Bhatti HN (2020) A review of advances in engineered composite materials popular for wastewater treatment. J Environ Chem Eng 8:104073. https://doi.org/10.1016/j.jece.2020.104073
15. Gernon MD, Wu M, Buszta T, Janney P (1999) Environmental benefits of methanesulfonic acid: comparative properties and advantages. Green Chem 1:127–140. https://doi.org/10.1039/a900157c
16. Guglielmi N (1972) Kinetics of the deposition of inert particles from electrolytic baths. J Electrochem Soc 119:1009–1012. https://doi.org/10.1149/1.2404383
17. Guillard C, Puzenat E, Lachheb H, Houas A, Herrmann JM (2005) Why inorganic salts decrease the TiO_2 photocatalytic efficiency. Int J Photoenergy 7:1–9. https://doi.org/10.1155/S1110662X05000012
18. Hamlaoui Y, Pedraza F, Remazeilles C, Cohendoz S, Rebere C, Tifouti L, Creus J (2009) Cathodic electrodeposition of cerium-based oxides on carbon steel from concentrated cerium nitrate solutions. Part I. Electrochemical and analytical characterisation. Mater Chem Phys 113:650–657. https://doi.org/10.1016/j.matchemphys.2008.08.027

19. Hamlaoui Y, Tifouti L, Remazeilles C, Pedraza F (2010) Cathodic electrodeposition of cerium based oxides on carbon steel from concentrated cerium nitrate. Part II: influence of electrodeposition parameters and of the addition of PEG. Mater Chem Phys 120:172–180. https://doi.org/10.1016/j.matchemphys.2009.10.042
20. Hou F, Wang W, Guo H (2006) Effect of the dispersibility of ZrO_2 nanoparticles in Ni–ZrO_2 electroplated nanocomposite coatings on the mechanical properties of nanocomposite coatings. Appl Surf Sci 252:3812–3817. https://doi.org/10.1016/j.apsusc.2005.05.076
21. Ito S, Deguchi T, Imai K, Iwasaki M, Tada H (1999) Preparation of highly photocatalytic nanocomposite films consisting of TiO_2 particles and Zn electrodeposited on steel. Electrochem Solid-State Lett 2(9):440–442. https://doi.org/10.1149/1.1390864
22. Kanakaraju D, Glass BD, Oelgemöller M (2014) Titanium dioxide photocatalysis for pharmaceutical wastewater treatment. Environ Chem Lett 12:27–47. https://doi.org/10.1007/s10311-013-0428-0
23. Khysh V, Luk'yanenko T, Shmychkova O, Amadelli R, Velichenko A (2017) Electrodeposition of composite PbO_2–TiO_2 materials from colloidal methanesulfonate electrolytes. J Solid State Electrochem 21:537–544. https://doi.org/10.1007/s10008-016-3394-1
24. Kurian M (2020) Cerium oxide based materials for water treatment—A review. J Environ Chem Eng 8:104439. https://doi.org/10.1016/j.jece.2020.104439
25. Lachheb H, Puzenat E, Houas A, Ksibi M, Elaloui E, Guillard C, Herrmann JM (2002) Photocatalytic degradation of various types of dyes (Alizarin S, Crocein Orange G, Methyl Red, Congo Red, Methylene Blue) in water by UV-irradiated titania. Appl Catal B 39:75–90. https://doi.org/10.1016/S0926-3373(02)00078-4
26. Laohhasurayotin K, Pookboonmee S (2013) Multifunctional properties of Ag/TiO_2/bamboo charcoal composites: preparation and examination through several characterization methods. Appl Surf Sci 282:236–244. https://doi.org/10.1016/j.apsusc.2013.05.110
27. Lazar MA, Varghese S, Nair SS (2012) Photocatalytic water treatment by titanium dioxide: recent updates. Catalysts 2:572–601. https://doi.org/10.3390/catal2040572
28. Low CTJ, Wills RGA, Walsh FC (2006) Electrodeposition of composite coatings containing nanoparticles in a metal deposit. Surf Coat Technol 201:371–383. https://doi.org/10.1016/j.surfcoat.2005.11.123
29. Luan X, Wang Y (2014) Preparation and photocatalytic activity of Ag/bamboo-type TiO_2 nanotube composite electrodes for methylene blue degradation. Mater Sci Semicond Process 25:43–51. https://doi.org/10.1016/j.mssp.2013.10.023
30. Luo S, Xiao Y, Yang L, Liu C, Su F, Li Y, Cai Q, Zeng G (2011) Simultaneous detoxification of hexavalent chromium and acid orange 7 by a novel Au/TiO_2 heterojunction composite nanotube arrays. Sep Purif Technol 79:85–91. https://doi.org/10.1016/j.seppur.2011.03.019
31. Magagnin L, Bernasconi R, Ieffa S, Diamanti MV, Pezzoli D, Candiani G, Pedeferri MP (2013) Photocatalytic and antimicrobial coatings by electrodeposition of silver/TiO_2 nano-composites. ECS Trans 45(8):1–6. https://doi.org/10.1149/04508.0001ecst
32. Mahlambi MM, Ngila CJ, Mamba BB (2015) Recent developments in environmental photocatalytic degradation of organic pollutants: the case of titanium dioxide nanoparticles—a review. J Nanomater 2015:790173. https://doi.org/10.1155/2015/790173
33. Miyake M, Takahashi A, Hirato T (2017) Electrodeposition and anodization of Al–TiO_2 composite coatings for enhanced photocatalytic activity. Int J Electrochem Sci 12:2344–2352. https://doi.org/10.20964/2017.03.41
34. Mohajeri S, Dolati A, Ghorbani M (2017) The photoinduced activity of Ni–TiO_2/TiO_2 multilayer nanocomposites synthesized by pulse electrodeposition technique. Int J Electrochem Sci 12:5121–5141. https://doi.org/10.20964/2017.06.50
35. Mudhoo A, Paliya S, Goswami P, Singh M, Lofrano G, Carotenuto M, Carraturo F, Libralato G, Guida M, Usman M, Kumar S (2020) Fabrication, functionalization and performance of doped photocatalysts for dye degradation and mineralization: a review. Environ Chem Lett 18:1825–1903. https://doi.org/10.1007/s10311-020-01045-2
36. Nakata K, Fujishima A (2012) TiO_2 photocatalysis: design and applications. J Photochem Photobiol C 13:169–189. https://doi.org/10.1016/j.jphotochemrev.2012.06.001

37. Nasirian M, Mehrvar M (2016) Modification of TiO_2 to enhance photocatalytic degradation of organics in aqueous solutions. J Environ Chem Eng 4:4072–4082. https://doi.org/10.1016/j.jece.2016.08.008
38. Ohtani B, Prieto-Mahaney OO, Li D, Abe R (2010) What is Degussa (Evonik) P25? Crystalline composition analysis, reconstruction from isolated pure particles and photocatalytic activity test. J Photochem Photobiol A 216:179–182. https://doi.org/10.1016/j.jphotochem.2010.07.024
39. Oturan MA, Aaron JJ (2014) Advanced oxidation processes in water/wastewater treatment: principles and applications. A review. Crit Rev Env Sci Technol 44:2577–2641. https://doi.org/10.1080/10643389.2013.829765
40. Patel PS, Bandre N, Saraf A, Ruparelia JP (2013) Electro-catalytic materials (electrode materials) in electrochemical wastewater treatment. Proc Eng 51:430–435. https://doi.org/10.1016/j.proeng.2013.01.060
41. Pelaez M, Nolan NT, Pillai SC, Seery MK, Falaras P, Kontos AG, Dunlop PSM, Hamilton JWJ, Byme JA, O'Shea K, Entezari MH, Dionisiou DD (2012) A review on the visible light active titanium dioxide photocatalysts for environmental applications. Appl Catal B 125:331–349. https://doi.org/10.1016/j.apcatb.2012.05.036
42. Pifferi V, Spadavecchia F, Cappelletti G, Paoli EA, Bianchi CL, Falciola L (2013) Electrodeposited nano-titania films for photocatalytic Cr(VI) reduction. Catal Today 209:8–12. https://doi.org/10.1016/j.cattod.2012.08.031
43. Protsenko VS, Kityk AA, Vasil'eva EA, Tsurkan AV, Danilov FI (2019) Electrodeposition of composite coatings as a method for immobilizing TiO_2 photocatalyst. In: Inamuddin SG, Kumar A, Lichtfouse E, Asiri A (eds) Nanophotocatalysis and environmental applications. Environmental Chemistry for a Sustainable World, vol 29. Springer Nature Switzerland AG, pp 263–301. doi: https://doi.org/10.1007/978-3-030-10609-6_10
44. Protsenko VS, Tsurkan AV, Vasil'eva EA, Baskevich AS, Korniy SA, Cheipesh TO, Danilov FI (2018) Fabrication and characterization of multifunctional Fe/TiO_2 composite coatings. Mater Res Bull 100:32–41. https://doi.org/10.1016/j.materresbull.2017.11.051
45. Protsenko VS, Vasil'eva EA, Smenova IV, Baskevich AS, Danilenko IA, Konstantinova TE, Danilov FI (2015) Electrodeposition of Fe and composite Fe/ZrO_2 coatings from a methanesulfonate bath. Surf Eng Appl Electrochem 51:65–75. https://doi.org/10.3103/S1068375515010123
46. Protsenko VS, Vasil'eva EA, Smenova IV, Danilov FI (2014) Electrodeposition of iron/titania composite coatings from methanesulfonate electrolyte. Russ J Appl Chem 87:283–288. https://doi.org/10.1134/S1070427214030069
47. Protsenko VS, Vasil'eva EA, Tsurkan AV, Danilov FI (2017a) Electrodeposition of electrocatalytic and photocatalytic Fe/TiO_2 composite coatings using methanesulfonate electrolytes. In: Jacobs T (ed) Electrospinning and electroplating: fundamentals, methods and applications. Nova Science Publishers Inc., New York, pp 177–226
48. Protsenko VS, Vasil'eva EA, Tsurkan AV, Kityk AA, Korniy SA, Danilov FI (2017b) Fe/TiO_2 composite coatings modified by ceria layer: electrochemical synthesis using environmentally friendly methanesulfonate electrolytes and application as photocatalysts for organic dyes degradation. J Environ Chem Eng 5:136–146. https://doi.org/10.1016/j.jece.2016.11.034
49. Rahmawati F, Wahyuningsih S, Irianti D (2014) The photocatalytic activity of SiO_2-TiO_2/graphite and its composite with silver and silver oxide. Bull Chem React Eng Catal 9(1):45–52. https://doi.org/10.9767/bcrec.9.1.5374.45-52
50. Rajeshwar K, de Tacconi NR, Chenthamarakshan CR (2001) Semiconductor-based composite materials: preparation, properties, and performance. Chem Mater 13:2765–2782. https://doi.org/10.1021/cm010254z
51. Rosolymou E, Spanou S, Zanella C, Tsoukleris DS, Köhler S, Leisner P, Pavlatou EA (2020) Electrodeposition of photocatalytic Sn–Ni matrix composite coatings embedded with doped TiO_2 particles. Coatings 10:775. https://doi.org/10.3390/coatings10080775
52. Saravanan A, Kumar PS, Vo DV, Yaashikaa PR, Karishma S, Jeevanantham S, Gayathri B, Bharathi VD (2021) Photocatalysis for removal of environmental pollutants and fuel

production: a review. Environ Chem Lett 19:441–463. https://doi.org/10.1007/s10311-020-01077-8
53. Shan AY, Ghazi TIM, Rashid SA (2010) Immobilisation of titanium dioxide onto supporting materials in heterogeneous photocatalysis: a review. Appl Catal A 389:1–8. https://doi.org/10.1016/j.apcata.2010.08.053
54. Sonawane RS, Kale BB, Dongare MK (2004) Preparation and photo-catalytic activity of Fe–TiO_2 thin films prepared by sol–gel dip coating. Mater Chem Phys 85:52–57. https://doi.org/10.1016/j.matchemphys.2003.12.007
55. Spanou S, Kontos AI, Siokou A, Kontos AG, Vaenas N, Falaras P, Pavlatou EA (2013) Self cleaning behaviour of Ni/nano-TiO_2 metal matrix composites. Electrochim Acta 105:324–332. https://doi.org/10.1016/j.electacta.2013.04.174
56. Stoychev D (2013) Corrosion protective ability of electrodeposited ceria layers. J Solid State Electrochem 17:497–509. https://doi.org/10.1007/s10008-012-1937-7
57. Vasil'eva EA, Smenova IV, Protsenko VS, Konstantinova TE, Danilov FI (2013) Electrodeposition of hard iron–zirconia dioxide composite coatings from a methanesulfonate electrolyte. Russ J Appl Chem 86:1735–1740. https://doi.org/10.1134/S1070427213110177
58. Vasil'eva EA, Tsurkan AV, Protsenko VS, Danilov FI (2016) Electrodeposition of composite Fe–TiO_2 coatings from methanesulfonate electrolyte. Prot Met Phys Chem Surf 52:532–537. https://doi.org/10.1134/S2070205116030278
59. Vasudevan S, Oturan MA (2014) Electrochemistry: as cause and cure in water pollution—an overview. Environ Chem Lett 12:97–108. https://doi.org/10.1007/s10311-013-0434-2
60. Walsh FC, Ponce de Leon C (2014) A review of the electrodeposition of metal matrix composite coatings by inclusion of particles in a metal layer: an established and diversifying technology. Trans Inst Met Finish 92:83–98. https://doi.org/10.1179/0020296713Z.000000000161
61. Walsh FC, Ponce de León C (2014) Versatile electrochemical coatings and surface layers from aqueous methanesulfonic acid. Surf Coat Technol 25:676–697. https://doi.org/10.1016/j.surfcoat.2014.10.010

Printed in the United States
by Baker & Taylor Publisher Services